THE ROUTLEDGE COMPANION TO INTERNATIONAL HOUSING MARKETS

Edited by Magnus Andersson, Peter Palm, Helena Bohman, Béatrice Balivet, and Omokolade Akinsomi

Routledge
Taylor & Francis Group
LONDON AND NEW YORK

Cover image: Eigo Tateishi

First published 2022
by Routledge
4 Park Square, Milton Park, Abingdon, Oxon OX14 4RN

and by Routledge
605 Third Avenue, New York, NY 10158

Routledge is an imprint of the Taylor & Francis Group, an informa business

British Library Cataloguing-in-Publication Data
A catalogue record for this book is available from the British Library

Library of Congress Cataloging-in-Publication Data
Names: Akinsomi, Omokolade, editor. | Andersson, Magnus (College teacher), editor. | Balivet, Béatrice, editor. | Bohman, Helena, editor. | Palm, Peter (College teacher), editor.
Title: The Routledge companion to international housing markets / edited by Omokolade Akinsomi, Magnus Andersson, Béatrice Balivet, Helena Bohman, and Peter Palm.
Description: Milton Park, Abingdon, Oxon ; New York, NY : Routledge, 2022. | Includes bibliographical references and index.
Identifiers: LCCN 2021058184 (print) | LCCN 2021058185 (ebook)
Subjects: LCSH: Housing development. | Housing policy. | Cities and towns—Study and teaching.
Classification: LCC HD1375 .R66 2022 (print) | LCC HD1375 (ebook) | DDC 333.33—dc23/eng/20211202
LC record available at https://lccn.loc.gov/2021058184
LC ebook record available at https://lccn.loc.gov/2021058185

ISBN: 978-0-367-64664-6 (hbk)
ISBN: 978-0-367-67231-7 (pbk)
ISBN: 978-1-003-13038-3 (ebk)

DOI: 10.1201/9781003130383

Typeset in Bembo
by Apex CoVantage, LLC

CONTENTS

FIGURES

Participating countries map

Civil law countries map

PhD in construction and real estate economics. His research focuses on the real estate market and real estate valuation.

Helena Bohman is a senior lecturer in economics in the Department of Urban Studies, Malmö University, Sweden. Her research focuses on urban and regional development, with a special focus on issues concerning real estate and transport.

Béatrice Balivet is a senior lecturer in private law in the Faculty of Law department at Jean Moulin University Lyon 3, France. Her research focuses are property law and real estate law, co-ownership, and construction.

Omokolade Akinsomi is an associate professor in real estate finance and investments in the School of Construction Economics and Management at the University of the Witwatersrand, South Africa. He is the president of the African Real Estate Society (2021–2023) and a member of the Royal Institution of Chartered Surveyors (MRICS). His research focuses on real estate investment trust, price discovery in commercial real estate, real estate portfolio management, housing, and proptech.

THE ROUTLEDGE COMPANION TO INTERNATIONAL HOUSING MARKETS

There is a lack of international comparative housing studies, possibly because it requires thorough knowledge of the real estate market in question. This book brings together scholars with knowledge of different national markets in order to facilitate comparisons for real estate and housing and urban studies scholars and practitioners. By studying international markets using new data as well as new analysis of existing data, the chapters in this book present insights into the institutional constraints on national housing markets. Specifically, the contributors seek to examine the role of institutional factors and their influence on transaction costs in these markets. Exhibiting a diverse range of geographical, legal, and economic perspectives, the countries are grouped together based on legal institutional similarities, and each group includes an introduction and a conclusion highlighting similarities and differences from the institutional perspective.

The book is divided into three parts:

Part I sets the theoretical context and the reasons for writing a book focusing on national housing markets.

Part II presents national markets from the perspective of the transaction process and covers Europe, North and South America, East Asia, the Pacific, and Africa.

Part III contains conclusions with a critical discussion on how to compare national housing markets and a reflection on future directions of housing markets in an increasingly competitive international environment.

The Routledge Companion to International Housing Markets is essential reading for academics and professionals in housing studies, real estate, economics, and urban studies.

Magnus Andersson is an associate professor in economic geography in the Department of Urban Studies and senior lecturer at the Institute for Urban Research at Malmö University, Sweden. His research focuses on socio-economic development in contexts with limited access to official statistics. He has since 2010 participated as researcher and adviser for the United Nations' work to implement its Sustainable Development Goals.

Peter Palm is a senior lecturer in economics in the Department of Urban Studies, Malmö University, Sweden, and chairman of the Division of Property Valuation, Sweden. He holds a

Nordic law countries map

Common law countries map

TABLES

ABOUT THE AUTHORS

Evgenia Afanaseva is an appraiser in the Institute of Valuation of Property and Financial Activity, Russian Federation, and holds a PhD in cadastral valuation of land with subsurface rights. Her research has a main focus on the real estate market and real estate market valuation, including mass (cadastral) valuation.

Enrico Al Mureden holds multiple teaching courses such as civil law in the Department of Legal Studies of the University of Bologna; family law in the Department of Legal Studies of the University of Bologna based in Ravenna; private law, held entirely in English, at the School of Economics, Management and Statistics of the University of Bologna; product safety, product liability, and automotive, held entirely in English, within the Advanced Automotive Engineering Degree course at the Motorvehicle University of Emilia-Romagna (MUNER), based in the Department of Engineering Enzo Ferrari of the University of Modena and Reggio Emilia.

Wilfred K. Anim-Odame is a chartered valuer and land administration expert with over 30 years of experience. He is a senior technical advisor on land administration, urban management, housing and human settlement at the National Development Planning Commission in Ghana. Prior to that, he spent six years as the chief executive officer of the Ghana Lands Commission and as the director of the Land Valuation Division between 2010 and 2012, where he acted as the valuer-general for Ghana. He is a fellow of the Ghana Institution of Surveyors and a member of the Royal Institution of Chartered Surveyors. He holds a PhD in finance with a specialization in real estate finance and investment from the prestigious Cass Business School, City University, London, UK. He is an author and a scholar of the Commonwealth Secretariat.

Stéphane Berre is an associate professor in private law at Lyon 3 University, France. His research focuses on civil law notary, taxes matters, and rural law.

Yousra Chaaban is a lecturer at the Faculty of Law Ain Shams University, Egypt, and a PhD holder from the Jean Moulin University Lyon 3, France. Her research focuses mainly on comparative, international, and private law. She also carries out various activities in women's rights and public society.

Wong Woei Chyuan is an associate professor at the School of Economics, Finance and Banking, Universiti Utara Malaysia. He received his PhD from National University of Singapore. His research interests include REITs, housing markets, and corporate finance.

Joseph Djogbenou is a professor at the Abomey-Calavi University, Republic of Benin, and chairman of Centre for Research and Study in Law and Judicial Institutions in Africa (CREDIJ) and president of the constitutional court of Benin.

Emeka Eleh is a senior partner at Ubosi Eleh + Co., a firm of estate surveyors and valuers based in Nigeria. He is a past president of the Nigerian Institution of Estate Surveyors and Valuers. He is also a fellow of the Royal Institution of Chartered Surveyors. He is passionate about land reform and the establishment of appropriate land rights infrastructure in developing countries as well as issues of affordable housing.

Yaëll Emerich is a full professor at McGill, Faculty of Law. Her main research focuses on property from a comparative law perspective and on social issues related to property.

Emmanuel Kofi Gavu is a lecturer at the Department of Land Economy, KNUST Kumasi, Ghana. He is a board member of the African Real Estate Society (AfRES), chair of the Future Leaders of the African Real Estate Society (FLAfRES), and member of the Ghana Institution of Surveyors (GhIS). His research focuses on application of geographical information system in urban management, real estate, and housing market analysis.

Peter Gladoić Håkansson is an associate professor of economic history in the Department of Society, Culture and Identity, Malmö University, Sweden. His main research interest concerns labour market entry and mobility, the role of institutions, and social capital.

Christine Helbach is a real estate appraiser for the city of Frankfurt am Main with a focus on the valuation of public and fiscal properties. She has an engineer's degree in geodesy from the Technical University of Darmstadt, Germany, and is a member of the expert committees for property valuation in the cities of Frankfurt am Main and Offenbach am Main.

Walaaeldeen M. Ibrahim is an associate professor in private law at Zayed University, UAE, and an expert in private law at the National Center for Social and Criminological Research, Egypt. His research focuses on international private law, business law, real estate, and construction.

Ola Jingryd, Doctor of Laws (LL.D.), is a senior lecturer in Law and Real Estate Science at Malmö University, Sweden. His research focuses on legal and interdisciplinary aspects of real estate and the housing market, particularly the conveyance of real estate and the professionals involved in such transactions.

Raphael Kieti is a registered valuer and land economist. His research focuses on property valuation, land, and housing markets. He holds a PhD in land economics and is currently a lecturer in the Department of Real Estate and the Department of Construction Management and Quantity Surveying at the University of Nairobi, Kenya.

Kimitaka Aoki is a Japanese architect who holds a PhD in urban planning and design from the University of Tokyo, Japan. His current research explores methodologies for solving urban challenges in dense wooden urban areas of Tokyo.

Jasenka Kranjčević is an assistant professor and senior researcher in the field of architecture and urbanism at the Institute for Tourism, Zagreb, Croatia. Her research focuses on spatial planning and architecture in contexts of sustainable tourism and heritage.

Rebecca Lewis is associate professor in the School of Planning, Public Policy and Management at the University of Oregon, United States, and co-director of the Institute for Policy Research and Engagement. Her research focuses on land use policy, housing, and governance.

Sylwia Lindqvist holds a PhD in infrastructure with a specialization in real estate economics and is a senior lecturer in real estate science in the Department of Urban Studies at Malmö University, Sweden. Her research primarily deals with real estate brokerage and housing in an international perspective.

Peter Mburu holds a Doctorate of Laws degree from the University of Groningen, Netherlands. He is an employee of the Ministry of Lands in Kenya as land registrar. He is also an adjunct lecturer at the faculty of law in the University of Nairobi. His research interests include conveyancing law and the law of land registration.

Muhammad Najib Razali is an associate professor with the Faculty of Built Environment and Surveying, Universiti Teknologi Malaysia (UTM), Malaysia. He is currently the director of real estate, Faculty of Built Environment and Surveying, UTM. He is also a research fellow at the Centre for Environmental Sustainability and Water Security and a board member of the Pacific Rim Real Estate Society. His research and writing interests include property investment analysis in developing countries, property portfolio management, the international property market, big data, blockchain in real estate, and business valuation.

Olusegun Olanrele holds a PhD in real estate development and environment with a specialization in property investment performance analysis. His research has focused on REIT and housing affordability.

Imane Oualji is a professor in private law at the Faculty of Legal and Political Sciences at Hassan 1st University, Morocco. Her research focuses on business law, property law, and securities law.

Robert Parker is director of strategy and technical solutions at the Institute for Policy Research and Engagement in the School of Planning, Public Policy and Management at the University of Oregon, United States. His research focuses on growth management, housing, and economic development.

Ana Paula Corrêa Patiño is a lawyer and holds both a master's and a PhD in civil law from Universidad de São Paolo and is professor of civil law at the Faculty of Law at the Centro Universitário Fundação Armando Álvares Penteado, Brazil.

Nattapong Puttanapong is an assistant professor at Faculty of Economics, Thammasat University, Thailand. His research interests are in the areas of economic modelling, spatial econometrics, and socioeconomic disparities. He was awarded with the Royal Thai Government Scholarship, through which he obtained his MPA, MS and PhD degrees from Cornell University, United States.

Kirsikka Riekkinen is an assistant professor in land management at the Real Estate Economics Research Group at Aalto University, Finland. The position is partly funded by the National Land Survey of Finland. Her research focuses on developing land administration systems and land administration tools.

Elena Sánchez Jordán is full professor of civil law at the University of La Laguna, Tenerife, Spain. Her main research interests focus on consumer protection and in property law, particularly in connection to urban planning and land registration.

Sandra Schilling is working at the municipal land surveying office Frankfurt am Main with a focus on the real estate market development in Frankfurt. She graduated from the international master programme 'Urban Agglomerations' at Frankfurt University of Applied Sciences, Germany, and Newcastle upon Tyne, Great Britain, with a specialization in real estate valuation.

Julian Sidoli studied at University College London and was called to the Bar of England and Wales by the Honourable Society of the Inner Temple. As well as practising at the Chancery Bar, he lectures in property law for the Open University, United Kingdom. He has published widely in property law and related fields.

Chaing Sinath is a supreme court magistrate and associate professor in private law at the Faculty of Law, Royal University of Law and Economics, Cambodia. Her research focuses on civil law and property law.

Neuv Socheata is a notary public and associate professor in private law at the Faculty of Law, Royal University of Law and Economics, Cambodia. Her research focuses on civil law, property law, and real estate law.

Jill St George is a lecturer in law at the Open University, United Kingdom. Her teaching includes land law, torts and family law. Her main research interests are focused in the Caribbean, particularly Barbados, looking at CARICOM, human trafficking, and human rights.

Marimo Taderera is a sessional lecturer for commercial real estate investments at the University of the Witwatersrand, South Africa, and is a transaction manager at Cushman & Wakefield | Broll in South Africa. His research focuses on real estate finance.

Eigo Tateishi is a PhD candidate in urban studies at Malmö University, Sweden. His primary research interest is post-urban governance of big cities under an extreme demographic decline such as a sustained population loss and aging population.

Fabian Thiel is professor for land valuation in the Department of Architecture, Civil Engineering and Geomatics at Frankfurt University of Applied Sciences, Germany. He studied law and geography and researches and teaches in the fields of property and planning law, land valuation, social land policy, and legal geography.

Saija Toivonen is an assistant professor in real estate economics at the Department of Built Environment at Aalto University, Finland, with a wide range of topics concerning the real estate market. She is also the chair of the Finnish Association for Real Estate Valuation.

Alison Tshangana is the head of research and market intelligence at the Centre for Affordable Housing Finance in Africa (CAHF), an NGO based in Johannesburg, South Africa. Her areas of interest include human settlement policies and programmes, inter-governmental finance, and public expenditure analysis.

Karina R. Villacura is a PhD student within the research group Studies in Housing and Welfare at the Department of Urban Studies at Malmö University, Sweden. Her research focuses on the housing mobility of young families with children.

Nikolay Volovich, Doctor of Economics, is a professor at the Financial University under the Government of the Russian Federation. He holds a PhD in valuation of industrial land. His research focuses on valuation of real estate, including real estate located abroad, and land acquisition.

Clive Warren is an adjunct professor in property economics at Bond University, Australia. He has a PhD in strategic property management and is editor of Emerald Publishing's international journal *Property Management*.

PREFACE

The importance of real estate and real estate transactions to the modern economy, both from a private and an institutional perspective, cannot be overrated. Sales of private homes is greatly important for a nation, both in terms of the large values that are transferred and from a perspective of mobility, in terms of enabling moving from one labour market to another. For these transactions to work accordingly, institutions like property rights and cadastral offices need to be in place. As noted later in the book, these can be arranged differently and are to a large extent dependent on the legal tradition within each country.

This book consists of the expert knowledge from 46 researchers covering 27 countries presenting and analyzing the conveyance process from their respective country. Each chapter shows the cross-country variations regarding the conveying process in terms of methods of sale, real estate agency arrangements, and transaction costs. This is examined in 27 national residential markets at varying stages of maturity and within different legal traditions and systems. An important aspect of the book has been the coherent presentation of data and other information complied by the authors. For this, the editors are grateful for the cartographic design and artwork created by Eigo Tateishi.

The theoretical framework of institutions, property rights, and transaction costs frames each of the chapters. This is a novel way to compare the different national conveyance processes, while at the same time giving an understanding of the differences in legal systems and traditions.

The research was carried out from autumn 2019 until the end of 2021. The authors were invited through the editors' international research network, focusing on not only the output itself but also the bridging of different parts of the world and different academic disciplines. Writing this book during the pandemic has been challenging, as we were not able to meet physically for workshops as initially planned. But with timely and efficient Zoom conferences, the project evolved in both time- and cost-effective manners.

Many thanks go out to everyone who contributed, from the authors to those who strengthened the data with expert knowledge and information. Also, to Janet Feenstra, the language editor, who has had to put up with our disparate writing and was always asked to have the editing done by yesterday. Further thanks go to the International Office at Malmö University, the Language Editing Group at Malmö University, the Institute for Urban Research at Malmö University, and CFFF for financial contributions to this book.

Magnus Andersson, Peter Palm, Helena Bohman,
Béatrice Balivet, and Omokolade Akinsomi
November 26, 2021

PART I

1

HOUSING MARKETS IN A GLOBALIZED ECONOMY

Magnus Andersson

Real estate constitutes a major part of individual or household wealth, and it is also an important driver of the economy. This is because every activity in the real estate sector requires space (land) for its setting and operation. The housing sector is important in national economies mainly because the way housing markets function is closely interlinked with many elements of the macroeconomy. This functioning of housing markets can be analyzed through standard models based on neoclassical economics. In these models, housing is approached as a one-dimensional homogeneous product, and the focus lies predominantly on the relationship of housing with the wider urban structure rather than on the operation of housing markets. First, in these models, the price of housing is determined by demand, as the supply of housing responds to changes in demand conditions. Given this, these models do not take into account the dimension that housing, like all real estate, is a heterogeneous commodity with complex and distinctive characteristics. Second, housing is geographically fixed. Therefore, as real estate cannot be moved, it is the households that need to relocate. This means that the availability or absence of housing in a location will influence the location decisions of households and therefore generate mobility. In conjunction with mobility, it is essential to consider not only market outcomes, for example, selling or buying a home, but also the conveyance process on the actual housing market of interest. It is through this process that real estate values are specified, information about the object is generated and made available on the market, legal interests are defined and enforced, uses are allocated, and development and redevelopment are stimulated. Developing a profound understanding of this process necessitates consideration of the institutional characteristics of the housing market. Furthermore, the spatial fixity of housing is why housing development gives rise to major planning debates, as there can be mismatches between the demand and supply of housing in different locations.

Housing markets in a globalized world

The housing market is a series of interconnected, location-specific submarkets on three different scales: international, regional, or urban. Analyses conducted on these spatial scales give rise to a variety of analytical and policy issues, for instance, how the study of international housing markets focuses primarily on the global interaction of housing markets. Many empirical studies

DOI: 10.1201/9781003130383-2

explore the dynamic link between housing prices and vital macroeconomic variables (Adams & Füss, 2010; Aizenman & Jinjarak, 2009; Favilukis, Ludvigson, & Van Nieuwerburgh, 2017).

The study of regional housing markets highlights how the immovability of housing determines the housing market as regional yet provides significant differences in the level of housing market development in different regions. These studies, however, focus mainly on house price diffusion among different cities or regions, for example, Duran and Özdoğan (2020) explored the dynamics of regional housing prices in Turkey by including cross-regional determinants and economic and demographic variables together with geographic and cultural variables. Their findings show that seaside density and cultural activity density are the significant climate and cultural variables seldom seen in the literature. Lean and Smyth (2013) found that, in Malaysia, house prices divert from the most developed states to the less developed states. On a higher geographical scale, Gupta, André, and Gil-Alana (2015) examined the co-movement of house prices in the Euro area using a fractional co-integration approach. They found that co-integration does exist between some European countries, and the Euro area is co-integrated with Belgium, Germany, and France.

However, limited research has been devoted to an understanding of the housing market and to its relationship with the actors and institutions within the market by comparing different national housing markets. In attempting to do so, these complex characteristics of housing, combined with the high transaction costs that generally apply to the property market (legal fees, real estate agent fees, search costs), constitute sources of inefficiency in the functioning of the housing market. Traditional neoclassical space/access models do not consider these complexities. It is important to fill this knowledge gap, particularly as the world has experienced an unprecedented house price boom, not only in terms of magnitude and duration but also of synchronization across countries. Since the 2001 recession in the United States, house prices have been disconnected from the business cycle, with low interest rates and mortgage market innovations driving housing demand and mortgage equity. This development then imploded, with household debt reaching record levels in many economies. In some markets, sustained house price increases led to the belief that the upward trend in the housing market could go on forever. After prices started to fall in in the United States around the end of 2006, the subprime mortgage market collapsed, causing a deep financial crisis (Duca, Muellbauer, & Murphy, 2010). Overheated housing markets were adjusted dramatically, with large price falls. In recent years, the integration of increasingly deregulated international financial markets and general business cycle links have led to the co-movement of house prices across countries globally (Agnello & Schuknecht, 2011). However, strong similarities in house price fluctuations across different countries are not solely a characteristic of the recent house price boom: studies of house price data for the period 1970–1992 from OECD countries show that house price dynamics are interdependent, and there is no evidence supporting the existence of an international house price cycle (Englund & Ioannides, 1997). The degree of international housing market synchronization has been researched, with results showing that even though individual countries may exhibit differences in levels of house price growth, there are significant ties between housing markets of big European countries and the United States (Kim & Renaud, 2009). The effects of the recent housing boom and subsequent bust have been a clear illustration of this interconnectedness, in the context of an increasingly globalized economy and a time of financial liberalization and internationalization. These spillover effects of the housing bust that spread to many countries provided policy makers with strong evidence that prudential and forward-looking regulation had to be developed to reduce the risk in housing markets. Here, the consideration of the institutional characteristics of the housing market becomes pivotal.

A recent study by Devaney, Livingstone, McAllister, and Nanda (2017) analyzes international variations in transaction processes and costs in commercial real estate markets. Globally, transactions involving development land sales account for a large proportion of market activity in commercial real estate markets and generate substantial fee income for real estate brokers. In this context, the results indicate a nuanced and intricate pattern of market conventions rather than any simple binary distinction between a universal, best-practice model and locally contingent models. The authors highlight the growing interest in comparative transaction processes and the operation of brokerage markets but also note few comparative investigations of such processes in an academic context. Focusing on transaction processes in housing markets, this book is situated in an area of growing interest: the transaction process of housing conveyances in a period of international integration.

The aim of this book is to analyze the transaction process and map the transaction costs in 27 countries around the world. Moreover, many of the country chapters include a regional overview of the national market by using the spatial pattern of regional price differences. The book provides an understanding of how similar, yet different, the various countries' housing markets function, in terms of the transaction process. It also provides a comprehensive scheme where transaction costs, on the national level, are comparable between countries. In order to facilitate comparisons for scholars and practitioners of real estate markets, the analyses require a general framework. This book addresses international housing markets by studying national markets and mapping transaction processes for housing conveyances. National markets' unique institutional constraints can serve as examples of why more integration is needed between different housing markets.

What role do institutions have on the transaction process of housing conveyances? And what are the levels of transaction costs in terms of money and time? These are the questions posed at the level of national housing markets. This book adds to the literature on housing markets by focusing on many national markets. Understanding national housing markets can facilitate movements and integration between countries. Also, the national market chapters provide regional views on each country's housing market.

The book is divided into three sections: Part I sets the theoretical context and the background reasons for writing a book focusing on national housing markets, given that commercial property markets are increasingly globalized by investment flows and have gained more attention from a global perspective. Also, housing markets are characterized by international mobility, as labour markets are increasingly becoming international. We address how institutional constraints serve as obstacles for more integration between national housing markets, and as a result, other markets. Part II presents national markets from the perspective of the transaction process. The context involves Europe, North and South America, East Asia and the Pacific, and Africa. The selection of countries is based on a diverse range of geographical, legal, and economic perspectives. The countries are grouped together using legal institutional similarities based on the three main legal systems: civil law, Nordic law and common law.

Each country group is introduced by way of explaining and discussing these institutional similarities. This will also facilitate the international comparison of the respective countries' legal systems. The final section, Part III, discusses how to compare national housing markets and gives some conclusions, with a reflection on the current status of more integrated national markets in today's global arena.

References

Adams, Z., & Füss, R. (2010). Macroeconomic determinants of international housing markets. *Journal of Housing Economics*, *19*(1), 38–50. doi:10.1016/j.jhe.2009.10.005

Agnello, L., & Schuknecht, L. (2011). Booms and busts in housing markets: Determinants and implications. *Journal of Housing Economics*, *20*(3), 171–190. doi:10.1016/j.jhe.2011.04.001

Aizenman, J., & Jinjarak, Y. (2009). Current account patterns and national real estate markets. *Journal of Urban Economics*, *66*(2), 75–89. doi:10.1016/j.jue.2009.05.002

Devaney, S., Livingstone, N., McAllister, P., & Nanda, A. (2017). Institutional convergence in real estate markets: Acomparative study of brokerage models and transaction costs. *Journal of Real Estate Literature*, *25*(1), 169–188. doi:10.1080/10835547.2017.12090446

Duca, J. V., Muellbauer, J., & Murphy, A. (2010). Housing markets and the financial crisis of 2007–2009: Lessons for the future. *Journal of Financial Stability*, *6*(4), 203–217. doi:10.1016/j.jfs.2010.05.002

Duran, H. E., & Özdoğan, H. (2020, July). Asymmetries across regional housing markets in Turkey. *Journal of Economic Asymmetries*, *22*. doi:10.1016/j.jeca.2020.e00178

Englund, P., & Ioannides, Y. M. (1997). House price dynamics: An international empirical perspective. *Journal of Housing Economics*, *6*(2), 119–136. doi:10.1006/jhec.1997.0210

Favilukis, J., Ludvigson, S. C., & Van Nieuwerburgh, S. (2017). The macroeconomic effects of housing wealth, housing finance, and limited risk sharing in general equilibrium. *Journal of Political Economy*, *125*(1), 140–223. doi:10.1086/689606

Gupta, R., André, C., & Gil-Alana, L. (2015). Comovement in Euro area housing prices: A fractional cointegration approach. *Urban Studies*, *52*(16), 3123–3143. doi:10.1177/0042098014555629

Kim, K. H., & Renaud, B. (2009). The global house price boom and its unwinding: An analysis and a commentary. *Housing Studies*, *24*(1), 7–24. doi:10.1080/02673030802550128

Lean, H. H., & Smyth, R. (2013). Regional house prices and the ripple effect in Malaysia. *Urban Studies*, *50*(5), 895–922. doi:10.1177/0042098012459582

2
A THEORETICAL FRAMEWORK FOR HOUSING CONVEYANCES ON NATIONAL MARKETS

Peter Gladoić Håkansson and Peter Palm

1. Introduction

As the property market becomes more global, so too have certain differences between national property markets become apparent. For example, the transaction processes of buying or selling look very different around the world. The transaction costs also differ depending on location, and this in turn influences residential mobility (see among others Caldera Sanchez & Andrews, 2011). Several studies on transaction costs and residential mobility have been conducted. One example is from van Ommeren and Leuvensteijn (2005), who found that a one percentage point increase in transaction costs reduces residential mobility by about 8% in the Netherlands. It is safe to say that all existing housing markets are to some extent regulated by state policies, and these policies, rules, and laws affect the transaction costs of moving. Therefore, it seems well motivated to take a closer look at the institutions that matter in national property markets, particularly those that determine the transaction costs. What are institutions? Why are they important? What do they do? And why do they differ?

2. Institutions

'Old' and 'new' institutional economics

The role of institutions has long been acknowledged within economic theory. Joseph Schumpeter (1883–1950), John Commons (1862–1945), and Thorstein Veblen (1857–1929) were among the first to acknowledge and emphasize the role of institutions. They criticized the neoclassical idea that the market economy is an independent and self-regulating system that strives towards equilibrium (Papageorgiou, Ioannis Katselidis, & Michaelides, 2013). For both Commons and Veblen, property and property rights are of central importance for the market economy to work. For Commons, contracts and property are fundamental, stating that 'hence, ownership becomes the foundation of institutional economics' (Commons, 1934, p. 5). Institutions are also important in Schumpeter's work, as they are responsible for the emergence of certain behaviours thus affecting individual behaviour. However, Schumpeter differs in his claim

DOI: 10.1201/9781003130383-3

that institutions are a necessary but not essential condition for robust capitalism (Schumpeter, [1942], 1992; Swedberg, 2002).

This – what we may call 'Old' Institutional Economics – in many ways, was forgotten in the 1960s and 1970s when neoclassical economic theory triumphed. However, from the mid-1980s, institutions once again became interesting to economists, and today, the institutional idea has merged with neoclassical perspectives and tools (Williamson, 2000). It is this 'merge' that distinguishes the 'new' from the 'old'.

So, what does this mean? In New Institutional Economics, institutional conditions are essential for a market economy to flourish (the role of institutions), but traditional economic factors, like incentives and changes in relative prices, also change institutions (the development of institutions). The role of institutions is to reduce uncertainty; therefore, for a market economy to flourish, it needs safe and secure property rights and stable rules of the game. Without secure property rights, a private investor will be unwilling to invest in real capital or property. For instance, if there is a risk that the property will be confiscated or heavily taxed, the incentive of the investment is reduced proportionally to the increased risk of confiscation. Therefore, uncertainty is a risk that must be considered when making a long-term investment.

Why do institutions seem to be effective, just, and secure in some countries, while the opposite in others? And how can institutions be reformed? These seem to be the Nobel Prize questions above all. The difficulties of reforming institutions are interlinked with how institutions evolve. As we will discuss in the coming chapters, institutions are intertwined with our belief systems, culture, and history and have developed over a long time, but institutions do change and can be reformed. We will come back to property rights and institutional change but start off by discussing how to define institutions.

The definition of institutions

It is safe to say that today there is a joint view among most institutionalists that institutions can be divided into those that are formal (laws, rules, etc.) and those that are informal (norms, religion, etc.). The institutional framework of formal and informal institutions lies imbedded in a human environment that consists of language, physical artefacts, and beliefs, and as mentioned, they reduce uncertainty. North puts it the following way:

> The structure we impose on our lives to reduce uncertainty is an accumulation of prescriptions and proscriptions together with the artefacts that have evolved as a part of this accumulation. The result is a complex mix of formal and informal constraints. These constraints are imbedded in language, physical artefacts, and beliefs that together define patterns of human interaction.
>
> (North, 2005, p. 1)

Institutions are thus defined as constraints, and both formal and informal institutions constrain our actions. By doing so, uncertainty is reduced. For example, we know that the buyer is going to pay because they signed the contract. The courts and legal system will act if the buyer does not pay, and the buyer is aware of this. The existence of a legal system makes both the buyer and the seller act trustworthy. A trustworthy legal system constrains the individual and reduces the number of alternatives, which, again, reduces uncertainty. In this way, formal (a just legal system) and informal (trust) institutions are interconnected.

Today, institutionalists often make a discrepancy between institutions and organizations (see e.g. North, 1990, 2005; Williamson, 2000), and the phrase, 'institutions are the rules of the game, while organizations are the players of the game' reflects this. North pointed out that institutions involve the constraints of human behaviour, while organizations are 'groups of individuals bound together by some common objectives' (North, 2005, p. 60). Trust (or reliability) is obviously a constraint. With this norm in society, defection in an agreement will be a less likely alternative – it is simply something you do not do. It can be considered a constraint for the actor who may gain from defection, but, of course, for the other party, it is a constraint that reduces uncertainty. For society in general, this norm lowers transaction costs and stimulates economic activity, which leads us to the next section on property rights and transaction costs.

3. Property rights and the mystery of capital

What is the difference between an asset and capital? What makes an asset capital? As Hernando de Soto (2000) points out, central to classical economists such as Adam Smith and Karl Marx is that capital gives a return, or rather, *it has the potential to give a return*. Therefore, capital is not money – far from it. Money may facilitate transactions that allow us to buy and sell things (e.g. capital), and capital is most often valued in money. For an asset to become capital, it needs a legal system that guarantees that the owner is (1) allowed to keep it, (2) allowed to keep the returns from it, and (3) allowed to sell it. This is why property rights are crucial. The reasons for this are quite straightforward: it is difficult for an asset to be productive if it does not belong to something or someone (de Soto, 2000).

The problems with weak property rights have been demonstrated many times (see e.g. North, 1990). If a society experiences weak property rights, meaning that the owner of the property may lose it in the future, the value of the property will be low. No one wants to buy something they cannot keep. Real estate has a long economic lifespan, and the value of the real estate is calculated from this remaining lifespan. Consequently, if there is a risk of losing the property, the value will decrease.

de Soto (2000) calls property without secure property rights 'dead capital'. It is an asset, but it does not have the property rights that embed it, and by this, it cannot give a return – it cannot be used as productive capital. It cannot be sold, as no one wants to buy something they do not know if they can keep. It cannot be mortgaged, as the bank cannot use it as security. It will not be used productively (e.g. for letting or production) because no one will put the effort into something when they do not know if they will be able to keep the returns on it.

According to de Soto, what is needed is not only property law but also a property system. The law itself is not enough; therefore, a complete system that can also enforce the law is needed. A property system is not simply one law: it is thousands of laws, rules, systems, and organizations that work jointly. The property system secures property rights and increases transparency, and as a result, trust is strengthened. When the system works and there is transparency so that everyone can see if someone defects, trust is easier to obtain.

Transparency may be obtained by a transparent and open cadastral, but this is not enough for a property system. Property rights needs to be guaranteed by an effective legal system, and this legal system needs to be able to enforce the law. In addition, financial markets must acknowledge mortgages as a financial tool, and city planning must be predictable and also consider property rights. With a property system, the house not only is a used good, a home to live in, but also becomes an economic concept, as in, an investment with economic potential that could be used as a financial asset.

Further, the property system integrates disparate information about the property into one transparent system. In an informal system, there is information, but it may be knowledge kept by specific persons (renovators, previous owners, extension builders, etc.). To get hold of this information may take time and effort, which is an example of a transaction cost. By putting this information into one transparent system, transaction costs decrease.

By transforming people's homes into economic concepts, people become economic actors responsible for capital with potential. In an informal system, people are dependent on neighbours and local agreements to protect the property and the property right. Who owns the property? If it is not written in a contract that is secured by property law, but only acknowledged by custom and tradition, the individual becomes dependent on the village, the tribe, or some other collective that can guarantee the ownership. The property system makes people responsible individuals, independent of neighbours to protect the property. Engaged and responsible individuals become more engaged in the property when they know they can keep it and capitalize from the investment; as a result, the incentive to invest in the property is higher.

4. Institutional reform: can we change institutions?

Institutions derive from historical heritage and are encapsulated and coded with information of past experiences and therefore are historically context specific. According to North (1990, 2005), norms derive from human beliefs and belief systems. The beliefs, in turn, derive from what is normally called 'culture'. Culture can be defined as 'the intergenerational transfer of norms, values and ideas' (North, 2005, p. 50). This means that over many generations, a population will learn more than a single individual will ever have had the possibility to do in their lifetime. Earlier generations' learning is transmitted to present generations as culture through belief systems. A prerequisite for building cultures is human genetics, or more specifically, the nature of the human consciousness and cognition. It was this cooperation, or human interaction, that created culture and the wide variety of cultures in the world. Consequently, it is not possible to reform the institutional framework overnight because it is embedded in our belief systems. Beliefs are passed down from parents to child, and institutions look the way they do as a result of historic traditions.

Is it then totally impossible to reform institutions? Is it all deterministic? Not really, but we have to be accurate when defining which factors are important. North uses both endogenous variables and exogenous variables to explain institutional change. Using endogenous variables, the model becomes evolutionary (see Figure 2.1).

As pointed out earlier, institutions are the rules of the game and the organizations are the players. In this model, the organizational framework is set up within the institutional framework. The opportunities provided by the institutional framework determine what kind of organizations will emerge. The organizations, on the other hand, will try to alter or change the rules of the game, as in, the institutions that benefit them the most. The ongoing interaction between the organizations and the institutional framework shapes institutional change. However, both exogenous as well as endogenous variables can be the driving force behind institutional change.

In Figure 2.1, North's idea of institutional change has been interpreted with a descriptive model. An endogenous variable could be competition between the organizations, and competition leads to the actors needing more knowledge because new and more knowledge may be a way to gain a comparative advantage. This drive for new knowledge among the actors leads to not only innovation but also a change in belief systems. Our beliefs and belief systems are, among other characteristics, a function of our perceptions, and our perceptions are a function of our cultural heritage. The potential profits depend in turn on the institutional framework, both informal and formal, but potential profits also affect people's will to gain new knowledge. As North points out, if piracy is seen as potentially profitable, then people will educate themselves

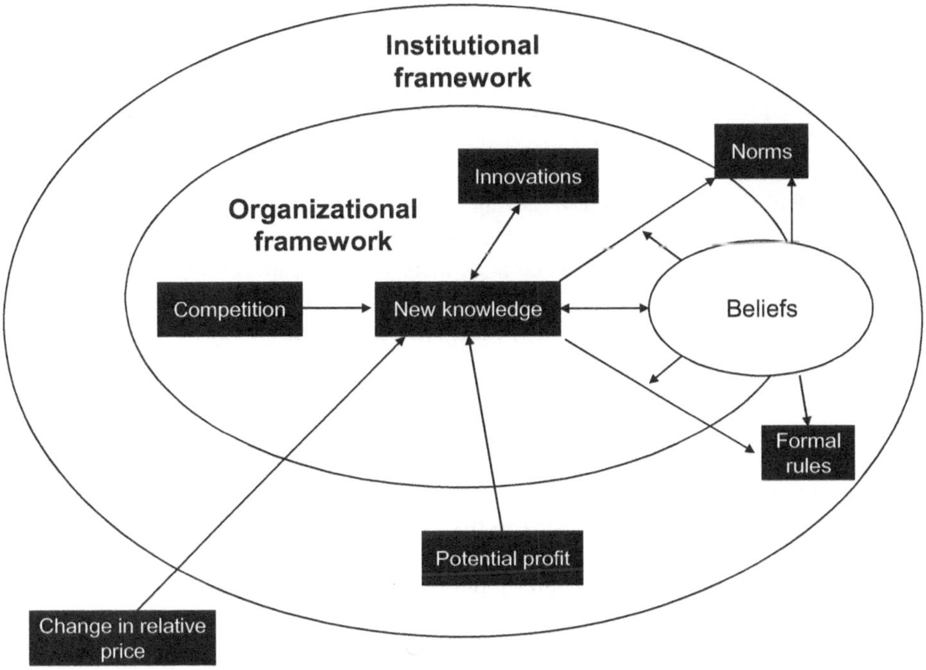

Figure 2.1 A descriptive model of institutional change

Source: Author

to become pirates. Given, of course, that norms and laws allow it (or at least that the risk of being punished, formally and informally, does not outweigh the potential benefits).

An exogenous variable in the model is a change in relative prices. However, changing relative prices can also be an endogenous variable (as shown earlier). An exogenous change in relative prices could come from, for example, demographic change or innovation. A change in relative prices changes the relations between the production factors and thus becomes an incentive for the actors to gain more and new knowledge.

To conclude, institutional change is a dynamic process, which begins as inherited culture, knowledge, and beliefs. The institutions are also inherited and rest on a foundation of historical tradition. The process of change also depends on historically determined trajectories. The evolutionary processes depend on a number of endogenous variables that develop within the model. Exogenous variables, like changes in relative prices, war, and oil crises, influence the model and change the institutions through knowledge accumulation.

5. Transaction costs

Transaction cost theory highlights how exchange is accompanied by several different costs. Costs vary depending on the type of transaction, which parties are involved, and any potential regulations. However, there is no consensus on how transaction costs should be defined, for example, Coase (1937) defines it as the cost of using the price mechanism; Demsetz (1968) as the cost associated with the transfer, capture, and protection of rights; and North (1990) as the cost of measuring the value of that which is being exchanged as well as the cost of monitoring the agreements.

The theoretical body of transaction costs takes its starting point in price as the coordination mechanism for transactions, where the housing market is no exception (Ball, Lizieri, & MacGregor, 1998). The theory gives that seller and buyer both meet on the market and that the price will coordinate their preferences. However, the fixed location of housing, as outlined in Chapter 3, will disrupt this theoretical starting point. Also, emotions may come into play when houses are traded, thus disrupting the model (Evans, 1983).

In relation to exchanges on the market, Benham and Benham (2000) define it as all costs associated with a transfer, capture, and protection of rights. This can be reformulated as all costs that can occur when a task is performed. Furthermore, North (1990) categorizes transaction costs as 'market costs', such as legal fees and taxes, and 'non-market costs', such as time. The latter type of transaction cost is higher when information is not easily accessible or activities must comply with rules and regulations (also called compliance costs). An example of this type of transaction cost includes the cost of time spent obtaining permits and so forth (North, 1990). Costs that go through the market are (often) measurable. However, information costs and time costs are hard to measure, and it is these costs that complicate the assessment of the total transaction cost (see North, 1990).

The nature of the analysis is that you should strive for lower transaction costs. This is converted into how the market exchange will try to minimize transaction costs (Williamson, 1975). However, there is also a societal need for institutions that monitor markets and safeguard the interest of all parties (see e.g. 'property rights' earlier in this chapter). A market without such institutions would have lower transaction costs but at the expense of security and may result in that property becoming 'dead capital'. Therefore, a lack of security is also a transaction cost. We merely transform a market transaction cost (measurable) to a non-market transaction cost. However, the market transaction cost is easier to observe and measure and thereby we might perceive the transaction costs as lower.

As mentioned, an effective and transparent property system with a transparent market and strong property rights reduces transaction costs. In a transparent property system, the information cost is low, because information about the property is easier to obtain. Although there may be information about a property only a few persons will know about (and it is impossible to create systems that reveal *all* information), major information (owners, major renovations, taxation, mortgages, etc.) will be included in a transparent system. Further, with secure property rights, the risk of misappropriation is less, and this also lowers transaction costs. Therefore, a property system simplifies the transaction, and by this, transaction costs drop.

6. Conclusions

In conclusion, institutions involving both formal and informal activities surround the property market. However, the formal aspect of institutions is particularly important for property conveyances, as homes to a large extent have been transformed into economic concepts. Without written contracts, property law, and strong property rights with a land register to protect the individual's ownership, houses become dead capital. This means, among many things, that it will be difficult to borrow money to buy a house, and in the long term, less mobility on the housing market and labour markets. Basically, people stay where they have a home because if a market for homes did not exist, it would not be possible to find a home elsewhere.

Property rights and formal institutions come with transaction costs in terms of rigorous processes and taxes as well as different fees. These transaction costs can be argued as necessary as they do not add any particular obstructs for the market and conveyance process. The absence of property rights and formal institutions would add transaction costs of a different kind, which would be much more harmful for the market and conveyance process. Therefore, transaction costs and institutions can be viewed as a double-edged sword. On one hand, high transaction costs affect the market, leading to fewer transactions, and by this, less mobility. On the other hand, it costs money to uphold an institutional framework (and the organizations that guarantee the institutional framework). Without effective institutions that can guarantee property rights, property will be dead capital, which limits the property market.

We can always question to what extent a formal procedure or level of tax is efficient or could be organized in a different way to lower the transaction costs. However, in the end, it comes down to the efficiency of the institutional framework, and most importantly, the efficiency of governance.

References

Ball, M., Lizieri, C., & MacGregor, B. D. (1998). *The economics of commercial property markets*. London: Routledge.

Benham, A., & Benham, L. (2000). Measuring the cost of exchange. In C. Menard (Ed.), *Institutions, contracts and organizations* (pp. 367–375). Cheltenham: Edward Elgar.

Caldera Sanchez, A., & Andrews, D. (2011). *To move or not to move: What drives residential mobility rates in the OECD?* OECD Economics Department, Working Paper 846. Paris: OECD.

Coase, R. H. (1937). The nature of the firm. *Economica*, *4*(3), 386–405.

Commons, J. R. (1934). *Institutional economics: Its place in political economy*. New York: Macmillan.

Demsetz, H. (1968). The cost of transacting. *The Quarterly Journal of Economics*, *82*(1), 33–53.

de Soto, H. (2000). *The mystery of capital: Why capitalism triumphs in the West and fails everywhere else*. New York: Basic Books.

Evans, A. W. (1983). The determination of the price of land. *Urban Studies*, *20*, 1219–1229.

North, D. C. (1990). *Institutions, institutional change and economic performance*. Cambridge: Press Syndicatee of the University of Cambridge.

North, D. C. (2005). *Understanding the process of economic change*. Princeton and Oxford: Princeton University Press.

Ommeren, J. V., & Leuvensteijn, M. V. (2005). New evidence of the effect of transaction costs on residential mobility. *Journal of Regional Science*, *45*(45), 681–702.

Papageorgiou, T., Ioannis Katselidis, I., & Michaelides, P. G. (2013). Schumpeter, commons, and veblen on institutions. *The American Journal of Economics and Sociology*, *72*(5), 1232.

Schumpeter, J. A. ([1942] 1992). *Capitalism, socialism and democracy*. London: Routledge.

Swedberg, R. (2002). The economic sociology of capitalism: Weber and Schumpeter. *Journal of Classical Sociology*, *2*(3), 227–255.

Williamson, O. E. (1975). *Markets and hierarchies: Analysis and antitrust implications: A study of internal organisation*. New York: The Free Press.

Williamson, O. E. (2000). The new institutional economics: Taking stock, looking ahead. *Journal of Economic Literature*, 595–613.

3

A CONCEPTUAL MODEL OF CONVEYANCES ON THE HOUSING MARKET

Peter Palm, Béatrice Balivet, Ola Jingryd and Julian Sidoli

When studying real estate markets, many make assumptions – explicit or implicit – of a competitive market. However, we should mention that there are several differences. The first important difference between real estate markets and the perfect competition theory is that the real estate market can be divided into a number of sectors: office, retail, industrial, and residential. The residential sector can then be divided into several subsectors, for example, those defined by tenure rights or those that are privately owned or rented. Also, these two subsectors dependent on tenure right can be divided into further subsectors, as the privately owned market is often divided between real estate (as small houses) and apartments (multi-owner properties). These, in turn, can be further divided into subsectors, where small plots of real estate land upon which the houses are situated can be owned (freehold) or leased (leasehold). All these subsectors within residential real estate solely refer to differences in tenure rights, and this calls for a different approach. To include institutional prerequisites is therefore needed to enable a thorough analysis of the residential real estate market.

The fact that the residential market consists of different subsections is important, as we are not studying a well-defined market but rather a series of subsectors with different characteristics that are nevertheless interrelated. Furthermore, within these subsectors, the individual residences will not be identical, given that the residential market consists of houses built in different times with different construction methods, sizes, and so on. Even two residential houses built during the same time period using the same design and construction will always be at a different location. And these differences in location, even within a street, can be important, as this could affect the view, garden position, sunlight, and so on. Therefore, we conclude that the residential market's heterogeneity is defined by:

- legal/regulatory differences
- physical differences
- locational differences

This chapter outlines how the specific characteristics of the real estate market set the overall rules governing the market. Certain prerequisites are often what make the national housing market national, with some added, rather unique, characteristics. This chapter concludes with a conceptual model of how the conveyance process has different stages. This conceptual model can then be used as foundation for comparative studies of national conveyances.

 DOI: 10.1201/9781003130383-4

The remainder of the chapter is structured as follows: First, a review of different legal systems will be presented. Second, a short exposition of the specific characteristics of the real estate market will be given. Third, and last, the conceptual model of conveyances on the housing market will be outlined and presented.

1. Three legal families

Ubi societas ibi jus. Every society is organized according to specific rules that each member must respect. Each law is in fact a system, and as such, there is a multiplicity of legal systems, one for each given society at any given moment. Without denying the relativity and multiplicity of systems, in terms of the content of the rules between States and within each State, certain consistent features can be noted when observing, not the content of the legal rules, but rather the techniques used to draw them up, state them, classify them, or interpret them. From these structuring elements underlying the legal rules, families of law are deduced. The two main ones are the Romano-Germanic family and common law, as well as a group of so-called mixed systems.

Civil law

The civil law family, which prevails in continental Europe and has proliferated in other parts of the world, particularly Latin America, has two main branches: the Napoleonic (or Romanistic) and the Germanic. Where Napoleonic law originates in France and the *Code Civil* (or *Code Napoléon*) of 1804, Germanic law originates in Germany, and its predecessor was the Holy Roman Empire. The great Code of Germanic law is the *Bürgerliches Gesetzbuch* (BGB) of 1896, which was preceded by the Bavarian civil code of 1756 and the General Land Law for the Prussian States of 1794. While both branches of civil law rest on a heritage of Roman, Canon, commercial, and customary law, the reception of Roman law took place in different times historically. The *Code Napoléon* and the BGB also differ in many ways; for instance, where the former merges the old *droit écrit* of Southern France and the *Coutume de Paris*, with the ideas of the Enlightenment, the latter is an intricate work of jurisprudence and a systematization of late Roman law (Zweigert & Kötz, 1998, pp. 74–179; Gordley & von Mehren, 2006, pp. 18–64; Bogdan, 2003, pp. 147–172).

A general trait of civil law is the focus on statutory law over legal precedents. Another common trait of the civil law family is the prominence of the public notary, whose intervention is mandatory in several types of transactions, including real estate conveyances.

Within this family, there are differences. A distinction can be made between:

- Napoleonic (Romanistic) civil law countries, with influence from the Napoleonic Code, even if this code, adopted in 1804, has lost its attractiveness (France; Spain; the former French colonies, particularly in South and Central America, with Brazil and Chile; and French-speaking Africa, with Benin, Morocco, and Egypt)
- Germanic civil law countries influenced by the German Civil Code of 1900: the BGB *Bürgerliches Gesetzbuch* (Germany, Croatia, and some far Eastern countries such as Japan)

Nordic law

The Nordic family comprises Denmark, Iceland, Norway, Sweden, and Finland. They are historically closely related and interconnected since before the Kingdoms of Denmark, Norway, and Sweden were even founded. Scandinavian law was originally an offshoot of Germanic law and remained largely so for many centuries, barring elements of Canon law in some areas of the law. Pan-European ideas such as the Enlightenment and its effect on most areas of the law can be seen in the legal development from the eighteenth century onwards. At the end of the nineteenth century, the Scandinavian countries cooperated closely with respect to the law of obligations, resulting *inter alia* in national statutes upon the entering of contracts that is virtually identical and still in use today (Zweigert & Kötz, 1998, pp. 276–285; Ekeberg, 1939).

On a methodological level, Nordic law is mainly focused on statutory law, although judicial precedents from courts of higher instance are a central source of laws. Of particular note is the prominent role of *travaux préparatoires*.

The Nordic countries do not have civil law notaries, and therefore, no such intervention exists in real estate conveyances. While the exact rules on how to accomplish legally binding conveyances differ between the Nordic countries, simplicity and pragmatism are common traits of the respective legal regimes.

Common law

Common law is the dominant legal system in the English-speaking world and in many former colonies of the British Empire. It also features in a number of countries with hybrid legal systems, including Scotland, South Africa, and India. Its central characteristic feature is the fact that decisions are made by judges using the system of judicial precedents that bind lower courts in matters where the facts align. This is called the *doctrine of stare decisis* (lit. 'let the decision stand').

In some respects, the common law approach can be said to be inductive, working up from the factual matrix to 'discover' the law rather than working deductively downwards from a codified body of law. Only the very senior courts of each jurisdiction bind other courts. Also, judgements are typically far longer and more discursive than in other legal systems, and in the senior appellate courts, judges can give separate judgements that sometimes do not fall in line with the majority view. Other notable features of the common law system include the fact that judges are invariably drawn from senior lawyers and the adversarial system, where barristers present their case to a judge who does not usually take an inquisitorial role.

Mixed systems

Beyond this classical presentation of the three main legal families, the reality is often more complex, with several legal systems applied simultaneously within the same country. Mixed systems may cover different realities. It may also be a concurrent application of:

- several civil law systems: Napoleonic and Germanic, as in Poland, etc.
- the civil law system and common law system, as in South Africa, Thailand, etc.
- the civil law system and the religious law system, as in UAE, etc.
- the common law system and the religious system, as in Malaysia, Nigeria, etc.

Also important to remember is that many countries can have differences in how the legal family is influenced, depending on which part of the legal regulatory is considered. For example, in many countries, family law can be heavily influenced by the religious law system, but in other aspects, like that of real estate law, the civil or common law system dominates. All countries participating to this study are shown in Figure 3.1.

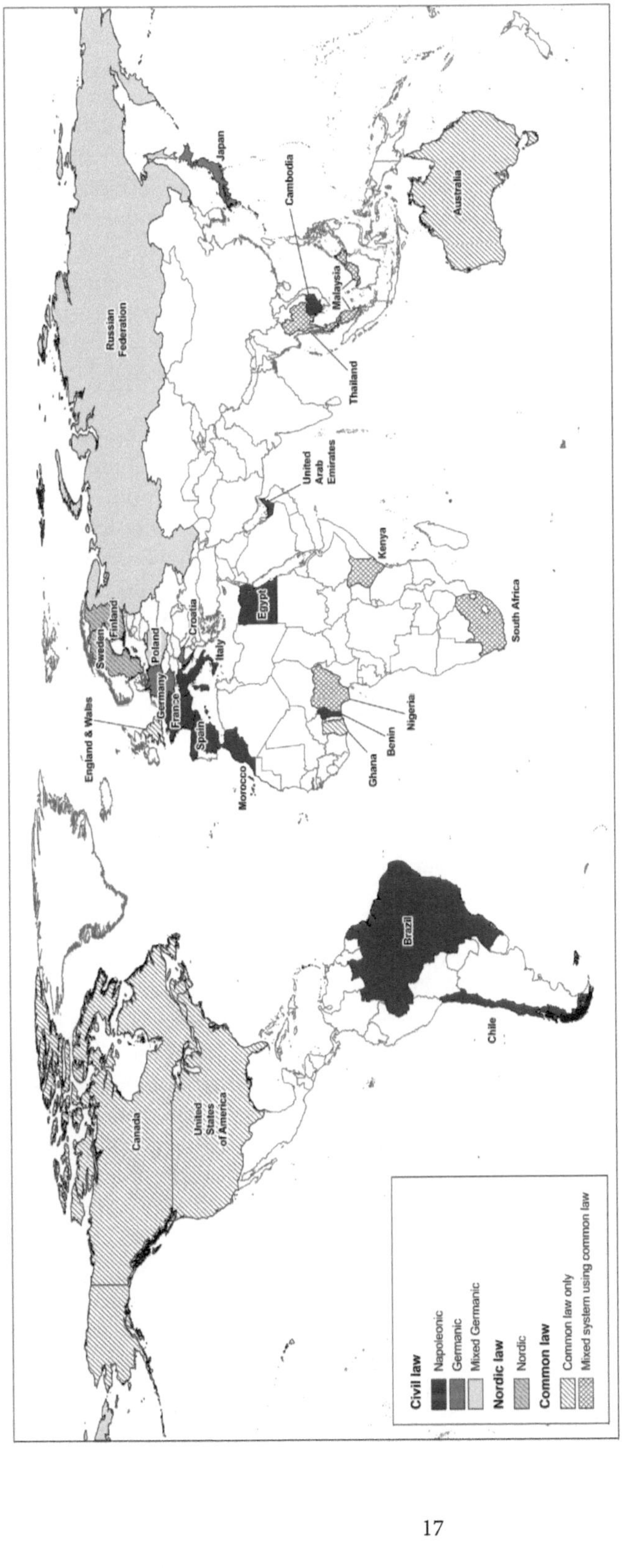

Figure 3.1 World map showing the participating countries

Source: Author

2. The real estate market and its characteristics

The real estate market has certain characteristics that distinguish it from other markets. These different characteristics, in turn, have implications on how the real estate market functions.

Heterogeneity: All real estate is heterogeneous. Two parcels cannot be identical, only similar to each other in terms of building style, building year, and size. Regardless of how similar to each other, they will at the very least differ in terms of location – the plots on which they stand. Each unit is thus unique, and therefore, two units cannot be considered exact substitutes for each other.

Fixed location: Real estate is immobile, with each property fixed in location. This distinguishes real estate from other goods. And unlike other commodities, it cannot be moved to locations of high demand or from locations with low demand. In turn, this has implications for the efficiency on the market and how individual properties may be used. Furthermore, the fixed location also implies a great sensitivity to changes on the local market and sensitivity for externalities that will play a significant role on the price of the property.

Large values: Real estate is traded at large values. In relation to average incomes, the prices of housing are relatively high. Therefore, the majority of purchases depend on the credit market for the purchase of a property. Therefore, the cost and availability of credit directly impacts on the demand of property. But cost and the availability of credit also have an impact on the supply side. And given that new construction is also dependent on the availability of credit, the pricing of new construction is directly influenced by the cost of credit.

Durability: Real estate is more durable than most other commodity types. We can see that most real estate is old, and new properties are developed to last for many years to come. This also implies constraints on the market, from making marginal changes in the number of properties in relation to demanded and supplied in response to marginal price changes, which, in turn, will affect the market efficiency.

Government intervention: The fact that the real estate market consists of durable goods, at fixed locations and involving large values, means government intervention is inevitable. For example, real estate is an easy target for taxes and fees, as you cannot move your property. Furthermore, government intervention is necessitated by political interest, not least in terms of the question of housing. This intervention may be direct or indirect, and directed towards either the demand or supply side or both. It can take the form of rent control for housing or the subsidy of certain buildings. Another governmental instrument on the real estate market is planning control. This directly determines the amount of land available, the use of available land, and the use of buildings. Overall, government intervention in real estate markets influences the market efficiency.

Inelasticity of supply and demand: New development of real estate takes time, and the supply can, therefore, in the short term, be characterized as fixed. Furthermore, regarding supply, there are constraints to how the market can react to price changes. Due to the legal procedures, government intervention, and the relatively long construction period required, there will be a lagged response to the market demand. On the other hand, the demand for housing is not very responsive to price changes, given that housing as such is a necessity for all and no substitutes exist. The implication of this inelasticity of supply and demand of housing is that the market does not efficiently adjust to prices in the same way an efficient market would.

Transaction costs: In addition to the relatively high price of housing, participants on the real estate market incur relatively high transaction costs. These costs may be in the form of money or time. Compared to other commodities, housing is related to the high cost of transfer. In addition to the price of the house, various statutory fees and taxes, fees to professionals such as real estate agents, notaries, and inspectors will arise. Also, it is worth considering that the time taken to complete a given transaction may take months.

All in all, these specific characteristics of the real estate market add to obstacles on the market and affect the efficiency of the real estate market's function.

3. A conceptual model of conveyances

Market activity can be seen on a timeline – from when the seller decides to put their home up for sale or the future buyer decides to look for a new home to when the new owner, the buyer, takes over control of the home, in terms of the legal, economic, and practical matters. Although one can argue that every transaction is unique and has its specific prerequisites and features, we will try to normalize the process, including the different steps that are 'normally' included, and categorize them according to when on the timeline they 'normally' take place.

This normalized conveyance process is divided into four phases by Ferlan, Sumrada, and Mattsson (2007): marketing activities, pre-contract, contract, and registration. In this book, we have developed this model to include a fifth phase – the post-transaction phase – because not all transactions and the associated costs end with the registration of the deed. These five phases include different activities, depending on not only the national context but also the complexity of the transaction, and basically, how the seller, and sometimes the buyer, chooses to go through the conveyance. In this book, we present some national markets that do not feature all of the stages. These are often dependent on the legal system and how it is put in practice in the respective country; for example, there is no pre-contract phase in Sweden.

Marketing phase

This phase involves all the activities associated with the sellers' marketing and the buyers' searching. Marketing activities range from featuring the property on different channels, such as online platforms and newspapers, to negotiating the price of the property. It is also here that the estate agent often comes into play. In different countries and different legal contexts, the estate agents have different roles, but in most contexts, these are the activities that we often associate with the role of the estate agent.

There are several transaction costs associated with this phase. Both searching and matching costs arise in this phase, and not always solely between the seller and the buyer but also between, for example, the seller and the estate agent, in terms of uncertainties in relation to both the agency contract and observing and monitoring the agent's work.

During this phase, there can also be some fees and taxes associated with the process. However, these fees are most often more or less voluntary in terms of not being regulated in statutes, for example, the estate agent's fee or the fee for marketing the property on an online market platform or in the newspaper. These fees are voluntary, but if you want your property to be sold, you are more or less obliged to pay them.

Pre-contract phase

In some legal systems, a preliminary contract of sale is used as a means of binding the seller and buyer. It often clarifies the terms of the contract, including the conditions for a loan. Although this phase will add to the transaction costs, in terms of time and the need to contract legal competence, it will also reduce other transaction costs, such as uncertainty. For example, it is possible to avoid a situation where one party incurs expenses when the other party withdraws from the proceedings before they are completed.

Contract phase

This phase marks the finalization of the binding agreement. However, we must consider that the phase ranges from the time of the contract signing to when the property is taken in possession. The contract of sale is the final purchase document. The buyer and seller agree upon all conditions and sign the final sales contract. This is also when all binding commitments and conditions of the sale, such as financing, are agreed upon.

In this phase, mortgage activities are often seen as a part, but in some legal contexts, mortgages are treated separately. This could, for example, already have been dealt with in the pre-contract phase.

The common point is that, with the signing of the transaction, this phase marks the transaction as finally confirmed. Different legal systems have different prerequisites to confirm it as legally valid. In most contexts, it is when the purchase amount is handed over. This can also vary between countries and legal systems. One can imagine a down payment to be handed over in connection to the signing of the contract and the rest of the purchase sum to be paid when the property is taken in possession.

In this phase, different transaction costs come into play. The phase can involve different parties, both private advisers and officials, depending on both the cultural traditions and legal systems. This might both add to complexity and/or reduce risks.

Normally, this phase will also be associated with different fees and taxes. Also, it is here that we will see fees from both private and official organizations. For example, in one legal context, it is the bank that will charge the buyer for taking up new mortgages, but in another legal system, it is paid to the state through its land registry.

Registration phase

This phase comprises the activities related to the registration of the change of ownership in public land registry and/or other public registries. This registration in public registries serves two purposes: transparency and protection against third parties. After this registration, 'everyone' will know who the new owner is and sometimes also the purchase sum. At the same time, this will protect the new owner from third parties, as no one else can claim possession of the property. This phase will most often contain different fees and taxes from officials.

Post-transaction phase

After taking the property in possession and the purchase has been registered, the transaction can still be associated with costs. This phase will compile all the costs that are current. One example is when a profit from the sale is taxed. In Sweden, this happens the year after the sale of the property.

References

Bogdan, M. (2003). *Komparativ rättskunskap*. Stockholm: Norstedts Juridik.

Ekeberg, B. (1939). Det nordiska samarbetet på lagstiftningens område. *SvJT*, 5–13.

Gordley, J., & von Mehren, A. T. (2006). *An introduction to the comparative study of private law: Readings cases, materials*. Cambridge: Cambridge University Press.

Ferlan, M., Sumrada, R., & Mattsson, H. (2007). Modelling property transactions. In J. Zevenbergen, A. Frank, & E. Subkjær (Eds.), *Real property transactions procedures, transaction costs and models*. Amsterdam: IOS Press.

Zweigert, K., & Kötz, H. (1998). *An introduction to comparative law* (3rd ed.). Oxford: Oxford University Press.

PART II

Civil law countries

Russian Federation
Poland
Germany
France
Croatia
Spain
Italy
Morocco
Egypt
Japan
Cambodia
United Arab Emirates
Benin
Brazil
Chile
Civil law
Napoleonic
Germanic
Mixed Germanic

4
BENIN

Joseph Djogbenou

1. The Benin housing market

The real estate sector in Benin[1] is governed by three main laws according to the mode of occupancy: property and condominiums are governed by the land and property code, rentals are governed by the law on leasing, and rentals are further governed by either the uniform act on general commercial law if it is a professional activity or by the law on leasing for residential use.

We were not able to access the official statistics on the volume of transactions. In any case, these would be incomplete because of the informality of many of them despite the government's efforts to make all transactions official and to facilitate, if necessary, their execution. However, it can be safely said that transactions continue to increase with urbanization, the development of economic activity, and the realization of major public projects.

The official figures that are available do not reflect the reality of the cost of real estate transactions. However, they do show certain proportions. As a reference, in regard to minimum purchase prices, the acquisition values of buildings within the framework of the real estate capital gain tax indicate a variation in decreasing order of € 457 per square metre for the most expensive areas at € 0.01 per square metre in the most rural areas.

2. Regulatory framework

The land and federal code governs the transfer of real estate. If the property is acquired and transferred by succession, donation, purchase, will, exchange, accession, incorporation, or prescription, it is (along with any other legal effects of the bonds) mainly through the purchase that the transfers of ownership are carried out.

The sale is a synallagmatic contract, by which one party agrees to sell the building and the other to acquire it in return for payment of an agreed price. The law does not interfere in the setting of prices between the seller and the purchaser. However, the administration levies a real estate capital gain tax for each transaction between individuals. The setting of this tax is instituted by an order that references the acquisition values of per square metre for the specific geographical area.

DOI: 10.1201/9781003130383-6

The form of the convention is essential. When the building is registered, any deed of transfer of ownership must be overseen by a notarial deed or by a private deed filed in the rank of minutes of a notary. However, before the effective uniformity of the land regime, a certain duality remains. As a result, customary tenure lands are still subject to surrender without the use of a notary. However, the registration of the agreement in writing with the tax and land administration departments is required (Law No. 2013–1 of 2013). In some cases, the use of the notary is mandatory and exclusive. This is the case with real estate donation. This duality will end on 14 August 2023, when it will no longer be possible to make a transaction on a building located in the Republic of Benin without recourse to the office of a notary.

Leasing is a building operation that ultimately allows the tenant to acquire all or part of the leased property at an agreed price, taking into account at least part of the rent payments. This type of leasing may be for existing or future buildings but is nevertheless exclusively for the professional activity.

It is a synallagmatic contract that must be in written form. It is required to contain a description of the property and the characteristics that allow it to be identified, the lessee and the tenant the duration of the lease, the amount of the lease, the terms and conditions of settlement, and the terms of the purchase option. The other terms of the lease agreement are regulated by law, but the parties may modify or deviate from them through an explicit written agreement.

The residential lease is a lease of property by one of the parties, the lessor, for a price that the other party, the lessee, is obliged to pay. The former provides the lease to the latter for a specified period, determined in advance or not, for the use of a building for domestic purposes.

The residential lease agreement must be written and signed by both parties. It can be a standard form provided by the administration, which may be drawn up privately or by notarial deed. The duration of the length of the lease is determined at this time, but freely set by the parties. The rights and obligations of the parties are stipulated by law but can be supplemented by either party in additional clauses. In particular, large repairs are the responsibility of the lessor when they are urgent and necessary; however, the tenant can sometimes shoulder this responsibility, and if necessary, be authorized by the judge to carry out the repairs on behalf of the lessor. The taker executes the contract as a trustee by taking care of the premises and paying the rent according to the agreed terms.

Rental rates are also set by law. The maximum annual rent is set at 8% of the actual value of the apartment or rented building. The monthly rent cannot therefore exceed one-twelfth of this amount. A surety is generally required by the lessor when the lease is finalized, but it may not exceed three months' rent. This amount is refundable at the end of the lease.

The rent is reviewable according to the agreements of the parties in their lease agreement, or failing that, at the time of renewal. The increase resulting from a revision clause may not exceed 2% of the initial annual rent amount.

In finalizing the various real estate contracts, the use of a real estate agent is not mandatory. However, it is common for the parties to use the services of this professional. The real estate agent profession is regulated, although the regulations are not always enforced. Registration in the real estate agent register maintained by the Minister of Housing is a requisite to practice the profession. The real estate agent acts on the basis of a negotiated mandate of agreement parties in accordance with common law. This mandate specifies the conditions under which the real estate agent is entitled to receive funds and be held accountable as well as the conditions of the real estate agent's remuneration, with the specific naming of the person who bears this burden. The mandate is limited in time, on pain of nullification; however, the parties involved can renew it in writing at the end of its initial term. The commission is freely set between the parties but may not exceed 50% of the monthly rent when the agent limits themselves to connecting the parties for the conclusion of the lease agreement. The real estate agent may also be the manager of the rental unit but cannot charge more than 10% of the monthly rent as compensation for this benefit.

Co-ownership is governed by Benin law, although there is very little recourse to it in practice. The rules of co-ownership are set by default in the land and subsidiary code, with the possibility of any changes left to the parties. The co-ownership revolves around the duality between private and common areas, and condominiums are administered by a syndicate.

It is necessary to mention the specific cases of the sale by the State of buildings belonging to its private domain in the context of urbanization projects through the construction of so-called social housing intended for a cash sale, a public auction, a private rental, or a particular type of rental sale. In the latter case, the buildings are acquired by the buyers after payment of the last rent without payment of a residual amount as required for conventional rental sales. In this case, rents are therefore advances on the sale price. The rental period varies between 15 and 17 years.

Since 2016, social housing construction programmes have been in place. The construction of 20,000 F4-type housing units has been underway in 14 cities in Benin, including those for 12,960 individuals and 7,040 collectives. The distribution is done by site. Therefore, 12,049 dwellings are under construction, including 10,849 in Ouèdo, 250 in Porto-Novo, and 500 in Parakou.

3. Transfer of ownership

The transfer of ownership of buildings or land located in the Republic of Benin is relatively open to any natural or legal person of Benin nationality, although Benin law does not prevent the acquisition of land by non-nationals. However, in this case, certain conditions must be met for the completion of the transaction. First, it must be urban land, as opposed to rural land. In addition, the state of origin of the foreign buyer must be in favour of the acquisition of real estate by a Beninese in that country. The absence of reciprocity is an obstacle to the acquisition of a building in Benin by a non-national (Memo No. 002/2020/MEF/ANDF/RP/DOFT/JCP/SP of 2020 January 24).

Foreign chancelleries and international bodies with a headquarters agreement in Benin may also acquire a building in Benin, but are subject to reciprocity agreements, international treaties, or other international agreements.

Marketing phase

This phase is completely free. The parties can get in touch directly or do so through a real estate agent. In this case, the agent sets their remuneration, which is not regulated by law. Apart from residential leases, the remuneration of real estate agents is not regulated. In addition, the parties involved in a contract for the sale of real estate can sign a promise of sale before a notary. However, certain conditions must be fulfilled by the applicant.

It is necessary that the building or land subject to the transfer of ownership is not subject to any rights or charges. The transaction relating to a registered building calls for the intervention of a notary. In addition, all agreements relating to a building located in the Republic of Benin must be registered in the land registers in order to be enforceable.

The most common method of transferring ownership is the acquisition of a registered property by purchase sale. The procedure is conducted in two stages for registered buildings:

- *The verification*: The notary makes sure that the operation is possible. Therefore, they must verify the authenticity of the title. To do this, they can connect to the e-Notaire platform *(https://enotaire.andf.bj/)* and instantly access the main menu with the descriptive status module. This step is performed instantly online. The cost of the transaction is € 8 for authentication of the land title payable online on e-Notaire.

  The notary's emoluments amount to 1% of the venal value of the land.

  At this verification stage, a copy of the seller's land title and a copy of both the seller's and buyer's identity documents are required. The transfer of ownership is requested online by the notary at this time.
- *The processing of the transfer request*: The National Domain and Land Agency receives the request made by the notary for transfer and proceeds with its treatment.

  The operation lasts 72 hours.

  The completion of the transaction by the ANDF is determined by an appointment granted to the notary for the withdrawal of the title of property. The notary goes to this appointment with a copy of the existing title.

Contract phase

As mentioned, it is not yet obligatory for a notary to oversee hire-purchases and sales. However, they will be as soon as the related provisions of the Land Code are fully in place.

There is no obligation to pay a deposit. In general, the parties are free to define the terms of payment. If the sale is made before a notary, these terms can be agreed upon in advance with the notary, and the transfer of ownership can be immediate or delayed until the date of full payment.

The removal of the geographical status certificate originally issued by the Institut Géographique National (IGN) and the removal of the formality of pre-registration of land deeds as part of the land transfer procedure have accelerated the authentication process for land titles.

Registration phase

When the building is not registered, the transfer may relate to the presumptive title of ownership, including the customary detention certificate (CDA) and the recase certificate. This change takes place at the town hall, in the service of state affairs, although from 14 August 2023, these transfers to the town halls will no longer be possible.

Disposal of properties belonging to the private domain of the State

In addition to transactions between individuals, the particular case of real estate transfers and state-initiated rental sales procedures through social housing projects are to be noted in the context of urbanization and living environment planning policies.

The transfer of social housing by the State follows specific and strict formal procedures (Act No. 2018–12 of 2 July 2018). It is based on the provisions of Article 299 of the Land and Federal Code Act in the Republic of Benin: 'Land and real property in *the private domain of the State may be transferred for a fee or free or leased according to the rules of common law, assigned to a legal person under public law, granted in concession to individuals or legal entities under private law.*'

On the basis of this text, the government proceeds in various ways:

Public auction: When the government choses to sell buildings belonging to the private heritage of the State by public auction, a decree on the terms of transfer for a fee is taken. The physical sales operation is led by the Housing Fund Agency (AFH). A specification is also adopted but does not constitute an offer or a contractual document. Tender offers are made by interested parties to the AFH, and allocations are made for the highest bidders. This process is not subject to public procurement rules. Where the State has ensured the publicity of future sales transactions, including announcements and press releases, any individual or legal person will be able to submit their offer under the terms of the specifications according to the prescribed forms and deadlines. Any individual or corporation can bid on one or more dwellings and on different sites. Each candidate in their offer to purchase must meet the following conditions of nationality, legal capacity, and fiscal regularity.

Rental sale: The Government's Urbanization Programme 2016–2021 incorporates a plan for 20,000 social housing units in Ouèdo in Abomey-Calavi. The acquisition model here is that of leasing. After a period of payment of rent between 15 and 17 years, the tenant becomes the owner without further formalities. To qualify for the acquisition of these units, the applicant must file with the AFH, and it must contain the following items:

- Application according to the registration form available at AFH
- Two legalized copies of the ID card or passport
- One copy of the last three pay slips
- Work certificate specifying retirement date
- Employer commitment for downstream and salary confirmation (private sector employees)
- Two legalized copies of the birth certificate or equivalent judgement
- Registration number at the National Social Security Fund or National Fund for Retirement of Benin
- One copy of the employment contract
- A residence certificate
- Payday transfer to the Tribunal in the event that the monthly payment is greater than the transferable
- Receipt of payment to the bank of the supplementary contribution (in cases of those whose salary or number of years of retirement makes them ineligible)
- Subscription to death and job loss insurance policies

Housing costs vary depending on the layout of the building.

- F2: € 13,568
- F3: € 16,769
- F4: € 19,513

With regard to other types of real estate sale, the parties are free to decide when to initiate the registration procedure for the sale, in particular, the transfer, and they do so by using the services of a notary. First, the agreement is registered and then transferred to the land and estates departments. The costs are shown in Table 4.1.

Post-transaction phase

When the seller realizes a capital gain by selling their property, the administration collects a tax of 5% on the amount of the profit.

Table 4.1 Fees and taxes in the context of a transfer of real estate ownership in Benin

Phase	*Fees and taxes*	*Seller*	*Buyer*
Contract	Mortgage write-off	0.3% of the value of the warranty	-
Registration	Transfer of ownership and name change on land registers	-	< € 15,245 = 0.3% of the transaction amount
		-	€ 15,245–76,225 = € 46
		-	> € 76,225 = 0.5% of the transaction amount
	Mortgage	-	0.3% of the value of the warranty
	Depositing mentions to the opposition book on existing land titles	-	€ 76
Post-registration	Sales tax	-	-
Total (EUR for the model case price)		**-**	**€ 576**
Total (% share to the model case price)		**-**	**0.6%**
Excluded costs from the model case		Mortgage write-off: 0.3% of the value of the warranty	Mortgage 0.3% of the value of the warranty

Note. Model case price: € 100,000

Source: Author

4. Summarizing the ownership transfer process and transaction costs

To recapitulate the process of property transfer in Beninese law, we can say that it is essentially consensual, that is, that the agreement of the parties is sufficient and can be recognized by a judge should the need arise. However, it must be formalized by signing a sales agreement, registering the agreement, and finally, registering the transfer of the title deed with the administration concerned. No pre-contract phase is mandatory (Finance Act, 2020).

Note

1 1 Central African Franc = 0.0015 Euro (as of 1 October 2021).

References

Act No. 2018–12 of 2 July 2018, which includes the legal regime for the residential lease.

Finance Act 2020.

Law No. 2013–1 of 2013 August 14, regarding land and federal code in the Benin Republic amended and supplemented by Law No. 2017–15.

Memo No. 002/2020/MEF/ANDF/RP/DOFT/JCP/SP of 2020 January 24, removing the formality of pre-registration of land deeds as part of the land transfer procedure.

5

BRAZIL

Ana Paula Corrêa Patiño

1. The Brazilian housing market

To comprehend the Brazilian[1] real estate market, it is important to know that it is an enormous country, with 213,317,639 people as of October 2021. Statistics show that in 2010, there were 160,925,792 people living in urban areas and 29,830,007 in rural areas. These numbers certainly increased a lot in last decade, but the actual number is still unknown. There are several regular ways of dwelling tenure. The real estate market for homes constitutes of ownership (buying, donation, exchange, construct, adverse possession, inherit, auction purchase) and rental. However, there are many families who cannot access housing through these regular ways due to financial difficulties and end up taking up residence in other dwellings or making their own irregular constructions. This is how slums are created in urban or rural areas. Due to decades of the irregular occupation of dwellings, Brazilian slums can be considered as shanty towns, almost cities in themselves, with some even containing hospitals, schools, supermarkets, and stores. It is estimated that, in 2021, approximately 5,127,747 people live in housing that is irregularly built or occupied. The large number of people without a place to live, sleeping on the streets, or living in irregular dwellings obviously has an impact on the housing market and creates a social problem that demands a government solution. Even though Brazil can be considered a developing country, and the majority of the population does not have enough resources to buy houses, prices on the housing market are slowly increasing after a long period of decline, starting right after 2014 and 2015, when housing prices reached their highest level. As house prices are getting higher, so too are new constructions increasing. The best year for the construction of new houses and apartments was 2019 (Sistema de Gerenciamento da Habitação, n.d.). In 2020, as a result of restrictive measures caused by COVID-19, new real estate constructions saw a drastic decline followed by a rapid recovery at the second semester of the year. Although housing prices and new constructions are currently experiencing an upswing, real estate financial activity is in its worst period due to high interest rates from banks.

At the same time, interest rates for financial investments and stock exchanges are currently low, so investors currently prefer to buy real estate property, which is why the real estate market is experiencing its best moment in Brazil.

DOI: 10.1201/9781003130383-7

2. Regulatory framework

The need to reduce poverty and alleviate social problems led the Brazilian government to create some social aid programmes, including financial aid for poor individuals and families to buy houses and apartments. Some government programmes are worth mentioning, like FGTS (savings from employment contract), *Programa Minha Casa Minha Vida* (My House, My Life Programme – financial aid to buy or construct houses or apartments, 2009/2020) and *Programa Casa Verde e Amarela* (Green and Yellow House Programme – financial aid to buy or construct houses or apartments, created in 2021) (Governo do Brasil, n.d.).

The FGTS, conceived by law in 1966 and substantially modified in 1990 Federal Law (n°8.036/1990), is required to form a fund (savings) for employees with a regular work contract. Employers will deposit 8% of the employee's monthly salary into an account (FGTS), and employees can use the savings to build or purchase a house or apartment, among other possibilities. Social programmes designed to help families acquire a private home are responsible for ensuring that millions of people have a place to live, and it has certainly improved the housing market over the years. Even before 2009, with the *Minha Casa Minha Vida* programme (Federal Law n°11.977/2009), federal and state governments give financial aid to those who cannot acquire their own homes (Brazilian Federal Constitution, 1988). Since 2009, 5,335,419 dwellings have been built or bought by socioeconomically disadvantaged families, helped by the federal estate financing programmes known as *Minha Casa Minha Vida* and *Casa Verde e Amarela*.

The Brazilian real estate market is regulated by certain federal laws, with each one relating to a specific subject (ABRAINC, n.d.). Federal Law n°8.245/1991 (Rental Law) relates to real estate rental agreements of properties located in urban areas. The rental agreement is a simple contract and does not require any intervention whatsoever or any specific form. However, in order to ensure tenants' and landlords' basic rights, it is important to determine the terms and conditions in an agreement, according to the law. Federal Law n°10.257/2001, called the 'City Statute', refers to real estate transactions and deals with the development policy of urban areas and the social function of property, principles also incorporated into the Brazilian Constitution and Civil Code. This law created the city's Master Plan, which is mandatory for any city with more than 20,000 inhabitants. Federal Law n°10.406/2002 (Brazilian Civil Code), hereinafter referred as CC, regulates the real estate conveyance process, contract matters (e.g. real estate donation and purchase agreement), adverse possession, and inheritance matters, among other subjects. Other laws also regulate specific aspects of the housing market, such as registration and taxation, but the CC is perhaps the most important law concerning real estate transactions. As mentioned, there are many ways to secure dwelling tenure. However, there is only one way to acquire ownership of real estate property, which is to register the title at the Real Estate Registry Office. In addition to proving who the owner of the real estate property is, the title registration is the only way to convey real estate property between owners other than inheritance. Real estate transactions, such as auction purchase, donation, selling and buying, or even building a house, do not convey real estate property *per se*, which, as mentioned, requires title registration. The majority of housing market transactions in Brazil is the buying and selling of houses or apartments, made by written contract after direct negotiations between buyer and selling.

Marketing phase

Anyone can directly and freely sell real estate property, without the intervention of a public or a private agent. The real estate property owner can promote the sale of their house or apartment by any means. Brazilians traditionally prefer to deal with other people, mainly in real estate transactions, where an enormous amount of money is spent. However, lately it has become relatively common for real estate sales and rental agreements to be made directly online, intermediated by digital real estate companies. After visiting the property, the buyer or tenant can bid online, and if seller or landlord accepts the price and terms of the contract, an electronic contract is sent to both of parties to be signed. Although the intervention of a real estate agent is not mandatory, the majority of transactions are intermediated by an agent, normally hired by the seller. This is because it is easy and normally faster to sell or rent a house or apartment if it is advertised by a real estate agent. The real estate agent's duties are to advertise the property, approach the buyer and seller, or tenant and landlord, and provide the hirer with all the information about price, the terms of the contract, security as well as any risks involved in the transaction. The real estate agent is required to be certified by a Regional Consul of Realtors. The broker's remuneration is due when the sale or rental agreement is signed, even if buyer or tenant does not finalize the sale or lease. The fee paid to the real estate agent varies according to the nature of the deal, the property's price, and the customs of the society in which the negotiation takes place. Counsels of realtors usually stipulate a remuneration fee of between 5% and 10% of the price or the first rent. When starting and concluding the deal directly between parties, no remuneration will be due to the estate agent, but if the estate agent was hired exclusively, they have the right to receive full remuneration, even if deal is carried out without their mediation. At this point in the transaction, it is important to investigate whether the seller has any debts, as creditors could seize the seller's assets, including the property that buyer wants to purchase. Therefore, it is highly risky to buy real estate property from an indebted seller. In order to know if there are lawsuits or debt protests against seller, a buyer can request official departments to provide certificates with this information.

Contract phase

Regardless of whether it is a sale or a donation contract, the terms have to be written in a formal contract called *escritura* (deed). Parties of the contract must be capable or be represented by their legal representatives. There is no need to sign a pre-contract; however, due to the high prices of land, houses, and apartments, the parties often sign a preliminary contract, and then a few months later sign the final contract (deed). Though not mandatory, it is expected that buyer pays a certain sum to assure the signing of the final contract, normally 10% of the price. During the time between the pre-contract and the signing of the deed, the buyer can investigate the existence of lawsuits or debt protests against seller. The written contract must specify the price of the property, if it is a sale, the parties' names and full qualifications, a description of the property, and the terms of the transaction. Payment, deadlines, and possession date can be adjusted between the parties, and the only formality required is that the contract is written, so that the title *(escritura)* can be registered afterwards.

A written contract is not required for real estate rental; however, when parties opt for the written form of the contract and rent the property for at least 30 months, it is easier and faster to evict the tenant at the end of the contract.

Registration phase

The registration of the title – a written contract named 'sales deed' *(escritura)* – in the Real Estate Registry Office is a formal act required to convey real estate property to a buyer or the person who has received a donation or inheritance. Each property has its own registration, and all transactions referred to that property are included in this registration, which is public. Therefore, anyone can know the current owner of a real estate property. To register the property title, the Real Estate Registry Office charges a fee. Each Real Estate Registry Office in Brazil has its own fee table, which varies widely.

Post-transaction phase

Real estate ownership transfer is taxable in Brazil. The tax rate depends on whether the real estate property transfer is a sale contract, donation contract, or inheritance. If the ownership transfer results from donation contract or inheritance, whoever receives the property has to pay a state tax (ITCMD). The rate varies from state to state but cannot exceed 8% of the value of the property, according to the Brazilian Constitution. On the other hand, if real estate property was purchased, either contractually or at auction, or even exchanged, the tax is established by municipal law and can vary up to 4% on the value of the property. The city tax (ITBI) rate and incidence are controversial, as there are cities that charge up to 5%, and some municipal laws, at the same time, stipulate property prices above the market value. The auctioneer fee is 5% and paid by the buyer. If seller makes a profit when selling their property, it is charged a federal tax called a 'real estate profit' at a rate of 15% to 22.5%, depending on the difference between the value of the purchase and value of the sale of the property.

3. Summarizing the conveyance process and its transaction costs

Given that Brazil is a huge country with 26 states, one capital district, and more than 5,570 counties, is very difficult to estimate the conveyance process and its transactions costs. Every county in the country has its own table of fees regarding registration cost.

In addition, costs may vary if there is a real estate agent involved in the transaction, which is not mandatory. However, it is estimated that in São Paulo (Brazil's largest city), the transactions costs of purchasing a real estate property are a 4% state tax (ITCMD) and a variable fee stipulated by the Real Estate Registry Office (Associação dos Registradores Imobiliários de São Paulo, n.d.). It is possible to simulate the cost of a sale and a donation or inheritance of a property of R$ 1,000,000 (nearly € 150,000) according to Table 5.1.

Table 5.1 Fees and taxes during the conveyance process in Brazil

Phase	*Fees and taxes*	*Real estate sale*		*Donation/inheritance*	
		Buyer	*Seller*	*Grantee/heir*	*Donor*
Marketing	Real estate agent's fees (optional and variable)	-	6%	-	-
Contract	Auction fee	5%	-	-	-
Registration	Real Estate Registry Office	€ 503	-	€ 509	-
Post-registration	ITBI (city tax)	3%	-	-	-
	ITCMD (state tax)		-	4%	-
	Real estate profit tax (variable)	-	15–22.5%	-	-
	Sales tax	-	-	-	-
Total (EUR for the model case price)		**€ 8,503**	**€ 6,000**	**€ 4,509**	**-**
Total (% share to the model case price)		**8.5%**	**6%**	**4.5%**	**-**
Excluded costs from the model case		-	15–22.5% CGT	-	-

Note. Model case price: € 100,000

Source: Author

Note

1 1 Brazilian Real = 0.15 Euro (as of 1 October 2021).

References

ABRAINC. (n.d.). *Main page*. Retrieved November 20, 2021, from www.abrainc.org.br

Associação dos Registradores Imobiliários de São Paulo. (n.d.). *Main page*. Retrieved November 20, 2021, from www.arisp.com.br

Brazilian Civil Code (Law n° 10.406/2002).

Brazilian Federal Constitution 1988.

Fundação Instituto de Pesquisas Econômicas. (n.d.). *Main page*. Retrieved November 20, 2021, from www.fipe.org.br/

Governo do Brasil. (n.d.). *Serviços do Governo para você*. Retrieved November 20, 2021, from www.gov.br/pt-br

Sistema de Gerenciamento da Habitação. (n.d.). *Casa verde e amarela*. Retrieved November 20, 2021, from http://sishab.mdr.gov.br/

6
CAMBODIA

Chaing Sinath and Neuv Socheata

1. The Cambodian housing market

There are two main forms of Cambodian[1] real estate market for homes: rental and ownership. Real estate and housing construction is a core among the four economic pillars of Cambodia. An estimate of the Cambodian population in urban areas in 2014 was about 4.5 million and will continue to increase to 7.92 million by 2030. These data show that for the 15-year period from 2015 to 2030, the housing demand in urban areas is about 800,000 units, with an average of about 15,000 units per year needed in Phnom Penh, which is equivalent to 1.1 million units by 2030 (Ministry of Land Management, Urban Planning and Construction, 2020).

Cambodia does not publish house price statistics. Nevertheless, Cambodia's housing market enjoyed unprecedented price rises from 2004 to 2007, with property prices rising by about 25% to 40% annually. Residential prices in Phnom Penh soared to € 1,376/m^2 in 2007, from € 516/m^2 in 2006. In mid-2008, the bubble burst. The market started to recover in 2011, buoyed by strong economic growth and by the new foreign ownership law (National Institute of Statistics, Ministry of Planning, 2014). In Phnom Penh, the average condominium sales price more than doubled from 2012 to 2019, from € 1,256 to € 2,738/m^2. Since the end of March 2020, the outbreak of COVID-19 has affected the real estate market, with the effects notable in adjusted rental rates and sale prices. Since 2020, about 68% of condominium projects were suspended, while the rest experienced delays (Realestate.com, n.d.).

The housing prices continue to grow and surpass local incomes. Only 20% of Phnom Penh's population can afford housing based on current and projected income levels. The average value of Cambodia Annual Household Income per Capita data from 2019 is € 1,183,947. It is estimated that a person needs between 33 and 53 years to buy a house with an area of between 36 m^2 and 48 m^2 with estimated costs € 43,000–68,800 (Ministry of Land Management, Urban Planning and Construction, & United Nations Cambodia, 2020). In 2013, the Ministry of Planning indicated that 13.5% of the population are below the poverty line, which would equate to some 2.1 million people. The migrant workers from the rural exodus and new families have led to informal settlements in town. About 22% of rural-urban migrants are reported to be working in the garment sector, and 22% in small businesses. A majority of households (66.8%) in Cambodia occupy only one room, 24% of households have two rooms, and 6.3% have three rooms. Therefore, about 97% of Cambodian households have only up to three rooms. However, about

DOI: 10.1201/9781003130383-8

92,599 households (2.9%) enjoy having four rooms or more. The United Nations reported in 2015 that 55% of the urban population in Cambodia live in slum conditions. In response to this issue, the government is implementing a national housing policy adopted in 2014 to promote and support the development of all types of housing, especially affordable housing for government officers and low-income people. This type of home costs less than € 25,800 with a small monthly payments in long-term instalments. In 2020, five affordable housing projects provided 8,331 houses and 7,256 condominiums. According to the 2020 summary report of the Ministry of Land Management, Urban Planning and Construction, there were 513 construction projects nationwide compared with December 2019 when there were 442 projects. The distribution of 22 condominium buildings and 434 residential projects (1,964 houses, 2,494 shop houses, and eight condominium buildings) are under construction.

2. Regulatory framework

2.1 Regulatory framework of the land system in Cambodia

All Cambodian land title records were destroyed between 1975 and 1979. Later, in 1989, Cambodian citizens were granted the right to own their own homes. In 1991, there were regulations on ownership grants and the sale of houses to people living in Phnom Penh, upon issuance of a title of land possession and on establishment of the stamp duties. The first land law was adopted in 1992, and a newer law, which has been in place since 2001, created the land registry system. Many provisions were modified or removed by the 2007 civil code, which came into effect on 21 December 2011 following the implementation of the Law on the Implementation of Civil Code dated 31 May 2011.

There are three types of land registry system in Cambodia. First, the systematic land registration in 2000 is a national policy of the government to register all the immovable properties in Cambodia by requiring a certificate of ownership. Second, the sporadic land registration in 2002 means that an individual must make a request for the registration of any immovable properties to obtain a certificate of possession. Third is when the complementary land registration for the area has already been completed, but the owner has not cooperated with the land officer and needs to apply for a certificate of ownership. Due to the complication of the land registry system, ownership can be proven with one of three different forms of title: a certificate of title (commonly known as a 'hard title'), a letter of possession (commonly known as a 'soft title'), and a form of private ownership in a co-owned building (also known as a 'strata title').

2.2 Different types of ownership

2.2.1 Hard titles

Hard titles are ownership certificates provided by the Land Department of the Ministry of Land Management, Urban Planning and Construction and are the official recognition of property ownership in Cambodia. Hard titles refer to different types of titles as follows:

- *Land possessory rights and use title*: This is called 'chicken feather titles', and it refers to titles issued to owners before the enactment of the land law in 1992.
- *Immovable possessory right title*: This title was provided before the establishment of the systematic land registration system in 2002 and for all the land that is registered through the sporadic land registration.

- *Certificate of immovable ownership*: This certificate is provided after a systematic land registration and the complementary land registration. This hard title also includes the Land Management and Administration Project (LMAP) titles, which is considered the safest kind of land title because it uses GPS to allocate land plots. An LMAP title is the latest, most accurate, and securest form of hard title in Cambodia.

2.2.2 Soft titles

Soft titles are delivered at the district and commune levels only, not the national level. Although soft titles are not registered at the national level, they are still considered a possessory right to land.

2.2.3 Strata titles (foreign ownership of co-owned buildings)

A strata title was introduced in Sub-Decree No. 126 ANK.BK of Royal Government of Cambodia dated 12 August 2009 on the Management and Use of Co-Owned Buildings. This title refers to the certificate of ownership of an individual unit in a co-owned building, such as an apartment/condominium.

Cambodia's Constitution in 1993, land law in 2001, and civil code in 2007 prohibit any foreigners, either a natural person or legal entity, to own any immovable property. As a result, there are different arrangements for foreigners to own a house or condominium in Cambodia.

2.3 Foreign property ownership mechanisms

2.3.1 Strata title

Law No NS/RKM/0510/006 dated 24 May 2010 on providing foreigners with ownership rights in private units of co-owned buildings allow foreigners to own an apartment/condominium in order to encourage foreign investment in the local property market and simultaneously accommodate an expatriate population seeking ownership over residential and commercial properties in Cambodia. Foreigners are not allowed to purchase a grounded unit, and their ownership is limited to 70% of units in any co-owned buildings. This strata title applies only to new buildings built after 2010. The price of the condominium with the strata title is more than € 860/m^2, and without the strata title, it is around € 172/m^2 (i.e. old buildings built before 2010).

Foreigners are granted ownership through the strata title and cannot own the apartment without it. Given this, another arrangement is considered useful for foreigners who want to invest in immovable property in Cambodia.

2.3.2 Nominee structure

Due to the prohibition, foreigners can arrange to buy a house by using a Cambodian citizen as nominee to act as an owner on the properties' title deed. When a property is owned through a Cambodian nominee, foreigners should enter into a nominee agreement with the Cambodian citizen. The Cambodian nominees should also grant foreigners a long-term lease. A mortgage should be made and registered at the Cadastral Administration. The foreigners should be in possession of the original title. To prevent any misunderstandings in the event of an accident or death, the nominee should also prepare a will. This structure is very complicated and risky in terms of diversion of the law and needs to be very well prepared by the legal expert.

2.3.3 *Cambodian citizenship*

In 1996, the Law of Nationality made it possible for foreigners to apply for Cambodian citizenship by either marriage or nationalization.

2.3.4 *Land holding company*

The Investment Law allows only Cambodian natural persons or legal entities to own immovable properties in Cambodia. A majority shareholder with 51% shares or more must be a Cambodian natural person or a Cambodian legal entity, while 49% or less of the shares can be held by a foreigner or foreign entity. The internal shareholder agreement can stipulate the mechanisms by which to manage, vote, and share 100% of the benefits in favour of the international investor.

2.3.5 *Trust*

The Royal Government of Cambodia promulgated the Trust Law on 2 January 2019. A trust is a legal relationship in which a trustor shall transfer his or her own property or funds to a trustee through a trust letter certified by a notary public, lawyer, or local authority. The trust property shall be automatically transferred to a trustee to administer or dispose of such property. The identity of the beneficiary or a specific purpose of the trust shall be mentioned in the trust letter. The trust has to be registered at the Ministry of Economy and Finance (MEF) no later than three months after the date of the signature of the trust letter. Several trust companies provide these services to their foreigner clients who wish to invest in immovable property in Cambodia. To date, no trust letter had been registered at the MEF because the registration system is not fully functional.

3. The conveyance process

The Cambodian conveyance process is summarized in Figure 6.1.

Marketing phase

There is no mandatory intervention of an estate agent. Estate agents must be registered at the Ministry of Commerce, pay tax at the general tax department (GTD), and have a license from the MEF to be allowed to practice the profession. However, in practice, anyone is able to be an estate agent, as it is not strictly controlled by the government.

In Cambodia, estate agency contracts are not exclusive, and most of the time, there are no contracts. The agent is usually remunerated with 3% of the sale price, which is usually paid by the seller. The estate agent advertises the property on their agency's website or an independent online marketplace without any charges and brings the buyer to visit the property. Everyone in Cambodia can advertise their own property on an independent online marketplace.

Pre-contract phase

This phase is not mandatory. If both parties agree on the purchase price, the buyer can also hire a lawyer to proceed with due diligence to check whether there is a conflict over this property at the court or the mortgage at the local authority and the Cadastral Administration. Then they conclude a deposit contract in which the buyer generally makes a deposit between 20% and 30% of the purchase price.

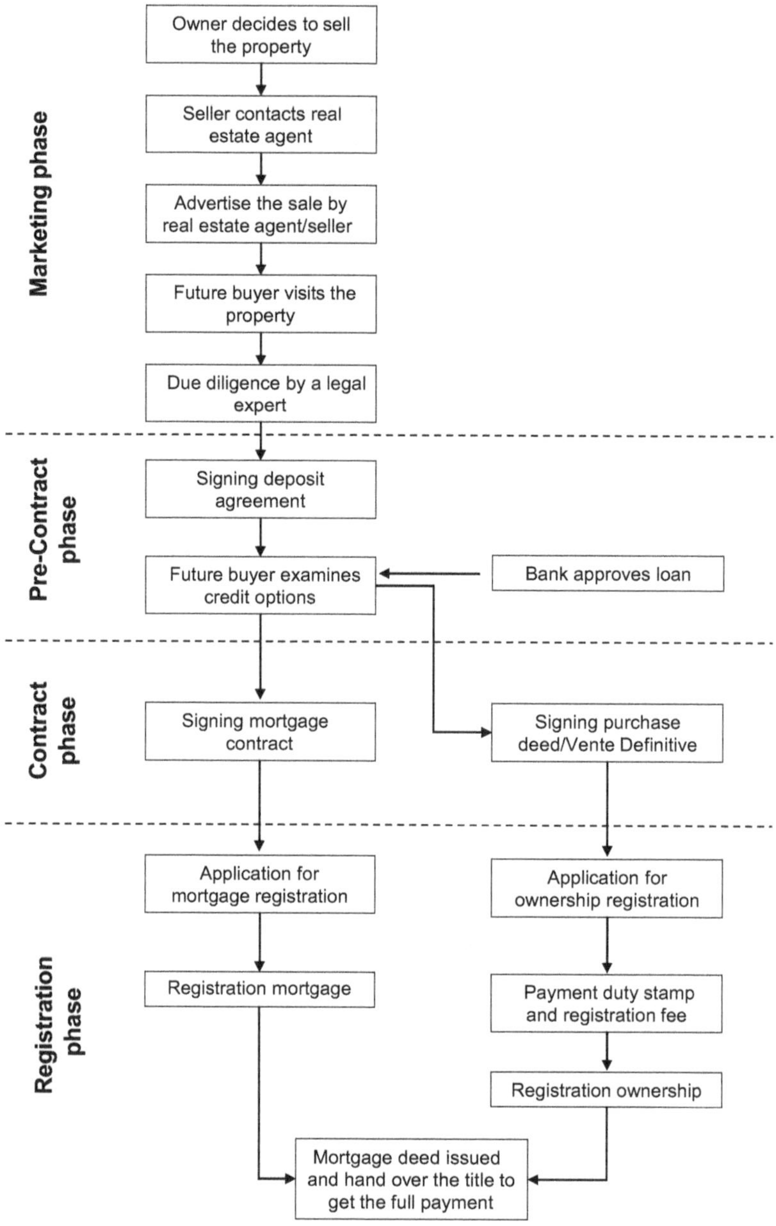

Figure 6.1 The conveyance process for real estate in Cambodia

Source: Author

The mode of payment varies with each situation. Upon the agreement of both parties, the deposit can be paid with a one-time transaction of 20% of the purchase price to proceed with the transfer of ownership, and then the remaining amount is paid after the title deeds are handed over to the buyer. The deposit can also be paid at the first time of 20% of the purchase price to discharge the mortgage at the Cadastral Administration, at the second time of 20% of the purchase price to proceed the transfer of ownership, at the third time of 20% of the purchase price when the seller paid the duty stamp, and at the final payment of the remaining purchase price after the handing over the title deeds to the buyer.

Upon the agreement of both parties, there is the possibility to open an escrow account at the bank to secure the transaction. This option could be useful for the important transaction in order to guarantee each step of the payment. During this period, the buyer examines their credit options. If the loan was not granted by the bank, the deposit can be reimbursed to the buyer.

Contract phase

The sale shall be in the form of a contract whereby the seller is obligated to transfer ownership to the buyer who engages to pay a purchase price. The sole agreement of both parties to transfer ownership of immovable property is not enough. The parties may require the execution of a notarized document, which refers to a document made by a notary public or a competent officer for the registration process.

However, in practice, in order to conclude the sale purchase agreement (SPA), both parties can involve a lawyer or public notary to prepare their contract or sign a template of sale contract at the local authority directly. Then, regarding the convey process of the hard title, both parties need to sign another deed called *Vente Definitive* at the Cadastral Administration.

Due to the land conflict issue in Cambodia, it is recommended to involve a lawyer or notary public to draft the SPA, where the important clauses are fully stated in the contract, such as the identity of the parties, the description of the object of the SPA, the purchase price, the form of the deposit and payment, the rights and the obligations of both parties, the guarantee of the seller, the applicable law, and the mechanism of the conflict resolution. This type of SPA is flexible according to the situation of each client. Detailed information about the sale is not fully mentioned in the sale contract at the local authority level, or in the *Vente Definitive* at the Cadastral Administration, which uses only the SPA template. This is not flexible and may lead to disputes generated from misunderstandings and misinterpretations.

Registration phase

The Cambodian Cadastral Administration records the transfer of ownership and issues a property title. Both parties are bound only by the authentic deed. Transfers of title by agreement pertaining to an immovable property shall come into effect only upon registration of the transfer of the right.

The Prakas No. 507 from the MEF dated 26 April 2017 on stamp duty tax collection determined the basis of registration taxable amount on transferring ownership or possession rights over immovable property. The minimum value of an immovable property is determined in accordance with its location and building characteristics. The basis of registration taxable amount is 4% of the value calculated according to the Prakas, or the value stipulated in a sale agreement if it is equal to or higher than such mentioned in the Prakas. One is prohibited to issue certificates of ownership of the property until the Property Transfer Tax has been paid. The charge to pay the tax is the buyer's responsibility, but in practice, the seller pays it in general or both parties can split the cost, upon agreement. There is a particular case regarding the payment of the duty stamp in accordance with the type of title. If this is the first transfer of soft title and the first registration at the GTD, there is no need to pay the duty stamp. This is how the government encourages the Cambodian population to register their transaction at the GTD.

If both transfers of soft and hard title have been conducted several times before 2 February 2016, they shall pay the duty stamp for the last transfer only.

The Prakas covers the transfers of all types of title deeds. The seller/buyer files an application with the Department of Land Management, Urbanization, and Construction (DLMUC), without which no binding mortgage can be accomplished, and pays the duty stamp at the GTD. Thereafter, the documents are transferred back to DLMUC to register the name of the buyer as the new owner of the title deeds and to register the mortgage if there is one. The remaining price shall be paid at the same time that the seller hands over the new title deeds to the buyer.

Post-transaction phase

The MEF issued Prakas No. 346 dated 1 April 2020 on capital gains tax on the profit, which is defined as a tax at a flat rate of 20% on capital gains on the sale or transfer of immovable property. However, the implementation of this regulation has been postponed due to the COVID-19 situation. Annual property taxes borne by the new owner are 0.1% of the property's market value in excess of Euro€ 25,000 (The Mission of the General Department of Taxation, n.d.a).

4. Summarizing the conveyance process and its transaction costs

The conveyance process and its transaction costs are outlined in Table 6.1. Ownership is conveyed upon registration at DLMUC, not with a sales contract. After the pre-contract phase with the deposit, both parties sign the deed. The seller/buyer fills out an application for the registration, pays the tax, obtains a title deed, and gets the full remaining payment (The Council for the Development of Cambodia, n.d.). For foreigners, the process of transfer of strata title is the same as for Cambodian people. However, regarding the transfer of soft title deed to foreigners,

Table 6.1 Fees and taxes during the conveyance process in Cambodia

Phase	*Fees and taxes*	*Ownership*	
		Seller	*Buyer*
Marketing	Real estate agent's fees	3%	-
Contract	Legal fees	€ 430–4,300	-
	Bank mortgages fees	-	1% of loan amount
Registration	Duty stamp	-	4% transfer tax based on criteria by the concerned land/cadastral office
	Properties registration fee	-	Hard title: € 129 Strata title: € 258 Soft title: € 10.75–86
	Application fee	-	€ 1.1
Post-registration	Capital gains tax	20%	-
	Sales tax	-	-
Total (EUR for the model case price)		**€ 3,430–7,300**	**€ 399–474**
Total (% share to the model case price)		**3.4–7.3%**	**0.4–0.5%**
Excluded costs from the model case		20% CGT	1% loan fee, 4% transfer tax

Note. Model case price: € 100,000

Source: Author

a different legal mechanism will be needed, such as the nominee, the citizenship, the trust and the land holding company, in which the role of a legal expert is very crucial in this case.

Note

1 1 Cambodian Riel = 0.0002 Euro (as of 1 October 2021).

References

The Council for the Development of Cambodia. (n.d.). *CDC-tax and accounting*. Retrieved November 20, 2021, from www.cambodiainvestment.gov.kh/investors-information/taxation-and-accounting.html

Ministry of Land Management, Urban Planning and Construction. (2020, December). *Summary report of the ministry of land management, urban planning and construction*. Phnom Penh: Author.

Ministry of Land Management, Urban Planning and Construction, & United Nations Cambodia. (2020). *Cambodia urban forum, smart sustainable and inclusive urban development*. Retrieved from https://urbandatabase.khmerstudies.org/get-data/NTcw/view

The Mission of the General Department of Taxation. (n.d.a). *Prakas no. 346 of ministry of economic and finance dated 1 April 2020 on capital gain tax*. Retrieved November 20, 2021, from www.tax.gov.kh/en/categories/lgZ8c05724474872

National Institute of Statistics, Ministry of Planning. (2014, March). *Cambodia inter-censal population survey 2013, housing and household amenities*. Retrieved from http://www.stat.go.jp/info/meetings/cambodia/pdf/c13ana09.pdf

Realestate.com. (n.d.). Cambodia real estate investment guide 2020. *Realestate.com*. Retrieved November 20, 2021, from www.realestate.com.kh/investment-guide/

Sub-Decree No. 126 ANK.BK of Royal Government of Cambodia dated 12 August 2009 on the Management and Use of Co-Owned Buildings.

7
CHILE

Helena Bohman and Karina R. Villacura

1. The Chilean housing market

The Chilean[1] housing market is often described as driven by two types of logics (Dohnke, Heinrichs, Kabisch, & Krellenberg, 2015). The first is a strong dependence on market forces, in which land markets have been privatized and development is driven by private sector and public-private partnerships. The second is a long tradition of housing subsidies directed towards providing affordable housing to low-income groups. Rather than provide rental housing, as many other countries do, the Chilean demand-oriented system has an explicit goal to stimulate ownership. Compared with other OECD countries, Chile has a relatively small share of rental housing, accounting for about 18% of the total housing stock, and relies mainly on owner-occupied housing (Salvi del Pero, 2016).

With an urbanization rate above 87% in 2017, Chile is one of the most urbanized countries in Latin America (World Development Indicators). Around 40% of the population lives in the capital region, *Región Metropolitana de Santiago*, implying that the capital region is at the very core of Chilean urbanization. In addition to the capital, much urbanization is concentrated to the adjacent fifth region, where the twin cities of Valparaíso and Viña del Mar are located. The strong urbanization trend not only in the capital region but also in other urban areas has led to densification and a tendency to build high-rise buildings. Criticism of this development has taken aim at how it has been detrimental to low-income urban households and spurred gentrification processes (López-Morales, 2016).

The property market is an important industry within the Chilean economy. When it comes to employment, the sector gives jobs to 9.4% (757,702 individuals) of the total labour workforce (8,027,786 individuals). A more detailed approach to employment in the sector shows that construction composes 8.6% of the workforce in which 93.5% are men. Real estate business has a more equal gender distribution; however, it is still a male-dominated area where 64.5% of the workers are men and 35.5% are women (INE, 2016).

Although the number of substandard dwellings has been reduced (Salvi del Pero, 2016), many households nevertheless live in informal housing, making it a widespread phenomenon. According to Celhay and Gil (2020), low-income households may choose informal housing because it offers better geographical and neighbourhood qualities than with subsidized housing, which is often in segregated and less secure neighbourhoods. Today, informal housing

DOI: 10.1201/9781003130383-9

comprises 81,643 households in 969 squatter settlements or shantytowns (TECHO, 2021) that have either reduced or informal access to services (running water, sewage, electricity). The number of families living in squatter settlements has increased by 79% during 2019–2020, arguably as a consequence of the COVID-19 pandemic (El Mostrador, 2021). Another important concept for understanding the Chilean property market and the role of the informal market is *allegados*, which is a concept used to describe households that live with relatives, often as a result of not being able to afford their own housing. Allegados can, for example, be adult children who live with their parents. In 2017, it was calculated that 739,603 individuals were affected by the so-called housing shortage, and 58% of these belonged to allegado families (Cámara Chilena de la Construcción, 2019a).

As shown in Figure 7.1, housing prices in Chile overall have increased since 2005. The international crisis in 2008 did not have an impact on overall prices, although there were some shorter periods with limited dips in the following years. Since 2011, the development has been steadily positive, with a few exceptions.

The high concentration of the population and businesses to the capital can perhaps explain the limited availability of regional data in Chile. However, data on living and housing standards can nevertheless give some indication of how the markets differ geographically. Previous studies have found that inequality within the country is often measured at an individual level, but large and persistent spatial inequalities are rarely addressed (Paredes, Iturra, & Lufin, 2016). Large differences in income may also be reflected in the property markets. As mentioned, the capital region contains much of the country's economic activity, and accordingly, income and property prices are high, although large differences exist within the city. However, two additional regions of the country also feature greater wealth. The mining region of Antofagasta in northern Chile has high salaries and property prices as a result of the local industries. Also, the very southern regions, including Magallanes, are rich in raw materials and can be considered well-off.

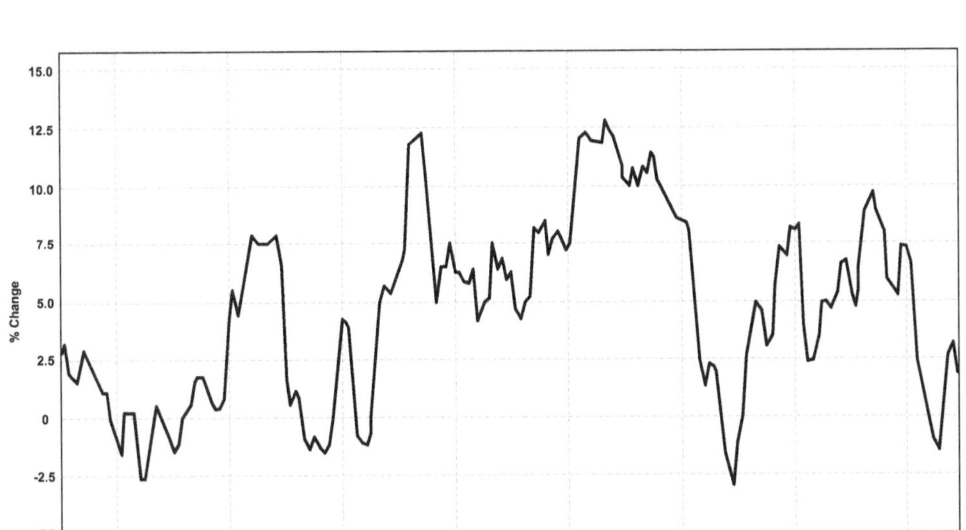

Figure 7.1 Change in housing prices, 2005–2020

Source: www.ceicdata.com (CEIC Data)

Large differences are also found within the regions, particularly the capital region. Santiago has witnessed a tremendous construction boom over the last decades with considerable densification. An example of this is with the construction of Gran Torre as part of the Costanera Center building complex, Chile's tallest building measuring 300 m, which was completed in 2013; however, the phenomenon of verticalization is widespread. As the name implies, verticalization is a strategy to densify the city, and planning has increasingly allowed higher buildings. From the 1990s, the inner *comunas* of Santiago, for example, began to change the local level master plans, relaxing their building codes and increasing the floor-to-area ratio. Consequently, the floor-to-area ratio of constructed buildings increased from roughly 6.5 to 9 between 2000 and 2010 (López-Morales, 2016). The tallest buildings tend to be located close to highways and in the city centre. Subcentres, often located around shopping malls, do not seem to have attracted the tallest buildings, although they also contribute to the densification of the local area (for a detailed analysis, see Caballero Villaroel, 2018). Figure 7.2 illustrates the capital region's price levels divided in commune levels indicating large differences between the commune with highest levels and the commune on the lower range.

Figure 7.3 shows the trends for a ten-year period, during which permits have increased slightly, and the construction of apartments reveals an overall positive trend. In contrast, the

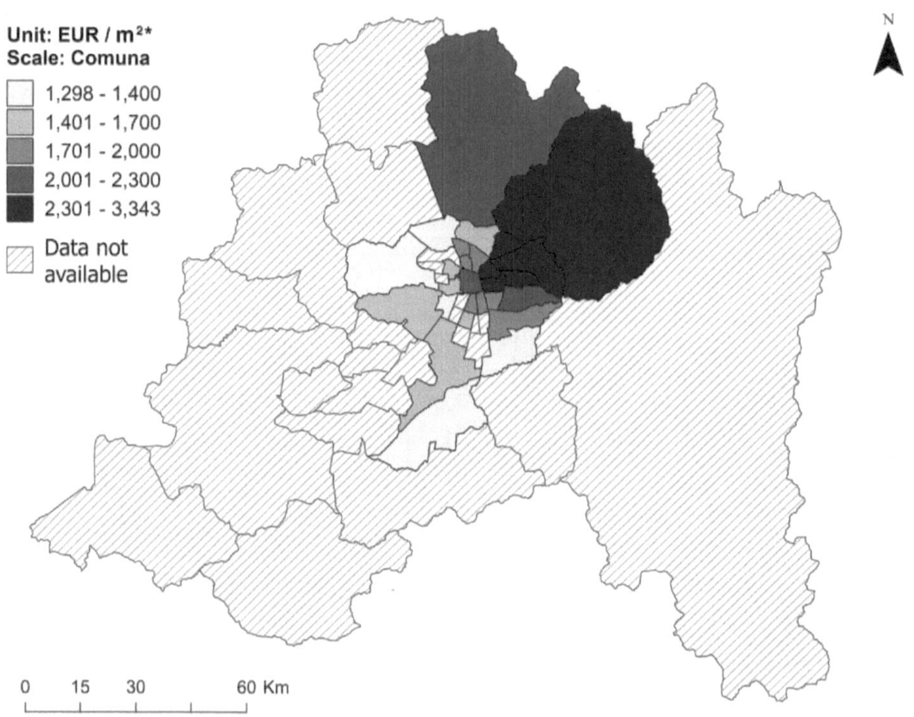

Figure 7.2 Average house price in the Santiago metropolitan region, 2019

Note. *1 Euro = 0.031 Unidad de Fomento (as of 2021 September)

Source: Biblioteca del Congreso Nacional de Chile (2019)

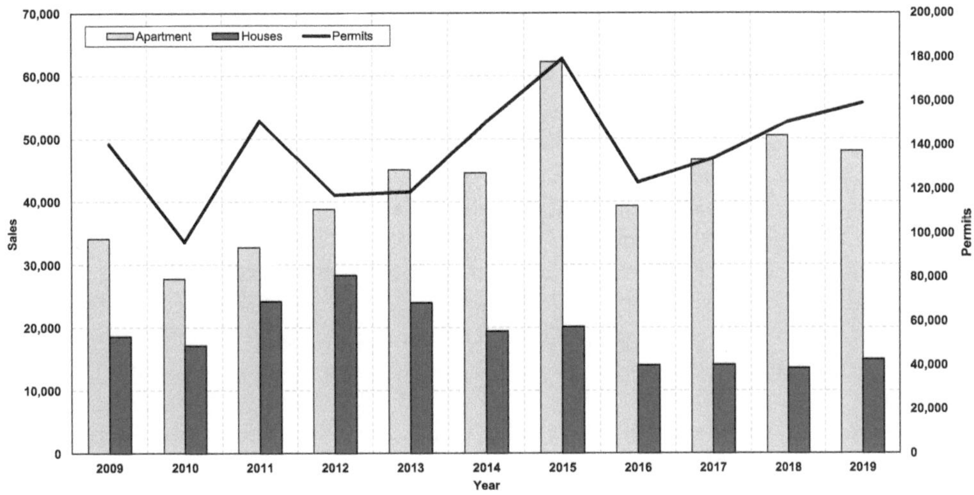

Figure 7.3 Sales by property type and construction permits, 2009–2019

Source: Own elaboration from CCHC (2021a, 2021b)

construction of small houses is decreasing, which can also be held up as an example of the overall trend of densification.

Criticism has been aimed at both oligopolistic market structures in the construction industry (López-Morales, 2016) and general access to housing at the national level. In 2019, the trade association CCHC *(Cámara Chilena de la Construcción)* focused on promoting the development of the construction industry and one of the crucial actors in the Chilean property market presented the report 'ìndice de Acceso a la Vivienda (price-to-income ratio)'.[2] This report argues that access to housing for Chilean households was 'severely unreachable' (CCHC, 2019b), even for families on a median income. The mean Chilean income by household, considered in the report, was 33.8 UF/month (€ 1,087 in 2021 prices), whereas the private housing market's mean price reached 3,089 UF (€ 99,373 in 2021 prices).

The situation was described as widespread in the country where the ratio for 'severely unreachable' (identified as 5) was surpassed at the national level, reaching 7.6. Other countries compared in the report had, for instance, a ratio of 6.5, as in the case of New Zealand, and 5.7 for Australia. The report also emphasized that this complex situation is evenly distributed throughout the country, for example, Santiago reached a ratio of 7.4, while other cities like Temuco had a ratio of 8.8, Great Concepción had a 7.7 ratio, and Gran Valparaíso had a ratio of 7.1. The report emphasizes affordability as one of the main challenges when it comes to access to housing that affects both the private and public sectors.

2. Regulatory framework

Property transactions are in many respects considered as any other goods, and the transactions are regulated through consumer laws. One difference, however, is the land register and cadastre. In a PhD thesis, Correa (2014) describes the historical evolution of the cadastral system in Chile as beginning as early as the 16th century. The study finds evidence of a land register from when the conquistadores arrived, as an attempt to secure returns on their work. Parallel to the ambitions of the conquistadores, the emerging state saw a need to locate national assets, and in the early nineteenth century, international experts were invited to perform cartographic studies. By the end of the nineteenth century and beginning of the twentieth century, public institutions were installed to provide the institutional foundations to develop a cadastre and taxation. Developing a cadastre is a complex and expensive process, but from a fiscal perspective, it is highly important, as it provides the foundation for developing a functioning property tax. Today, the cadastre is administered by the Ministry of National Assets.

According to Sánchez (2012), transaction costs are low when compared with other OECD countries, but it is not clear what is included in the calculations. There are no educational requirements to become a real estate agent in Chile, and real estate agencies need not comply with any special regulations, but rather they are subject to the same laws as any other commercial activity.

The tax reform implemented in 2014 implies that property transactions are now subject to value added tax (VAT), although there are several exceptions and few households are liable to pay VAT.

3. The conveyance process

Although the transaction process may differ somewhat, our research resulted in an emerging general pattern. To understand the process better, an interview with a real estate agent was conducted. However, the numbers and process described ahead should be interpreted as general examples rather than exact figures, as formal information is not available. Figure 7.4 provides an overview of the conveyance process.

Marketing phase

Before getting into the process, we will present the main actors in the process. Sellers do not need to contract a *real estate agent*, but if they do, the agent takes care of much of the administration of the sales. The market of real estate agents is not regulated; nevertheless, courses to become a real estate agents are offered by universities, institutes, and associations, and they keep their own registers of previous students. The diploma from this education thus serves more as a sign of quality and helps build trust. An important part of the agent's work is to provide a match and know which items are available on the market. Often, real estate agents are also active on the rental and commercial property markets. Another important person is the *notary*, who is required to witness copies and the signing of documents. The notary is thus involved in several steps of the transaction. Furthermore, a *lawyer* is often contracted by the buyer, or the buyer's bank, in order to verify the legal status of the property. The bank often also contracts a property appraiser to control the state and market value of the property. The costs for these services are then transferred to the buyer. Finally, the formal registry of the new owner is undertaken by the *conservador de bienes raices*, which could be translated as the 'property registrar'. The property registrar administers the inscription to the cadastre and has access to the information in it. It thus serves a very central function in the Chilean transaction system.

As mentioned, real estate agents often play an important role in the search process and can be used by buyers and sellers to reduce the time and effort spent on finding the right item or the right buyer. For example, agents collect information about the state of the property through

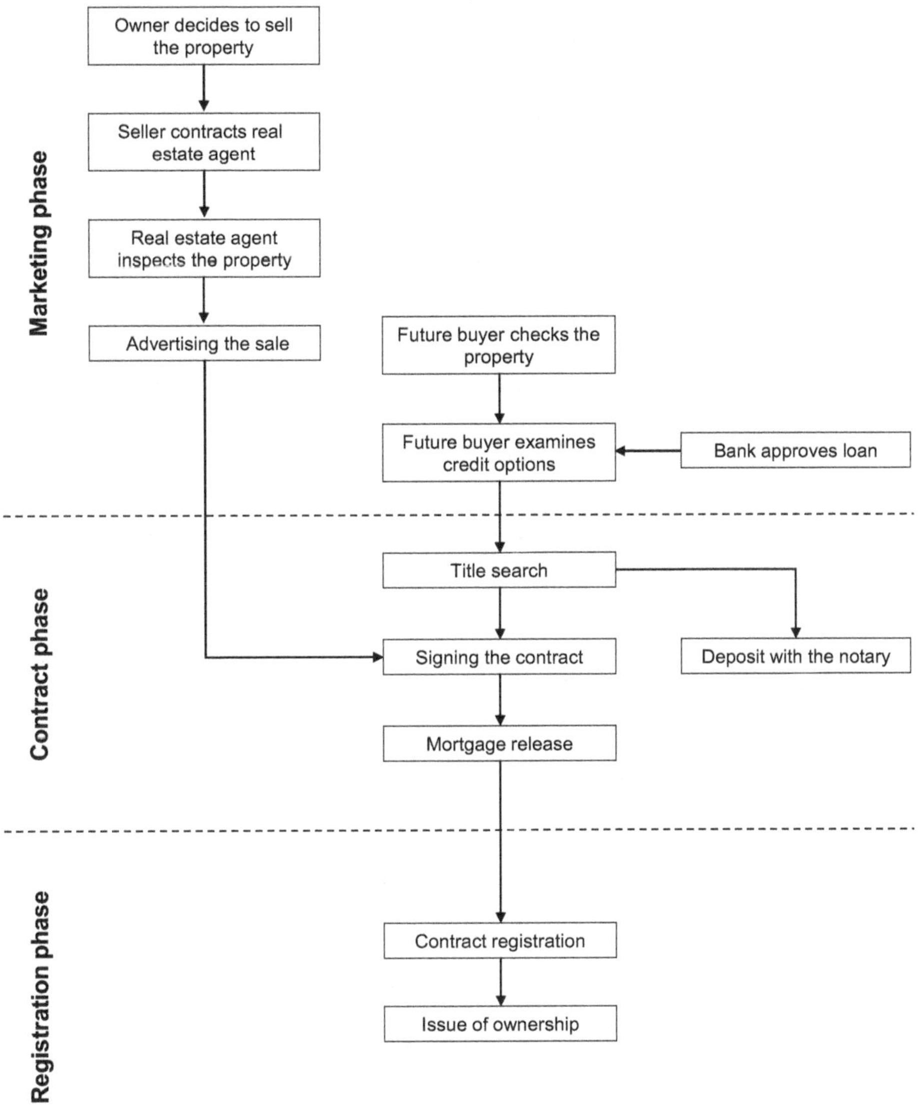

Figure 7.4 The conveyance process in Chile

Source: Author

different official agencies at the national level (Tax Agency, General Treasury) or at the local level (municipalities, service companies), thus ensuring the property is free of debts when the transaction is carried out. Agents charge 4% of the sale, which is split equally between the seller and the buyer; however, there is no formal contract between the agent and the seller. Marketing has moved online, and there are at present two common marketplaces, Portal Inmobiliario and TocToc, but there is also a number of smaller platforms. Other types of social media are also used often by buyers, sellers, and estate agents.

Pre-contract phase

In many instances, the buyer will write an offer to the seller, although this is legally non-binding. Before the actual signing of the sales, it is common to sign a document in which the potential buyer promises to buy the property before moving on to the formal administrative work. This contract can be written by a real estate agent but needs to be signed in the presence of a notary. Often, a down payment of 10% of the property value is paid at this time.

Contract phase

A first step of formalizing the sale is to verify the legal status of the property, which is done through a title search. The process seeks to make sure the seller is the rightful owner with the right to sell the property and that there are no liabilities attached to the properties that would prevent the sale from going through. This is often undertaken by a lawyer, contracted either by the buyer or by the buyer's bank when a mortgage is taken out. However, many of the documents involved are valid only for three months.

Properties are often sold expressing prices in *Unidad de Fomento*, which can be described as a non-circulating currency. Just like the Chilean *peso*, it is managed by the Central Bank of Chile and is an index providing an inflation-adjusted currency used for not only real estate but also cars and other types of capital goods.

The sales contract is signed in the presence of a notary, and the remaining 90% of the sale price is deposited with the notary until the sale has been finalized. In order for it to be legally binding, the formal document needs to be issued by the property registrar. Thereafter, the seller's mortgage is released by the bank.

Registration phase

The final step is the contract registration, in which the sale is registered by the property registrar *(conservador de bienes raíces)*. An ownership title is issued, confirming the formal validity of the sale. The property registrar has access to the cadastre system, which is necessary for the formal registration of the property.

Post-transaction phase

The potential post-contract costs seem primarily related to the failure to verify the property's legal status. Cases have been reported where the seller sells the property after already having previously sold it but before the property was registered, although it is not possible to verify the extent of such problems.

4. Summarizing the conveyance process and its transaction costs

The Chilean property market can be described as a deregulated market in many aspects. First, private ownership is the dominant tenure form, and therefore, transactions concern most Chileans at some point in their life. Second, while the role of the real estate agent is not regulated any more than other types of commodity markets, the role of the notary and the property registrar are. Most of the expenditures are proportional to the value of the real estate, although there are certain fixed costs concerning documents and similar. Table 7.1 outlines the fees and charges applicable to the sale of residential property.

Table 7.1 Fees and taxes during the conveyance process in Chile

Phase	*Fees and taxes*	*Seller*	*Buyer*
Marketing	Real estate agent	2% of property price	2% of property price
Contract	Lawyer's fee	-	App 1% of the property value
Registration	Stamp duty	-	Santiago, Valparaíso, Viña del Mar: 0.2%; rest of the country: 0.3%
	Registration	-	Generally around 0.10% of the property value, with a maximum charge of € 262
Post-registration	Sales tax	-	-
Total (EUR for the model case price)		**€ 2,000**	**€ 6,700 (Santiago, Valparaíso, Vina del Mar)–6,800 (Other areas)**
Total (% share to the model case price)		**2.0%**	**6.7–6.8%**
Excluded costs from the model case		-	Apply stamp duty = 0.3%

Note. Model case price: € 100,000

Source: Author

Notes

1 1 Chile Peso = 0.0010 Euro (as of 1 October 2021).

2 Price-to-income ratio is defined by the OECD as *the nominal house price index divided by the nominal disposable income per head and can be considered as a measure of affordability* (OECD, 2021).

References

Biblioteca del Congreso Nacional, Mapas vectoriales. (2019). Retrieved from https://www.bcn.cl/siit/mapas_vectoriales/index_html

Caballero Villaroel, C. I. (2018). *Análisis de la distribución espacial de la Edificación Residencial en Altura en el Gran Santiago: Año 2016, 166*. Retrieved from http://repositorio.uchile.cl/bitstream/handle/2250/153013/analisis-de-la-distribucion-espacial.pdf?sequence=1&isAllowed=y

Cámara Chilena de la Construcción. (2019a). *Balance Vivienda y Entorno Urbano*. Retrieved from www.cchc.cl/centro-de-informacion/estudios/estudios-balance-de-vivienda/

Cámara Chilena de la Construcción. (2019b). *Ìndice de Acceso a la Vivienda (Price income ratio)*. Retrieved from https://cchc.cl/uploads/archivos/archivos/INDICE_DE_ACCESO_A_LA_VIVIENDA_-_28_agosto_2019.pdf

Cámara Chilena de la Construcción. (2021a). *Permisos de Edificación*. Retrieved from https://cchc.cl/centro-de-informacion/indicadores/permisos-de-edificacion

Cámara Chilena de la Construcción. (2021b). *Mercado Inmobiliario – Ventas Nacional*. Retrieved from https://cchc.cl/centro-de-informacion/indicadores/mercado-inmobiliario-venta-nacional

Celhay, P. A., & Gil, D. (2020). The function and credibility of urban slums : Evidence on informal settlements and affordable housing in Chile. *Cities*, *99*(102605). doi:10.1016/j.cities.2020.102605

Correa, L. A. (2014). *Catastro de propiedad en Chile: Orígenes y evolución*. Universitat de Barcelona. Retrieved from http://hdl.handle.net/10803/145497

Dohnke, J., Heinrichs, D., Kabisch, S., & Krellenberg, K. (2015). Achieving a socio-spatial mix ? Prospects and limitations of social housing policy in Santiago de Chile. *Housing Studies*, *30*(6), 839–857. doi:10.1080/02673037.2014.982516

El Mostrador. (2021, March 25). Precariedad fuera de control: Estudio revela dramático aumento de 74% de familias viviendo en campamentos entre 2019 y 2021. *El Mostrador*. Retrieved from www.elmostrador.cl/noticias/multimedia/2021/03/25/precariedad-fuera-de-control-estudio-revela-dramatico-aumento-de-74-de-familias-viviendo-en-campamentos-entre-2019-y-2021/

INE. (2016). *Segmentación Horizontal del Mercado del Trabajo. Caracterización de la ocupación por sexo*. Instituto Nacional de Estadísticas (INE). Departamento de Estudios Sociales. Retrieved from https://historico-amu.ine.cl/genero/files/estadisticas/pdf/documentos/segmt_mercado_trabajo.pdf

López-Morales, E. (2016). Gentrification in Santiago, Chile: A property-led process of dispossession and exclusion dispossession and exclusion. *Urban Geography*, *37*(8), 1109–1131. doi:10.1080/02723638.2016.1149311

OECD. (2021). *Housing prices (indicator)*. Retrieved from https://data.oecd.org/price/housing-prices.htm

Paredes, D., Iturra, V., & Lufin, M. (2016). A spatial decomposition of income inequality in Chile. *Regional Studies*, *50*(5), 771–789. doi:10.1080/00343404.2014.933798

Salvi del Pero, A. (2016). Housing policy in Chile : A case study on two housing programmes for low-income households. *OECD Social, Employment and Migration Working Papers*, *173*. doi:10.1787/5jm2hzbnqq33-en

Sánchez, A. C. (2012). Building blocks for a better functioning housing market in Chile. *OECD Economics Department Working Papers*, *943*. doi:10.1787/5k9fj3hgsnvh-en

TECHO, Fundación Vivienda & CES. *Catastro Nacional de Campamentos 2020–2021 (Resumen Ejecutivo)*. Retrieved from https://ceschile.org/wp-content/uploads/2020/11/Informe%20Ejecutivo_Catastro%20Campamentos%202020-2021.pdf

8

EGYPT

Yousra Chaaban

1. The Egyptian housing market

The Egyptian[1] real estate market for homes is constituted of two forms: there are rentals, but ownership is the main form. The Egyptian state seeks to provide adequate housing for citizens with low and middle incomes that are suitable for their limited financial capabilities. Therefore, the state intervened recently to issue a law establishing the Social Housing Fund and real estate financing support.

Since 2005, property values in Egypt have increased by 13.7%, but this increase has not been stabile: it decreased by 0.4% in 2006 and by 0.6% in 2007. Then came the global financial crisis of 2008. By the end of that year, housing prices had fallen by 37%. However, the Egyptian real estate market recovered again in 2010; in that year, real estate prices rose by 10%. Despite the decline in the market witnessed during 2011, the real estate price index rose again by 9% in 2013 and by 11.4% in 2014. The year 2015 showed a decrease in house prices by 4.7%, but real estate prices rose once again by 24.7% in 2016 in response to the economic policies adopted by the government from floating of the Egyptian pound.

The volume of real estate financing granted during 2019 amounted to more than € 0.14 billion, with a growth rate of 18.2% compared with 2018, reaching € 0.12 billion. Globally, real estate sector revenues in Egypt represent 16% of the gross domestic product (Source: Investigate.me, n.d.). Experts believe that real estate investment in Egypt remains one of the safest and most profitable economic sectors in 2021.

In 2020, the Egyptian real estate market was badly affected by the coronavirus crisis. The value of the property map index reached the lowest value recorded during the last five years, and a decline in demand amounted to 29.59%. The Central Bank of Egypt, in March 2020, reduced interest rates by 300 basis points and set the lending rate at 10.25%. Despite this crisis, real estate remains second in terms of value traded in Egyptian pounds. Accordingly, June 2020 witnessed the largest growth in real estate demand in Egypt's history, which focused on monitoring the demand for real estate in the Egyptian market, as the demand increased by 20% with the return of public interest in buying real estate.

As some experts estimate, prices for residential units in luxury housing and resorts cost approximately € 0.08 million, while prices for residential units in popular housing vary between

DOI: 10.1201/9781003130383-10

€ 13,750 and € 38,500 due to the price increases in 2016; thereafter, the Egyptian pound has fluctuated compared with the dollar.

As estimated by experts, the real estate market is expected to continue growing, as it is one of the most important sectors in the country, which greatly affects both the macroeconomic performance and the performance of the Egyptian Stock Exchange (Digital Portal in Egypt, n.d.).

The Egyptian real estate market is going through a considerable *revolution* because of the various legislative amendments that have taken place and are taking place today (the data processed in this chapter are those updated until March 2021).

The real estate market is also experiencing an *evolution*, due to the huge projects that the real estate companies are currently working on in different regions in Egypt, in particular, the new capital of Egypt, also called 'The New Desert Capital' or the 'New Administrative Capital' (General Authority for Investment and Free Zones, n.d.).

In addition, the government tends to build fourth-generation cities, smart cities, and new cities for each governorate, for example, New October, Badr City, New Obour, Shorouk City, New Damietta, New Assiut, 10th of Ramadan City, New Sohag, New Beni Suef, and the New Alamein.

2. Regulatory framework

You can go through the conveyance process on your own, but it is advisable to be assisted by a lawyer. The most common way to transfer ownership is by using a sale agreement (see Articles 418 and following; and 932 and following of the Egyptian Civil Code). The ownership is not transferred automatically; instead, you have to go through the governmental property registration authority in order to transfer it (see point of registration phase later in this chapter).

In addition, legal advisors know the special legislations in relation to the ownership of real property such as Law No. 114 of 1946 concerning registration and its amendments; Law No. 27 of 2018, concerning the real estate registration in new urbanism and its regulations; and Law No. 196 of 2008 concerning the tax on real estate and its amendments. New legislation concerning the New Unified Building was adopted in 2021 amending the Law No. 119 of 2008. This will allow the establishment of projects for public benefit outside the urban space and the organization of the urbanization.

Today, the Egyptian Parliament and the Egyptian government are studying a set of laws that aspire to eliminate slums and building violations (see Law No. 17 of 2019 concerning reconciliation in construction violations) and other legislation aiming to regulate the relationship between the owner and the tenant, to regulate the work environment for real estate developers, and to avoid crises in the real estate market.

It is important to point out that Law No. 230 of 1996, regulating the ownership by foreigners of buildings and land plots, governs the ownership of property by natural foreign persons for residential purposes. There are restrictions on such ownership. Non-Egyptians cannot buy more than two properties with a total of 4,000 m^2. The purpose of the purchase must be for the buyer's own use or for members of his family. Foreign ownership is prohibited, especially in the border areas, and the North and South Sinai governorates. The ownership in these areas is regulated, allowing some exceptions by many laws and decrees, but making a special case for Sharm el-Sheikh. The foreign buyer who acquires ownership of a building or vacant land is obliged to build on it within a period not exceeding the five years following the month of the purchase. Moreover, for non-Egyptians special rules are applied for acquiring and selling residential properties in Egypt.

3. The conveyance process

Marketing phase

In the search for a property in Egypt, you can refer to, first, the different agencies of the private real estate developers (the most common market). And, due to the coronavirus conditions, many of these companies have shifted to online marketing developing specific websites. Second, you can also refer to the governmental projects, but this depends on the proof of your resources. And for that, you will turn to several governmental bodies such as the Housing and Development Bank, the New Urban Communities Authority, and the newly established real estate finance fund. Third, banks in Egypt play an important role in offering several financing opportunities to buy a property.

Finally, you can also refer to a real estate agent intermediary: independently or through an agency. The regime of an intermediary has been regulated recently by the Ministry of Commerce and Industry Decision No. 38 of 2020, issuing controls to monitor real estate brokers regarding combating money laundering and terrorist financing. No person is allowed to practice the profession of brokerage unless they have obtained the required licenses from the competent authorities. The real estate broker's commission is divided into two stages: First, the broker gets a commission from the buyer when the broker accompanies the buyer to see the house or real estate. This commission is estimated to be on average between € 6 and € 17 in the case of a rental and from € 55 to € 275 in the case of ownership. Second, in the case of selling, the broker gets a percentage of the price of the property from the seller of the property, and this percentage ranges from 1.5% to 2.5% in the case of ownership. Note that the buyer and seller have their own brokers and are responsible only for the broker fee of the one they have hired. In the case of a rental, the value of the commission shall be estimated by one month of rent or what is agreed upon in the contract concluded between them.

Pre-contract and contract phase

Concerning the pre-contract phase, when it is about a sale's agreement, the general rules for the formation of the contract and its execution are applied. The Egyptian Civil Code governs and regulates the rights and obligations of both the seller and the buyer. These contractors must be able to give enlightened consent and must be capable to form a contract. As for the price, this is specified according to the agreement of the two contractors, depending on the segment of the housing market concerned: private sector, social sector, and especially the place and the district where the property is located.

The promise to sell and the down payment are possible for buying a real estate property (see Article 103 and following Egyptian Civil Code). This article cites that

> (1) A down payment made at the time a contract is entered into indicates that either party may repudiate the contract, unless the contract indicates otherwise. (2) If the person who made the down payment repudiates the contract, his down payment is forfeited. If the person who received the down payment repudiates the contract, he shall reimburse double the amount. The above applies even if no harm is caused by the repudiation.

Yet, practically, you can pass this step and go directly to the contract itself. The purchaser must be sufficiently familiar with the object sold, as mentioned in Article 419 of the Egyptian Civil

Code. This familiarity is considered sufficient if the contract contains a description of the object sold and its basic characteristics, such that it may be identified.

The contract phase is regulated by Egyptian Civil Law. As for the ownership of real estate, it will not be transferred unless the contract is registered, as mentioned in Article 204 of the Egyptian Code (See Law No. 114 of 1946 concerning the regulation of the registration of real estate and its amendments). The formation of the contract is valid, but the ownership will not transfer until registration.

It is important to mention that, during the contract phase and its registration, the Egyptian bar/syndicate takes a percentage of the apartment value. The fee rises to 1% in return for certification of the contract (Ministry of Justice – Real Estate Registration and Documentation Authority, n.d.). And for the real estate disposal tax, Article 42 of Law No. 196 of 2008 specifies a percentage of 2.5% on the total value of disposal of built real estate or land on which to build. The buyer is obligated to pay the tax within 30 days from the date of disposal, otherwise, a fine will be applied.

Registration phase

As stipulated by Egyptian law, when you buy a property (an apartment or land), you will not be the legal owner until registering the contract in the competent registration authority. The 'full registration' of the land or building is carried out in real estate registry to which the land or building belongs geographically. It is the recognized method to transfer real estate from the owner.

However, there is another usual way called 'signature of validity', although it is not the proper one to register a property. It is often used but does not provide security or protection. This way does not transfer ownership.

You can obtain 'full registration' in the following one of two ways. The first is by consensual registration, whereby the two parties aim for a direct registration in the real estate registry, or the buyer goes into the agreement on his own behalf and authorizes the registration. The second way is a type of registration that goes through the court. The judgement itself is registered in the real estate registry. In this latter type of registration, it is necessary to take into account the completion of all the procedures and documents required to obtain the correct judgement and enforcement from the first judgement. Otherwise, to postpone the sale for a second audience will force the party to pay 7% of the value of the apartment to the court.

The 'signature of validity' method removes restrictions. It is also not nearly as time-consuming as full registration. When a property is registered under the signature of validity, the buyer begins by obtaining what is called a 'negative certificate' from the registry. This document certifies that there are no outstanding debts or claims on the land. The tax authorities then produce a document showing what taxes are due on the property. The next step is to draft a contract that outlines the property size, purchase price, and so forth. In the final step, the buyer files a suit to legally certify the signature of the seller, which, in effect, completes the sale.

It is important to point out that, at the beginning of March 2021, the new legislative amendment will have come into force, standardizing or unifying the registration process in Egypt. This will keep the 'full registration' method as the only method of registration in Egypt, thus phasing out the 'signature validity' method. This result is that from today, the public registry authority will refuse to recognize primary contracts and contracts that were validly signed (following the second method of registration).

The following documents are usually required to complete the registration process: the title deed contract, the preliminary contract between the two parties, and an application for requesting the registration. If the property descriptions change, or in the case of the absence of an old list, a new survey record for the residential unit is needed. The Real Estate Registry's acceptance of the request for registration on the property is also required *(maqbul lilshahr)*, which involves a costly real estate certificate from real estate taxes if the property is old and a building license if the property is new.

In practical terms, you begin by submitting an application to the Real Estate Registration Office, which has attached to it the ownership data, tax assignment, and other supporting documents. Thereafter, it is important to make a survey determination list for the apartment and have it reviewed in the real estate registry. You should also obtain a site inspection by the Egyptian Surveying Authority and an approval of inspection report by the Measurement Department. At the same time, you must prepare a registration for the claim petition in the real estate registry. After this, the final contract is presented to the lawyer's syndicate for verification and ratification, and then the Real Estate Registry will approve the sale contract. It is the buyer who delivers the contract to the Real Estate Registry to make the contract legal and then the registration number is given.

As a point of interest, in 2020, the Egyptian Ministry of Justice decided to determine the real estate fees for registering built real estate and land space. Before this date, the fee was relative to the property value. To summarize, the fees were set as follows: € 28 for a property whose area does not exceed 100 m^2, € 55 for a property whose area does not exceed 200 m^2, € 83 for an area not exceeding 300 m^2, and € 110 for any property with an area exceeding 300 m^2. In addition, to facilitate the payment of some of these costs and to follow the digitalization phase, the Egyptian government has created a dedicated portal (for website, see the references at the end of this chapter) (Digital Portal in Egypt, n.d.).

However, these are not the only fees. There are also fees paid for the detection, identification, and expert inspection of the property; fees for maps and appraisal; the lawsuit fee; fees for contracts at the syndicate, and other legal fees. It is not possible to determine a fixed value for all fees, as every property has its own unique conditions.

Post-transaction phase and taxes

During the registration phase, we have highlighted the costs that this step in particular can entail. However, these are not all the costs that can be incurred, in particular, the tax on real estate. Concerning this, Article 12 of Law No. 196 of 2008 cites that

> the tax rate shall be (10%) of the annual rental value of taxable real estate, after excluding (30%) of this value for places used for residential purposes and (32%) for the percentage used for non-residential purposes, in exchange for all expenses incurred by him. The person charged with paying the tax, including maintenance expenses.

Article 18 of the same law gives same cases of tax exemption, such as the real estate unit that the taxpayer takes as a private residence for themself and their family and other cases. The real estate tax is due on the first of January of each year, and you can pay in two equal instalments: the first until the end of June and the second until the end of December of the same year (Egyptian Real Estate Tax Authority, n.d.).

4. Summarizing the conveyance process and its transaction costs

To summarize the process of buying property in Egypt, it is a process that is varied and complicated. It is varied from the point of view of marketing offers, as the sources of where to acquire property in Egypt are diverse and varied. The process is also complicated, particularly the registration process, which currently has many new rules and costs. As shown, the government is trying to put this process into some order by unifying different types of registration and facilitating procedures by creating a digital platform for real estate registration and the payment of taxes. The ownership is conveyed mainly with a sales contract. In the pre-contract phase, a down payment or a promise to sell the property can be made, but this is not an obligatory step. Being assisted by a lawyer is possible, but not required. There are many fees to pay, but for the moment, the existing regulations are not expensive.

Note

1 1 Egyptian Pound = 0.055 Euro (as of 1 October 2021).

References

Abdel Razzaq Al-Sanhouri, A. W. (2010). Part 9 reasons for earning ownership. In *Explaining civil law* (1st ed.). Cairo: Dar Al-Shorouk.

Baker Mackenzie. (2016). *Report on doing business in Egypt*. Retrieved from www.bakermckenzie.com/-/media/files/insight/publications/2016/03/bk_egypt_dbi_2016.pdf

Digital Portal in Egypt. (n.d.). *Your government transactions from one place*. Retrieved November 20, 2021, from https://digital.gov.eg

The Egyptian Civil Code (No. 131 of 1948).

Egyptian Real Estate Tax Authority. (n.d.). *Tax on built real estate*. Retrieved November 20, 2021, from www.rta.gov.eg/Home/BuildingTaxType

General Authority for Investment and Free Zones. (n.d.). *Real estate*. Retrieved November 20, 2021, from www.investinegypt.gov.eg/english/pages/sector.aspx?sectorid=84

Investigate.me. (n.d.). *Main page*. Retrieved November 20, 2021, from https://invest-gate.me

Ministry of Justice – Real Estate Registration and Documentation Authority. (n.d.). *Official website of the public registry office in Egypt*. Retrieved November 20, 2021, from https://rern.gov.eg

9
FRANCE

Béatrice Balivet and Stéphane Berre

1. The French housing market

The French real estate market for homes consists of two main forms: rental and ownership. Under the cover of a single meaning, the ownership of a house encompasses various situations. Usually, the transfer of ownership is instantaneously put into effect through a sale, or sometimes through a liberality. The owner then has a right of real property, which confers all uses of the good acquired. More marginally, a right of property can be progressively acquired within the framework of social homeownership schemes. The contract of tenant-ownership thus includes a period of rental use during which the holder of the right is a tenant, followed by an option time during which they decide whether to buy or not. Finally, the right of ownership may be temporary and concerns only buildings that have been constructed or renovated on the basis of a long-term lease. Also, a distinction is made between the ownership of land and the ownership of a building. The owners of a home own the house until the end of a long-term lease and also have a real right to the use of the land.

The housing market in metropolitan France is made up of 36.3 million homes, with an average increase of 1.1% per year since the 1980s (National Institute of Statistics and Economic Studies [INSEE], 2018). A total of 82% of the housing are primary residences occupied by an owner (57.6%), a tenant (39.9%), or a person with another status (2.5% housing in France on 1 January 2018, Insee Focus n° 126, October 2018). In 2020, the yearly volume of transactions exceeded 1 million. In contrast, a total of 13,000 sales of social housing in the social sector are below the target of 40,000 per year. In regard to housing prices, they have almost doubled between 1996 and 2016 (Index 2015 = 100), as shown in Figure 9.1.

A study carried out on 17 French metropoles indicates five phases of price evolution. Property prices increased sharply (+18% for collective housing and +14% for individual housing) between 1997 and 2005, followed by a phase of deceleration in 2006–2007, before the subprime crisis broke out in 2008, which led to a decrease in prices of up to 8% for old houses. This was then followed by a moderate decrease in prices between 2012 and 2015, before a new increase began in 2016. However, prices fluctuate between territories and even within a given territory. The median price for existing houses and flats is € 2,160/m^2 in 2020,[1] except for in Paris, where the median price for an existing property is € 11,040/m^2 in 2020 for a flat and € 12,100 for a house.[2] The strongest increases are located in the

DOI: 10.1201/9781003130383-11

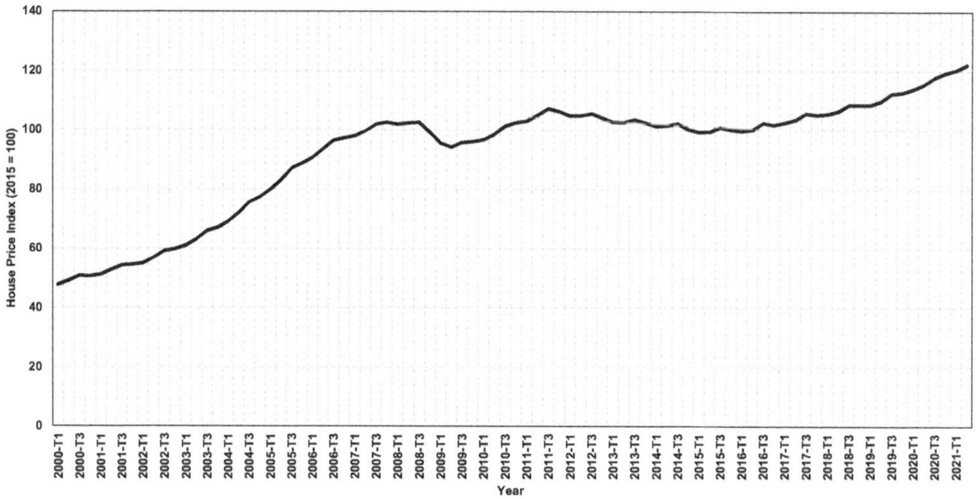

Figure 9.1 Yearly change in house prices in France

Note. House price index, new and second-hand – raw series – base 100 = annual average of year 2015

Source: immobilier.statistiques (n.d.)

Metropolis of Bordeaux (price evolution of up to 250% for houses) and the Metropolis of Lyon (+168%: Current currency, source: Notaries-INSEE, PERVAL and Insee indexes) as shown in Figure 9.2.

More recently, in the context of actions carried out on the 7–8 million thermal residences to fight against global warming, two phenomena have been observed. First, the French notaries annually measure the impact of a property's energy performance on its sale price. The 2020 study shows that the poor energy performance of a property has a negative effect on the value of the property, which can be as high as 12%, although this varies from one region to another. On other hand, the good performance of a property seems to have only a limited impact, at best 3%. Second, notaries are considering a change in property market due to environmental transition and the effects of COVID-19. In 2021, real estate projects are more likely to be carried out in the provinces or suburbs than in Paris.[3]

A new INSEE survey on housing conditions in France is currently in progress, and it will update the 2013 data (INSEE References, 2017 edition – Files – Housing stock). In the survey, the average size of housing in 2013 was 80/m^2. Half of them have three or four rooms. Existing housing, built before 1949, represents almost a third of the stock, and those built after 1999 represent 15%. Since 1999, more than 212,000 housing units built before 1949 and 194,000 built between 1949 and 1974 have been destroyed. However, it is the stock built between 1949 and 1974 that has been subjected to the most demolitions: 2.3% against 1.8% for stock built before 1949. Recent buildings are not evenly distributed over the metropolitan territory. The central towns tend to have housing built between 1949 and 1974, the peri-urban areas have housing mainly dating from 1975 to 1998, and 'multi-polarized' suburbs have housing built after 1999. In order to relaunch housing policy by creating a 'shock of supply', measures have been taken to remedy the lack of new housing and to build more and better homes in France, setting a target of 500,000 new homes per year. During 2020, 99,500 new homes were reserved

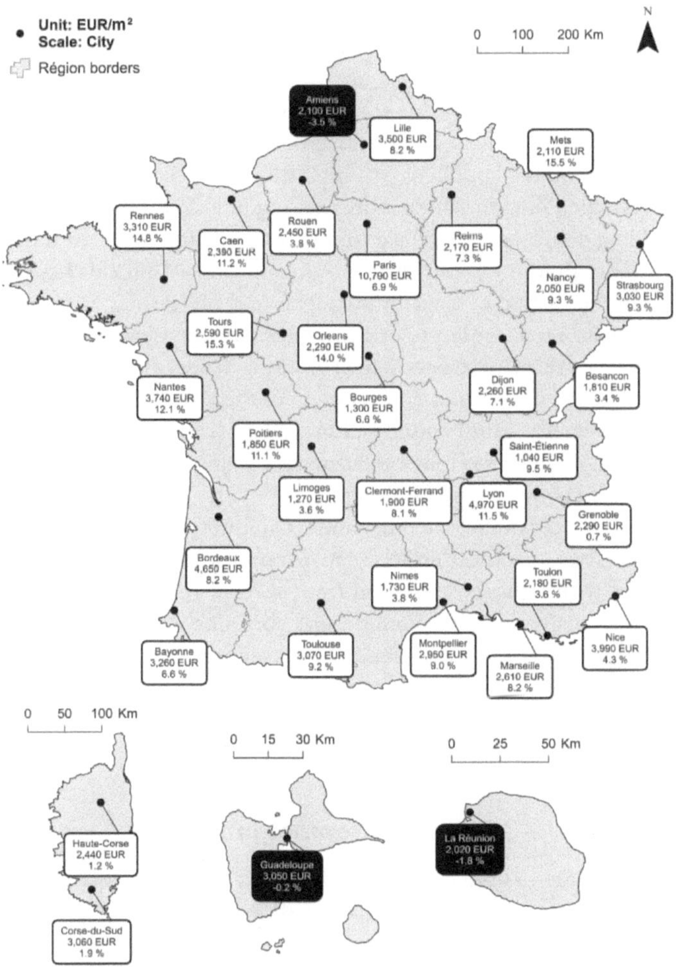

Figure 9.2 Change in apartment prices in France, 2019–2020

Source: Bases immobilières des Notaires de France

in metropolitan France, with a decrease of almost a quarter of the market compared with 2019, mainly due to the COVID-19 crisis.

2. Regulatory framework

The conveyance process for real estate appears simple, given that the sale contract is consensual in principle. According to Article 1583 of the Civil Code, a sale 'is perfected between parties, and ownership is automatically transferred by the buyer in relation to the seller as soon as the property and price are agreed, even when the property has not yet been delivered or the price paid'. A notarial act is thus not necessary for the contract to be valid, but in terms of the requirements for land registration, notaries have a monopoly in France on the completion of these formalities. In practice, however, a clause in the preliminary contract makes it obligatory to have recourse to a notary in order for the sale to be regularly concluded. The notary's role is important for guaranteeing the security of the transactions, as the legal framework for the transfer of a property may vary according to different parameters.

First, all property sales are governed by a corps of common rules, such as rules of the Civil Code on sales and rules on land publicity resulting from a 1955 decree; therefore, they are subject to specific rules depending on whether the housing unit is a condominium lot in an existing building or a building yet to be built, for example. Information given to the candidate for the acquisition of a condominium lot as early as possible in the sale process has been strengthened since 2014. They are informed of their future rights and obligations as co-owners, the seller's financial situation regarding to syndicate of co-owners, and the physical condition of the building (Article L. 721–2 of the Construction and Housing Code). In addition, the candidate for acquisition must be able to give enlightened consent to the purchase. Rules also apply to protect the syndicate of co-owners against unscrupulous sellers (e.g. willingness to sell while owing debts to the syndicate) or unscrupulous candidates for acquisition (e.g. the candidate is already a co-owner within the co-ownership and is in debt).

Second, when the candidate for the purchase is a consumer or a non-professional purchaser, they benefit in these two qualities, respectively, from protective rules of consumption law if they have a contract with a professional, and from specific rules in terms of loans (any sale is concluded under the legal suspensive condition of obtaining a loan) or the right of retraction provided for in Article L. 271–1 of the French Construction and Housing Code (cf. ahead).

Third, specific conditions, including resource limits for the buyer and price limits, depend on the segment of the housing market concerned: private sector, social sector, or (since 2014), the intermediate sector.

3. The conveyance process

The aim here is to describe the typical process of selling an existing house (i.e. completed more than five years ago) between private individuals, as shown in Figure 9.3.

In 2019, the average time it took for a house to sell was 82 days, with a minimum time of 28 days and a maximum time of 135 days (Real estate sales: average time by city in France 2019 | Statista).

Marketing phase

The house is put on the market:

- in 50% of cases through an estate agent whose activity is regulated by the law of 2 January 1970, known as Hoguet law.
- in 35% of cases directly between the seller and the buyer. This proportion is steadily increasing.
- in 15% of cases through the intermediary of other professionals, including notaries who, in certain regions (notably the west of France), are very active in property transactions. The notary's office is also developing new forms of marketing, such as an entirely dematerialized interactive property sale (Notaires de France, n.d.b).

When a house is put up for sale through a professional, a sales mandate must be signed in as many copies as there are parties. This mandate may be simple or exclusive; in the latter case,

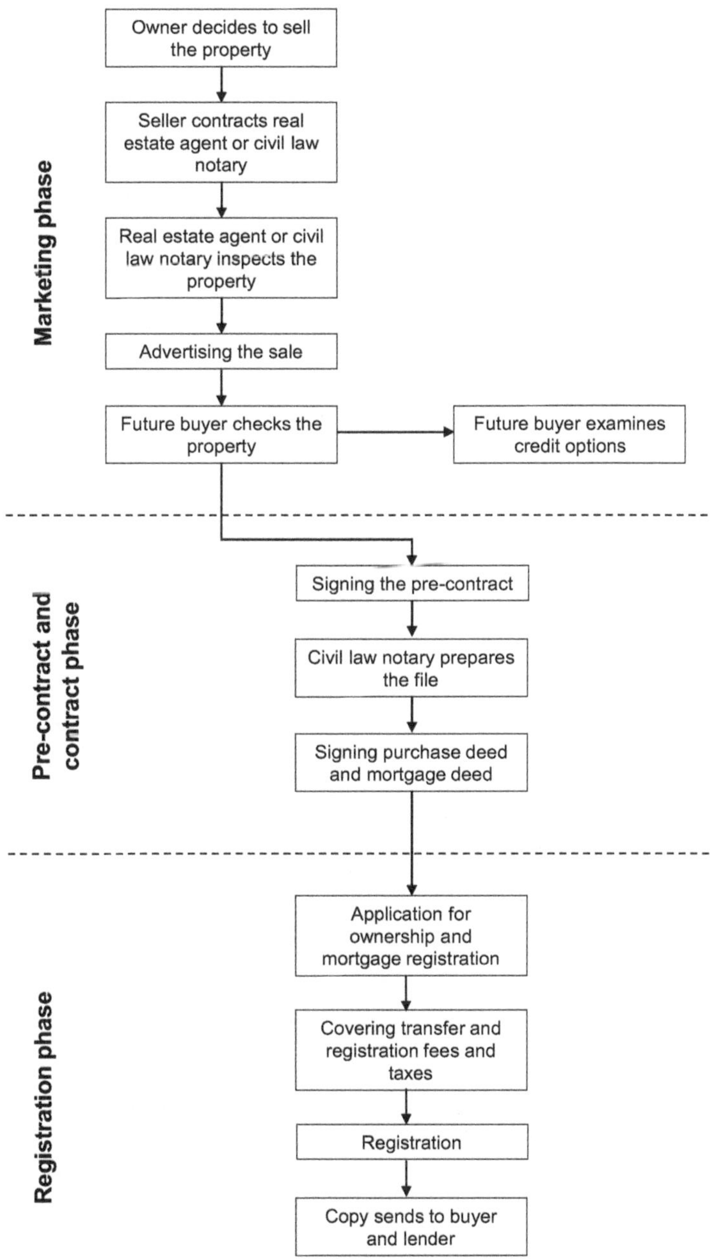

Figure 9.3 The conveyance process for real estate ownership in France

Source: Author

it is irrevocable for a period of three months. It must also set out the terms of the agent's fees. In most cases, this is between 3% and 9% of the sale price and is paid by seller. In all cases, the law regulates the content of the advertisement regardless of which media is used (window display, publication on the internet, or in a newspaper). In addition to a precise description of the property and, where applicable, the existence of co-owners, the advertisement must specify the asked sales price, the agent's fee, the party responsible for paying the agent's fee, and the property's energy performance.

Pre-contract phase

The pre-contract phase is not obligatory, but it is very common because, although the sale of a property is, by law, based on a consensual contract, its preparation has been gradually made more complex by legislation in order to ensure the protection of the purchaser. Despite its name, it is nevertheless a contract, with legal obligations for each of two parties. It can take the form of a unilateral or a synallagmatic promise to sell, and there are two contracts that have different consequences for buyer and seller. Both can be established by private act or by notarial act. The promise to sell is now becoming more common.

In the promise of sale, also known as a 'unilateral promise of sale', the owner promises the candidate buyer (called a beneficiary) to sell the property at a specific price. The owner thus gives the buyer an exclusive 'option' for a limited time (two to three months), during which the owner is not allowed to sell the property to another. Accordingly, the purchaser pays the seller an immobilization indemnity generally equal to 10% of sale price. If they decide to buy the property, this indemnity will be deducted from the total price to be paid. However, if the purchaser decides not to buy or does not express an acceptance of the sale within the option period, the indemnity will be retained by the owner as compensation. In order to be valid, the promise to sell must be registered with the Tax Authorities; the registration fee payable on this occasion by the purchaser is € 125. In 'compromise' or a synallagmatic promise to sell, both the seller and buyer promise to conclude the sale at a jointly determined price.

In legal terms, the synallagmatic promise of a sale is therefore equivalent to a contract of sale. If one of the two parties renounces the sale, the other can force them to carry it through and also demand damages. The signing of the pre-contract is accompanied by a payment by the purchaser of a sum between 5% and 10% of sale price. This is called a deposit and will be deducted from the purchase price when the notarial act is signed. This promise most often contains suspensive conditions which allow parties to regularize the final contract of sale only if certain events that are necessary for completion of transaction take place before the definitive act of sale. For example, it may be a pre-condition that the buyer obtains a loan to finance the purchase (this is a minimum of one month) or the buyer must carry out certain work on the property, or perhaps an urban planning easement is discovered.

Finally, it purges the ten-day right of retraction that a non-professional purchaser has. The exercise of this right does not have to be motivated and is completed by a simple registered letter with a receipt notice. In all cases, the sums paid by the purchaser must be returned in full.

Contract phase

The final sale must be concluded by notarial act. This formality is not required for the validity of the sale but rather to ensure that it can be blocked to third parties by its publication in the real estate register (Notaires de France, n.d.a). The notarian system also, indirectly, impose guarantees of reimbursement of real estate loans. Indeed, the contract establishing a mortgage must be made in notarial form. Similarly, in the case of registration by bank of its lender's privilege, the origin of funds and declaration of use must also be established by notarial act.

The costs of this contract of sale are to be assumed by purchaser. This principle is not a matter of public policy and is rarely derogated. The fee payable to the notary is determined by a national tariff set by public authorities and is proportional to the value of the property – generally 1% of the purchase price. In addition, there are taxes (in particular, land registration tax or VAT); as a general rule, these taxes amount to 5.81% of the purchase price. The buyer is required to pay the taxes, and the rate may be adjusted by the department where the property is located (immobilier.notaires, n.d.).

The seller, for their part, is subject to pay a tax on the capital gain they make; this tax is calculated at a rate of 19% plus social security contributions at a rate of 17.2%. Exemptions exist, for example, in the case of the sale of a property that has been held for more than 30 years or when it is the seller's principal residence. This tax is calculated by a notary and paid by the notary in the name of, and on behalf of, the seller.

The price is usually paid by the notary's accounting department, which hands it over to the seller after deducting taxes and costs, usually within 2 to 3 weeks.

Registration phase

Once the contract of sale has been signed, the notary is responsible for the so-called merged formality. In doing so, the notary carries out two separate formalities:

- *Civil formality*: that of land registration, which ensures that the sale is opposable to third parties.
- *Fiscal formality*: the registration formality, which is a fiscal formality during which the various taxes required by the sale are paid.

This formality is carried out in a dematerialized form using software developed by the public authority and notary office, Télé@cte. The notary who writes the act of sale sends all the information to the local land registry, which carries out regulatory checks before publishing the act. In this regard, a so-called legal security contribution is levied at a rate of 0.10% of the purchase price, and it is payable by purchaser.

The Land Registry Service, which currently manages the real estate file, is a State administration attached to the General Directorate of Public Finance. In addition to publishing contracts of sale, it is responsible for publishing all acts relating to real estate or real rights and for issuing, to any person who requests it, records that make it possible to identify the owner of a property and its successive mutations. A trial initiative is currently being conducted to give the notary's office direct access to this property file (Access to the file by notaries project).

4. Summarizing the conveyance process and its transaction costs

To summarize the process of selling property in France as illustrated in Table 9.1, it is a process that appears easy at first but may take longer than in other countries, particularly those of common law. The pre-contract phase has become increasingly important, as the applicable regulations have become more complex to ensure better information for the buyer. The intervention of a notary – a public officer appointed by State, and as such, an impartial advisor to each of parties – ensures the greatest degree of legal security possible. And this is carried out at a relatively low cost because, on average, the cost of a real estate transaction, all elements included (remuneration of real estate agent and notary, various taxes and duties) amounts to a little less than 10% of the purchase price.

Table 9.1 Fees and taxes during the conveyance process in France

Phase	*Fees and taxes*	*Real estate*	
		Seller	*Buyer*
Marketing	Real estate agent's fees	3–8% (of price)	-
	Energy performance certificate	€ 250	-
Contract	Notary fees	-	€ 0–6,500 = 3.945% € 6,500–17,000 = 1.627% € 17,000–60,000 = 1.085% € 60,000 and above = 0.814%
	Bank mortgages fees	-	2% (of mortgages)
Registration	Stamp tax	-	-
	Land registration fee	-	0.1% (of price)
Post-registration	Capital gains tax	36.2% (of profit)	-
	Legal fees	-	-
	Sales tax	-	5.09–5.81% (of price)
Total (EUR for the model case price)		**€ 3,250–8,250**	**€ 6,004–6,724**
Total (% share to the model case price)		**3.25–8.25%**	**6.0–6.7%**
Excluded costs from the model case		36.2% CGT	2% mortgages fee

Note. Model case price: € 100,000

Source: Author

Notes

1 Retrieved from www.immobilier.notaires.fr/en/home-prices?periodeReferences=1

2 Retrieved from https://immobilier.statistiques.notaires.fr/priximmobilier?typeLocalisation=DEPARTEMENT&codeInsee=75

3 Retrieved from www.notaires.fr/en/housing-tax-system/french-property-market/french-property-market-analysis

References

immobilier.notaires. (n.d.). *Real estate statistics from notaries*. Retrieved November 20, 2021, from www.immobilier.notaires.fr/en/home-prices?periodeReferences=1

immobilier.statistiques. (n.d.). *Prix au m² de l'immobilier*. Retrieved November 20, 2021, from https://immobilier.statistiques.notaires.fr/prix-immobilier?typeLocalisation=DEPARTEMENT&codeInsee=75

National Institute of Statistics and Economic Studies (INSEE). (2017). *Housing conditions in France INSEE references*. Retrieved from www.insee.fr/en/statistiques/2655734

National Institute of Statistics and Economic Studies (INSEE). (2018). *The housing stock in France on 1 January 2018 Insee Focus nr 126*. Retrieved from www.insee.fr/en/statistiques/3733235

Notaires de France. (n.d.a). *Main page*. Retrieved November 20, 2021, from www.notaires.fr/en

Notaires de France. (n.d.b). *French property: Analysis of the market*. Retrieved November 20, 2021, from www.notaires.fr/en/housing-tax-system/french-property-market/french-property-market-analysis

10
ITALY

Enrico Al Mureden

1. The Italian housing market

The Italian real estate market for homes consists of two main forms: rental and ownership. Historically, real estate is a common form of long-term investment policy for Italian families. According to the Italian National Institute of Statistics (ISTAT), 80% of Italians live in a house they own.

If we look at price growth, the Italian real estate market showed some resilience in view of the COVID-19 pandemic, as ISTAT preliminary data confirm (ISTAT, 2020, December). In fact, the growth trend of the house price index (IPAB) continues in a context of recovery in sales volumes that increased by 3.1% in the third quarter of 2020, as recorded by the Real Estate Market Observatory of the Revenue Service (Agenzia delle Entrate) for the residential sector. This is after a −27.2% in the second quarter attributable to the measures adopted to contain COVID-19 (largely lifted in the summer period), which drastically limited the possibility of stipulating notary deeds, especially in April (Lungarella, 2021). In particular, despite a decline in existing house prices, new houses prices have risen. On average, in the first three quarters of 2020, compared with the same period in 2019, house prices increased by 2.0%, with the prices of new houses recording +2.2% and those of existing houses (which account for more than 80% on the aggregate index) that grow by 2.0% (Bankitalia, 2021). The long-term evolution of IPAB is shown in Figure 10.1, giving the average price (solid line), new houses price (dotted line), and existing houses price (dashed line).

However, when looking at the real estate market in a dynamic perspective, we can see that total number of sales amounted to 157,126 in the first quarter of 2020 (−17.9% compared with the fourth quarter of 2019 and −17.7% on an annual basis) and to 149,764 in the second quarter (−17.3% compared with the previous quarter and −30.8% on an annual basis). The trend is negative in all geographical areas of the country (Center, −23.7%; South, −20.7%; Islands −19.9%; Northwest, −14.1%; Northeast, −10.7%; total Italy, −16.9%). A total of 95% of the deeds signed concern transfers of ownership of residential properties (142,265), 4.7% were for economic use (7,002), and 0.3% were for special use and timeshare (ISTAT, 2021a, p. 497).

If we look at construction works, we see that, in the second quarter of 2020, the residential sector experienced a sharp decrease in the number of homes and in the usable area ('walkable'

DOI: 10.1201/9781003130383-12

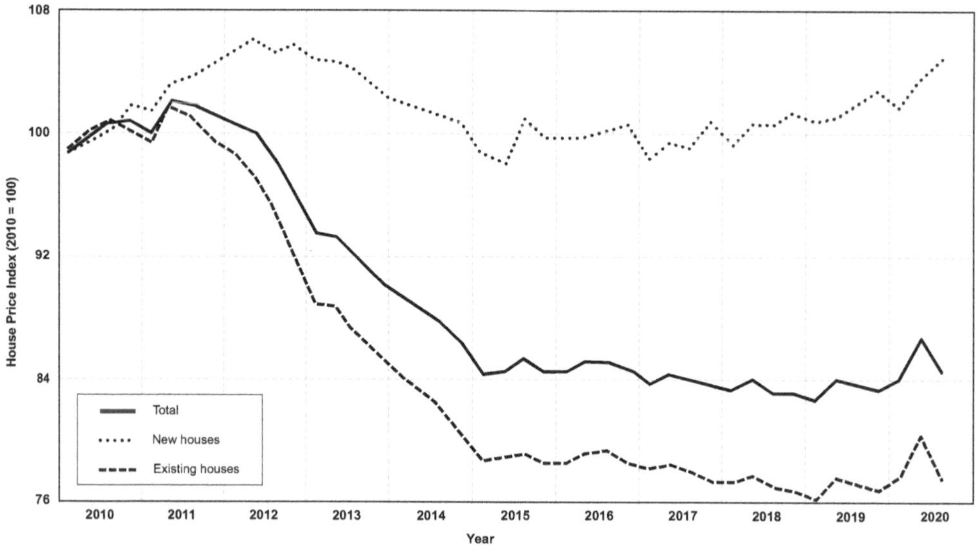

Figure 10.1 Evolution of the house price index, 2010–2020

Source: Author

area) is estimated. The surface area of non-residential buildings also fell sharply compared with the previous quarter (−15.5%). In the second quarter of 2020, the estimate of the number of homes in new residential buildings fell below the threshold of 10,000 units, the useful living area is around 912,000 m^2, while the non-residential number falls to about 1.9 million/m^2 (ISTAT, 2021b).

The spatial distribution of average sales price per region shows a clear north-south pattern as illustrated in Figure 10.2.

2. Regulatory framework

According to the Italian Civil Code (c.c.), the real estate sales deed must be concluded in writing (Art. 1350 c.c.); otherwise, it is deemed void. The sales contract is backed by a series of legal guarantees, such as the seller's guarantee for hidden defects (Art. 1490 c.c.). Book II of the c.c. also regulates the transmission of property upon death.

To ensure certainty and obtain legal knowledge from third parties, the deed is usually signed with the intervention of a notary serving as a public official, and then registered in the real estate public register (Art. 2643 c.c.). The notary who intervenes in the real estate deed is required to ascertain the traceability of the payment of the price declared in the contract, if the payment is made in their presence, and to fill in a profile of the parties for anti–money laundering purposes. Real estate deeds are also meticulously regulated by special laws in order to grant full compliance with tax requirements and urban regulations.

The sales contract is governed by the consensus principle (Art. 1376 c.c.); therefore, the right of ownership transfers when the agreement is reached between the parties and the contract is signed. In the practice of real estate bargaining, however, the parties often conclude a preliminary agreement (Art. 1351 c.c.) that may transfer possession of the house in advance to allow the party who intends to purchase it adapt it to their needs.

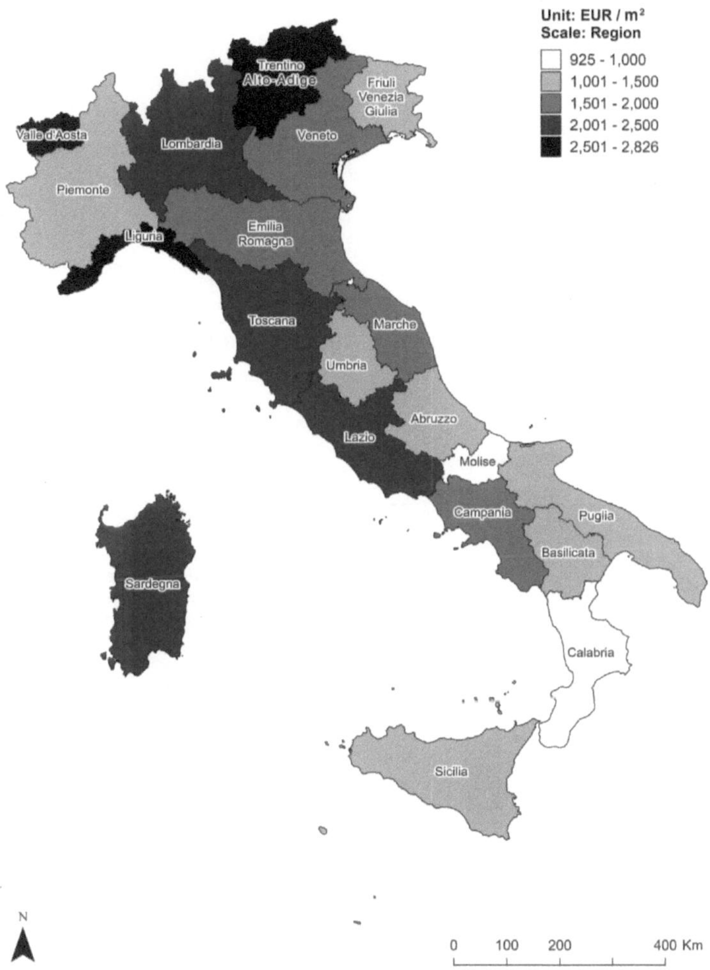

Figure 10.2 Average purchase prices of real estate, 2021

Source: Immobiliare.it (2021)

The owner has the right to use and dispose of the property (Art. 832 c.c.) within the limits imposed by law. For example, they can grant third parties limited rights on the property (called 'minor' property rights), such as usufruct (Art. 981 c.c.), right of use (Art. 1021 c.c.), and right of inhabitance (Art. 1022 c.c.). It can also attribute to third parties a personal right of use over the asset, which usually takes place through the rental (Art. 1571 c.c.) or lease (Art. 1615 c.c.) contract, which is governed by different rules.

3. The conveyance process

In this section, we examine the development of a typical real estate transaction in Figure 10.3. First, we outline the activity of real estate agencies, as they often play an important role as mediators in matching the supply and demand of houses. Second, we briefly discuss the preliminary agreements that are often carried out before the sales contract is finalized. Thereafter,

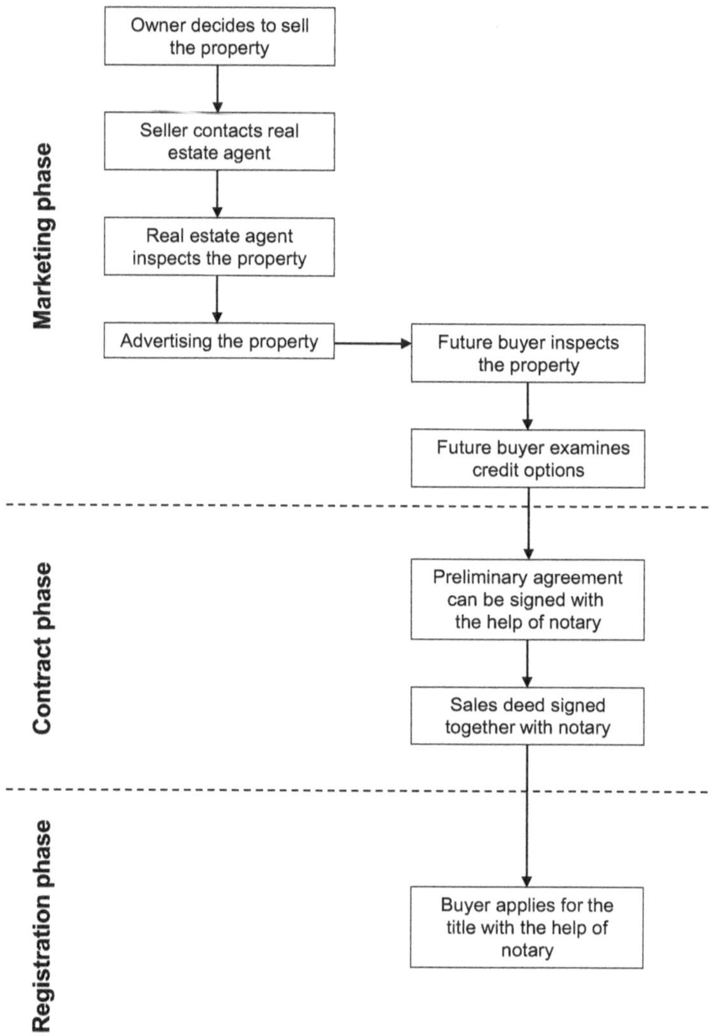

Figure 10.3 The conveyance process for real estate sales in Italy

Source: Author

we deepen the theme of the structure and content of contracts for the transfer of ownership of houses, while also examining the discipline of their transcription in public registers for publicity purposes. Finally, we explain some of the general principles regarding the taxation of real estate sales.

Marketing phase

Although it is not necessary, in practice, the meeting between demand and supply of houses is carried out with the intervention of a real estate broker (agent). Real estate brokerage is regulated both by the c.c. code and by a special law (Law No. 39/1989). The agent has the task of relating two or more parties for the conclusion of a deal (Art. 1754 c.c.), and, to carry out this business legitimately, they must be registered on a special list at the Chamber of Commerce after having attended a training course and passed an examination to ascertain professional aptitude and ability. The agent who works without being registered is subject to a fine and to civil and criminal liability and must return all the commissions collected by the parties.

It is usually the seller who appoints the real estate agent, but the latter's activity is also remunerated by the buyer, who turns to the agency for help in finding accommodation. The mediation contract can be concluded orally, but usually the parties sign a written mandate. Furthermore, the contract is always revocable.

The agent's right to remuneration (commission) is governed by Art. 1755 c.c., pursuant to which

> the mediator has the right to a commission from each of the parties if the deal is concluded because of his intervention. The extent of the commission and the proportion in which this must be borne by each of the parties, in the absence of an agreement, professional rates or uses, are equitably determined by the judge.

According to the most recent jurisprudence of the Italian Supreme Court, for the agent to accrue the right to the commission, it is not necessary that a sales contract is signed, but it is sufficient that the parties reach a preliminary agreement subject to compulsory execution. Usually, commissions vary from 1% to 5%, with an average 3% in metropolitan areas.

The price negotiation usually takes place through the submission of competing written offers. The bidder can place conditions on the offer, and the seller is not bound to accept the offer even if it is the best among those made. Sometimes, to guarantee the seriousness of the offer, the bidder is required to leave a sum of money on deposit to the agent, which will then be returned if the offer is not accepted.

During the marketing phase, the seller carries out a series of activities aimed at guaranteeing the urban and building regularity of the house, the compliance of the systems installed in it (e.g. electrical, heating, water, and sanitary systems) with the regulations in force at the time of their installation, as well as to assess the energy efficiency class of the home. The energy certificate must be provided to the buyer and attached to the sales contract.

Even the work of agents has been impacted by the COVID-19 pandemic. However, a recent survey by the Bank of Italy finds that the share of agencies that sold at least one property in the fourth quarter of 2020 rose again to levels close to those of 2019. Most agents reported that prices are substantially stable compared with the previous quarter, but one-third report a decrease in prices. Assessments of demand conditions and new listings worsened and the length of time on the market stabilized, while the average discount on asking prices increased slightly. After the outbreak of the pandemic, the average size of the houses sold increased.

Contract phase

In real estate bargaining in Italy, after the proposal is accepted, it seldom happens that the parties immediately conclude a definitive sales contract, which transfers the property on the basis of the consent exchanged between the parties. Usually, the parties sign a preliminary sales contract (Art. 1351 c.c.), which must have the same form (written under penalty of nullity) as the definitive sales contract that the parties will conclude later.

The preliminary contract has the sole effect of obliging the parties to conclude the future definitive sales contract. In the meantime, the parties usually cooperate by equipping the property with all the requisites required by law or by the purchaser and to carefully prepare the definitive sales contract, with the assistance of a notary. The preliminary sales contract is particularly appreciated and used in practice because it gives a certain amount of protection. In fact, in the event of a breach by one party, the other can obtain from a judge a sentence that replaces the definitive contract that the parties should have concluded and transfers (with constitutive effect) the ownership of the asset (Art. 2932 c.c.). Furthermore, the preliminary sale contract can be registered in the public real estate registers, which means that, for a limited time, the sale is blocked to any other third parties.

At the end of the market phase and of the phase covered by the preliminary agreement, the definitive sales contract requires the consent of the parties and records all the preliminary operations carried out by the parties in these phases. The final sale contract is drawn up with the assistance of a notary to give the contract the public form required for registration in the public real estate registers (Arts. 2643, 2644, 2657 c.c.). Furthermore, the real estate sale represents a significant event also from an administrative and fiscal point of view. The presence of the notary, therefore, also ensures that the parties comply with all the requirements imposed by law for the validity of the deed, for the administrative and fiscal regularity of the sale, and for the application of all the legal guarantees, especially those in the buyer's interest.

Registration phase

Real estate rights are subject to a registration regime, with the function of making the ownership of the right enforceable against third parties and of resolving any conflicts between subjects who consider themselves to be the holders of the same right. As a general principle, the registration of the sale has a 'declarative' effect (Art. 2644 c.c.); that is, it creates legal knowledge of the ownership of the right, allowing the holder to oppose it to third parties. For example, if the owner sells the same property in favour of two different subjects, the first to register the deed in the public real estate registers will prevail.

The system also requires the continuity of registrations (Art. 2650 c.c.) to verify all the steps that, since its construction, have brought the property through the various sales and into the ownership of the current owner.

As anticipated, the legal system also allows the registration of the preliminary sales contract, thus 'booking' the effects of the transcription of the definitive sales contract (or of the sentence that transfers the asset following the non-fulfilment of a party, pursuant to Art. 2932 c.c.). If the preliminary contract has been registered, the effects of the transcription of the definitive sales contract (i.e. the possibility of the buyer to 'prevail' over other subjects who have claims on the property) will not occur from the date of registration of the definitive sales contract (or of the sentence), but they will retroact to the date of registration of the preliminary contract, with the consequence that other registrations made in relation to that property cannot be opposed to the buyer after the registration of the preliminary contract (Art. 2645-*bis* c.c.).

The negotiation and registration costs are usually borne by the purchasing party, but the parties are free to reach a different agreement between themselves on the sharing of costs. Taxes, on the other hand, remain the responsibility of the person identified by the law as being obliged to pay them.

Before the sale, the seller bears most of the costs including a share of the real estate agency fee (usually their share is lower than the buyer's) and the costs needed to obtain the mandatory environmental certification (assessing the energy efficiency of the building). To avoid future controversies, the seller may also invest in a certification of compliance with urban and building regulations. In this phase, the buyer usually bears a higher share of the real estate agency costs. After the sale, the buyer bears the notary fee and pays a proportional register tax (9% or 2% of the sale value, depending on if the 'first house' exemption applies), a fixed mortgage tax, and a fixed cadastral tax (VAT does not apply to deals between private citizens but does apply when a company is involved), while the seller's capital gain is also subject to taxation.

References

Bankitalia. (2021, March). *Italian housing market survey: Short-term Outlook – 2020 Q4*. Retrieved from www.bancaditalia.it/pubblicazioni/sondaggio-abitazioni/2020-sondaggio-abitazioni/04/index.html

Immobiliare.it. (2021). *Regional sales prices*. Retrieved from https://www.immobiliare.it/

ISTAT. (2020, December). *Prezzi delle abitazioni*. Retrieved from www.istat.it/it/files//2020/12/CS-abitazioni-Q32020.pdf

ISTAT. (2021a, March). *Mercato immobiliare: Compravendite e mutui di fonte notarile*. Retrieved from www.istat.it/it/files//2021/03/Flash_Mercato-immobiliare_I-e-II-trim-2020PC.pdf

ISTAT. (2021b, January). *Permessi di costruire*. Retrieved from www.istat.it/it/files//2021/01/Permessi-di-costruire-II_trim_2020.pdf

Lungarella, R. (2021). *Effetto pandemia sul mercato della casa*. Retrieved from www.lavoce.info/archives/73129/effetto-pandemia-sul-mercato-della-casa/

11
MOROCCO

Imane Oualji

1. The Moroccan real estate market

In the Moroccan[1] economy, the real estate sector plays a key role with a direct contribution of 6.3% to the GDP and € 10 billion of added value. Also, housing occupies one of the three major items (non-food) of the household budget in Morocco. According to the national survey on household consumption and expenditure presented at the end of 2016, the High Commission for Planning (HCP) indicated that expenditure on housing increased by more than 60% between 2007 and 2014 and rose from € 219 to € 315. The share of this item in the overall budget increased from 20.3% to 23% during the same period.

Real estate remains the cornerstone of any asset management strategy. To attract foreign customers and investors, Morocco has launched many real estate projects. According to the annual report of the Bank Al-Maghrib (BAM), Morocco's central bank in 2018, the real estate market witnessed an increase in the number of transactions by 4.8%. As a result, sales rose by 3.3% for residential assets. In 2014, the housing stock reached 8.87 million housing units in Morocco, of which 69.8% (6.19 million) were in urban areas and 30.2% (2.68 million) in rural areas. Occupied dwellings represent 79.8% (7.07 million) of the national stock, vacant dwellings 12.3% (1.09 million), and secondary or seasonal dwellings 8.0% (706.1 million). In urban areas, these proportions are 74.8%, 15.9%, and 9.3% respectively.

The Moroccan housing market comes in two forms: ownership and rental. Nationally, the majority of households, 74%, are owners, while only 17.2% are tenants, with an average rent of € 109 (see Figure 11.1).

On this basis, three regions stand out with an average rent higher than the national average (€ 110). These are the three most dynamic regions economically (activity/employment) and where the standard of living is the highest: Tangier-Tetouan-Al Hoceima (€ 124), Rabat-Sale-Kenitra (€ 122), and Grand Casablanca-Settat (€ 115). It is also reported 34% of tenant households pay rent of between € 48 and € 96, more than 24% pay rent of between € 96 and € 144, and 26% pay a rent of € 144 and more.

The dwellings differ according to their type, their size (number of rooms and surface area), their architecture, and the mode of their occupation by the household. Nationally, more than half of households live in a modern Moroccan house, which is the highest proportion, followed by households that reside in rural-type housing. In urban areas, the distribution of households according to

 DOI: 10.1201/9781003130383-13

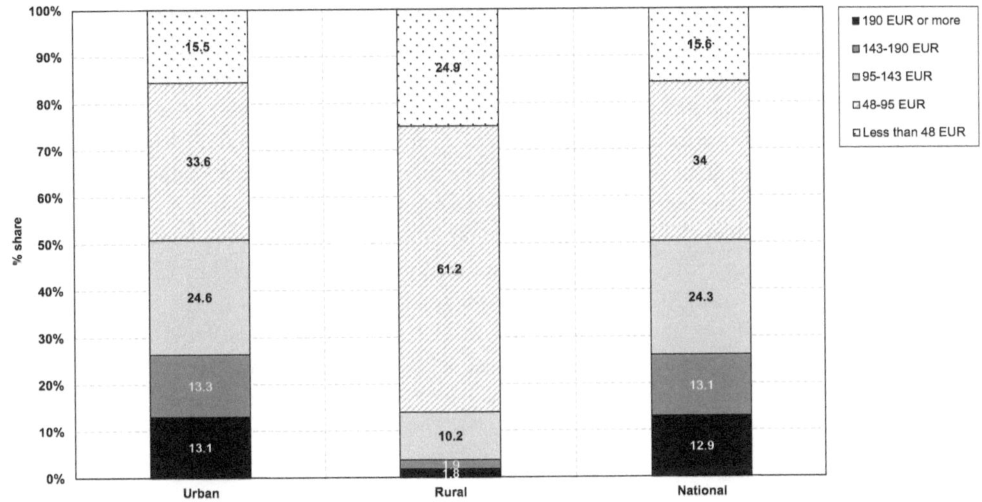

Figure 11.1 Rental costs – urban, rural, and national

Source: Author

the type of housing occupied shows an increase in the share of modern Moroccan houses (62.6% in 2004 to 65% in 2014) and the share of apartments (12.4% in 2004 to 17.5% in 2014). In contrast, during the same period, the share of households occupying traditional houses fell from 8.1% to 5.5% and that of households occupying basic housing or slums fell from 8.2% to 5.2%. In rural areas, although the predominant type of housing is 'rural-type housing' (64.1%), a significant increase in the share of modern Moroccan houses was recorded from 13.6% in 2004 to 25.9% in 2014.

2. Regulatory framework

Once the deed of sale is signed and registered, it is subject to an important formality, 'registration in the land books'. This is the registration for land conservation, on which the property sold depends. The goal is to ensure the transfer of ownership to the buyer and to register mortgages and charges that could encumber the property being sold. Notaries have a monopoly on registering in these books any deed relating to property subject to the status of co-ownership. The purpose of this procedure is to register the sale of a property with the land conservation service responsible for the property. The purchaser becomes, after registration of the deed of sale in the land books, the owner of the property with all the land charges that may encumber said property.

Housing policy

Morocco is facing profound changes caused by the accelerated and continuous process of social and economic transformation changes, resulting in an accelerated urbanization that directly impacts Moroccan citizens' way of life. The problem of the housing sector lies in the development of the supply market in line with the specificities of demand while improving the mechanisms of access to housing (property or rental) (The Center for Affordable Housing Finance in Africa Housing Finance in Africa, 2018).

Over the past 15 years, great efforts have been made to enable more Moroccans to own property. To do this, the authorities should act on the housing deficit. In this sense, one of the pillars of the national strategy in this area was social housing. Since 2007, developers and buyers have benefited from tax exemptions. This dynamism allows the production of nearly 260,000 housing units per year and the reduction of the housing deficit, which has fallen from 1.2 million units in 2002 to 400,000 currently.

The Moroccan government defines two types of affordable housing, called social housing, and these are differentiated by their maximum price. The first low total real estate value programme is a programme created by the 2008 finance law with a capped price of MAD 140,000 and housing with a living area of between 50 m^2 and 60 m^2. The programme targets people earning less than twice the minimum wage. The second type of social housing was launched by the 2010 finance law with a price capped at MAD 250,000 excluding tax. For this type of house, the surface area is between 50 m^2 and 80 m^2, and no maximum income is required for the acquisition (Ministry of Housing and Urban Policy, Department of Real Estate Development, 2016).

To facilitate access to housing, several actions have been carried out targeting the financing sector, which has several sources of mortgage loans. The average repayment term is between 15 and 25 years, with financing between 50% and 100% of the appraised value of the property. In March 2016, the Central Bank lowered its key interest rate by 25 basis points to 2.25%. On 19 June 2018, this interest rate remained unchanged. Mortgage rates have fallen steadily over the past five years; from 2012 to 2017, they went from 6% to 5% on average. The trend in home loans has slowed from 5.2% growth in 2016 with outstanding amounts of MAD 188.7 billion to 3.6% in 2017, with outstanding amounts of MAD 195.4 billion.

The financing of real estate is structured, for the most part, by recourse to bank credit for 51.2% of requests in urban areas (domiciled in the main regions of the country) and by family solidarity practices at 32.9% of requests in rural areas. The use of an intermediate practice combining family financing and bank credit represents 6.9% in urban areas and 10.5% in rural areas.

In Morocco, price levels vary greatly within the country as illustrated in Figure 11.2. In 2017, the real estate price index recorded an increase of 5% (compared with an increase of 1.3% in 2016)

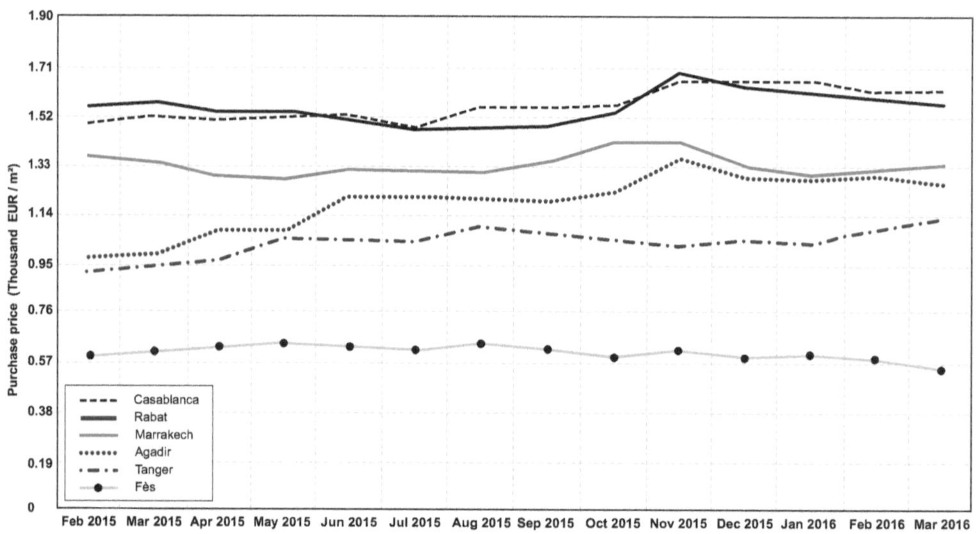

Figure 11.2 Evolution of the purchase price of housing by region

Note. *1 Euro = 10.48 Moroccan Dirham (as of 2021 October)

Source: Author

due to an increase of 4.9% for residential properties and 5.3% for land. However, the volume of transactions fell by 7.6% after rising 8.4% in 2016, with an 8.7% drop in residential property sales.

3. Ownership transfer process

In the so-called traditional regime, the legal basis for the ownership of *Melk* buildings is the fact of possession. According to Article 3 of the Code of Real Rights, this results from possession satisfying the legal conditions: the acquisition is based on peaceful possession under uninterrupted common knowledge for a period of ten years (towards third parties) and 40 years (towards family members).

The proof of the right of ownership of the unregistered property is effected by the presentation of a *moulkiya*, namely, an *adoulaire* act, by which 12 ordinary witnesses confirm the lawful possession of the person claiming the ownership of the property in front of two *Adouls* (Muslim law notaries). The transfer of ownership of unregistered properties occurs when there is consent between the parties (Articles 488 and 489 of the Dahir of Obligations and Contracts). The ownership cannot, however, be enforceable against third parties unless it is registered with the administration of Registration and Stamps. Any transaction relating to the real rights of unregistered property must be established by a notarial deed (drawn up either by the *Adouls* or by notaries or since Law No. 39–08 by lawyers approved by the Court of Cassation).

Regarding registered real estate properties characterized by publicity and the probative force of their registrations in the land register, the transfer occurs by registration in the land register (Law No. 14–07 on land registration), with a specific regime concerning co-ownership (Law No. 18–00) and future sales (Law No. 107–12).

Co-ownership regime

In Morocco, almost all apartments are subject to the condominium system, which applies to all co-owners and their tenants. In some special cases, houses may be subject to this legal regime.

Co-ownership of buildings and real estate complexes is governed by the real rights code and special provisions which set both the regime of the common portions as well as private portions.

On the other hand, the law sets mandatory rules for the use and management of buildings subject to the legal regime of co-ownership. These rules are gathered in the co-ownership regulations which contain both information on the building in general and on the intended use of the private parts of the property.

According to Law No. 18.00, any sale of co-ownership property shall be made by a notary who ensures, as a legal professional, the proper execution of the contract and confers, in addition to confidentiality, the necessary legal security.

'Off-plan' property sales

While the acquisition of registered and unregistered properties remains an important mode of land transfer, the 'off-plan' property sales remains very successful. Governed by Law No. 107–12 amending and supplementing Law No. 44–00, it provides mechanisms promoting investment in the real estate sector and encouraging access to housing. Law No. 107–12 relating to the 'off-plan' property sales includes:

- the obligation to abide by the law on pain of nullity of the contract, to deal with practices that are harmful to buyers and the law;

- the establishment of a new contract known as a 'reservation contract', under which the seller can obtain an advance from the purchaser before the preliminary contract is concluded by offering all the guarantees necessary to the purchaser for the security of the transaction;
- the establishment of the reservation contract or the preliminary contract of sale in the future state of completion by private deed at a certain date or by authentic deed, upon obtaining the building permit, not after the completion of the foundation work on the ground floor, as was the case with the former Law No. 44–00.

4. The conveyance process

The process is explained in Figure 11.3.

Marketing phase

In practice, this first phase of the transfer process most often involves a real estate agent. While as noted there is no mandatory intervention in the property transfer process, in practice, a significant number of transfers are carried out by real estate agents.

Real estate agents are one of the essential links in the real estate sector's value chain. Despite the importance of the real estate market and real estate agents in Morocco, this profession is not regulated and can be exercised by anyone and without any obligation of a diploma or authorization from the supervising ministry.

The latest survey carried out by the supervisory ministry stated the figure of 1,200 real estate agencies. Out of the 1,200 official agencies identified by the ministry, nearly 80% have only two employees at most, with a management rate of less than 25%.

In the absence of specific regulations, the real estate agent is first considered as a trader and must therefore be governed by the Commercial Code. He can also be a trader of a particular kind given that he can act as a broker who connects a buyer and a seller or an owner and a tenant. If this is the case, Articles 405 et seq. of the Commercial Code relating to brokerage must apply in the event of a dispute. The real estate agent can also play the role of an agent for his clients who assign him through a written mandate with certain assignments, such as the sale, rental, or management of the property. In this specific case, it is no longer the Commercial Code that regulates the profession, but the Dahir of Obligations and Contracts.

The market is trying to organize itself and the National Federation of Real Estate Agencies (FNAIM) has set the goal of supervising and promoting the profession. It is in this sense that a bill has been submitted to the Ministry of Housing to establish a legal framework that can protect professionals as well as consumers.

Contract phase

In practice, the sale of a building is preceded by the drafting of a sales agreement, also called a synallagmatic promise of sale, through which the seller and the buyer both agree to conclude the sale at a jointly determined price. Legally, the sales agreement is equivalent to a sale.

If one of the parties renounces the transaction, the other can force it to do so by legal means, and moreover, demand damages. The signing of the compromise could be accompanied by the payment by the purchaser of an advance of between approximately 5% and 10% of the sale price, which will be deducted from the price in the event of the final sale and during the signing of the notarial deed.

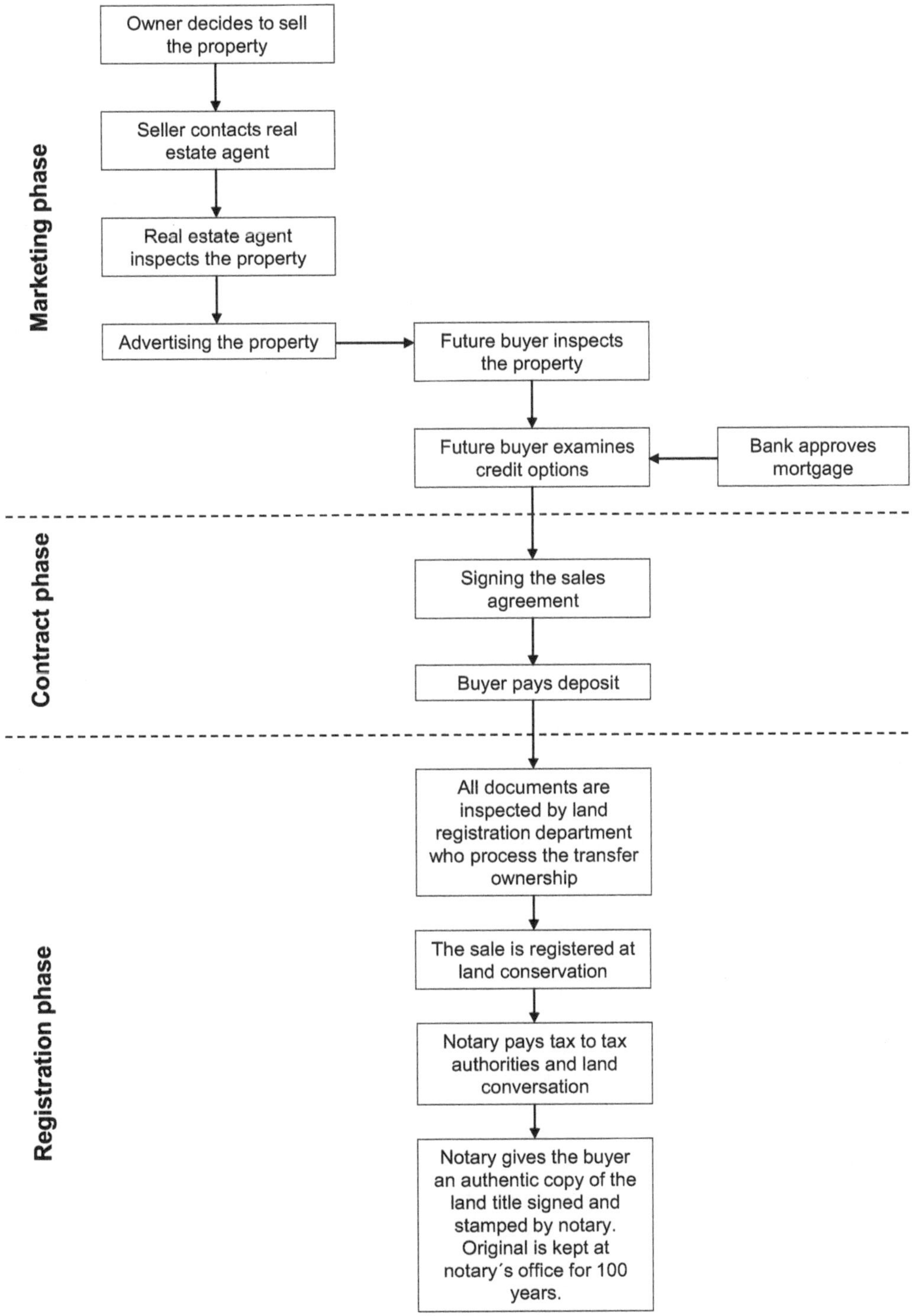

Figure 11.3 The conveyance process for real estate in Morocco

Source: Author

Like the promise to sell, the sales agreement is subject to the formality of registration.

Whether it is a promise to sell or a compromise, the buyer and seller can decide by mutual agreement to include a conditions precedent. This makes it possible to nullify the preliminary contract if certain events occur before the final sale.

In addition, a sales agreement may also contain a clause called an 'immobilization compensation', which allows the seller and/or the purchaser to waive the sale without cause by leaving the other party an agreed sum, which cannot exceed 10% of the sale price.

According to the Dahir of Obligations and Contracts, the seller of a building must inform the buyer. In general, the seller must send the buyer all the information they have regarding the property. The courts severely sanction any breach by the seller of this general obligation. The information to be provided is of a legal and technical nature. From a legal standpoint, the seller must, for example, inform the buyer of the existence of easements (of passage, view, the flow of water), mortgages or collateral (taken for the benefit of a bank to guarantee a loan), or other charges (pre-notation, preferential right, party wall, etc.). On a technical level, the seller of a condominium apartment has, for example, the obligation to provide the condominium regulations to the future purchaser.

Registration phase

After signing the sale, many subsequent formalities have to be fulfilled. The most important is registration and registration in land books. The former takes place at the registration department, and the latter in land conservation. Therefore, the notary pays the tax authorities and then the land conservation authorities fees fixed by law. Finally, the notary gives the buyer a special copy called an 'authentic copy', which is signed by the notary and then stamped. This document constitutes the origin of ownership, and the original of the deed will be kept by the notary in his office for 100 years.

The registration formality aims to acquire a fixed date for private contracts through their entry in a register known as the 'entry register' and to ensure the conservation of documents (Art. 126CGI 2008). The applicable tax rate is 4% of the sale price for built buildings and land intended to be built within seven years. However, the 2021 finance law provides for an exceptional total exemption for the acquisition of social housing and the acquisition of low-value property. Additionally, a reduced rate of 2% has been set out regarding properties for residential use and empty land intended for residential construction provided that the price does not exceed 4 million. This transitional measure is valid until 30 June 2021.

The amount of tax due must be paid within 30 days of signing the final deed. This rate varies depending on the nature of the transfer or the property transferred. It is important to note that this formality is a prerequisite to the registration of the transfer in the land register. After this formality, the buyer becomes the owner of the property sold. However, the seller will not automatically take possession of the sale price because they must definitively regularize their situation before the tax administration.

Once the deed of sale is signed and registered, it is subject to an important formality: 'registration in the land books'. This is the registration for land conservation, on which the property sold depends. The goal is to ensure the transfer of ownership to the buyer and to register mortgages and charges that could encumber the property being sold. Notaries have a monopoly on registering in these books any deed relating to property subject to the status of co-ownership. The purpose of this procedure is to register the sale of a property with the land conservation service responsible for the property. After registration of the deed of sale in the land books, the purchaser becomes the owner of the property with all the land charges that may encumber said property.

Post-transaction phase

The seller of real property is subject to the land profit tax (IR/PF). The IR/PF is obtained from the acquisition price of the property sold, to which a revaluation coefficient is applied to have the current value of the property. To this value are added the acquisition costs assessed at a flat rate of 15% of the acquisition price, investment expenses (with supporting documents), interest on loans and other financial costs to have the cost actualized or finalized. It is the difference between the discounted cost price and the sale price that makes it possible to have a taxable profit. This profit is multiplied by a rate of 20% to obtain the amount of tax payable. This amount cannot be less than 3% of the sale price, which represents the minimum collection even in the absence of profit. There are cases of exemptions from IR/PF which are defined in Article 63 of the General Tax Code (CGI).

After paying the IR/PF, the seller must obtain a tax discharge certifying that no tax debt of the administration remains in his regard on the property sold. After this document is produced, the notary can safely give the selling price.

Note

1 1 Moroccan Dirham (MAD) = 0.096 Euro (as of 1 October 2021).

References

The Center for Affordable Housing Finance in Africa Housing Finance in Africa. (2018, October). *Review of African housing finance markets*. Retrieved from https://housingfinanceafrica.org/

Ministry of Housing and Urban Policy, Department of Real Estate Development. (2016, March). *Study on the national survey on housing demand, summary report*. Ministry of Housing and Urban Policy, Department of Real Estate Development. Retrieved from https://mohua.gov.in/cms/report.php

12
SPAIN

Elena Sánchez Jordán

1. The Spanish housing market

The Spanish real estate market for homes has two main forms: rentals and ownership. Rental grants the lessee the enjoyment of the leased home during the period of time agreed upon in the lease, which gives a merely personal right, while ownership attributes the property over the home. The latter will be explained in detail in this chapter.

The Spanish real estate market and its housing market are ruled by an important number of acts and regulations. As Figure 12.1 shows, after the end of the 2008–2013 crisis, real estate prices increased quite quickly in Spain, that is, until the pandemic outbreak, which has cooled down both acquisitions and rental prices.

In Spain, around 80% of the homes are owned by the people who live in them, while approximately 15% are tenants, according to the data provided by the National Statistics Institute. This reveals great differences from other countries in Europe, where homes are enjoyed mainly through rental agreements. However, the Spanish situation may vary significantly after the COVID-19 pandemic, which has caused an important decrease in rental prices as well as an increase in both the supply and the demand for rental housing (idealista/news, accessible at https://bit.ly/3ig82T0). Nevertheless, in this chapter, we focus on homeownership, around which the housing market is articulated, strictly speaking.

The real estate sector has a fundamental weight in the Spanish economy: in 2018, it represented around 10.5% of the gross domestic product, while in 2005, it represented 7.1%, according to information contained in statistical websites (https://bit.ly/3o3DDZx).

The yearly transaction volume in Spain is between 400,000 and 500,000 (including all kind of homes, both second-hand and newly built). The market suffered a deep decrease from 2011 to 2015, according to data provided by the National Statistics Institute (www.ine.es/jaxiT3/Datos.htm?t=6155#!tabs-grafico).

The price levels are quite different within the country. The most expensive homes in 2020 were those located in Madrid, Catalonia, and the Balearic Islands (according to data provided by the National Statistics Institute, www.ine.es/jaxiT3/Datos.htm?t=25171#!tabs-grafico), while in February 2021, the most expensive homes were those in Madrid (€3,071/m^2), Balearic Islands and Basque Country (€ 2,800/m^2), and Catalonia (€ 2,508/m^2), according to the real estate website Fotocasa Life (www.fotocasa.es/blog/compraventa/conoce-el-precio-de-la-vivienda-en-venta-este-mes/).

DOI: 10.1201/9781003130383-14

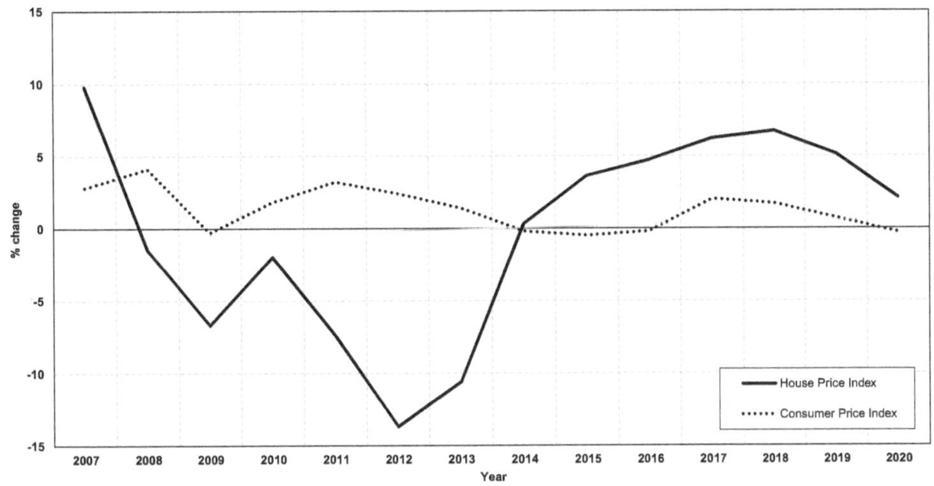

Figure 12.1 Yearly change in consumer price and house prices in Spain (in percent)

Source: National Statistics Institute

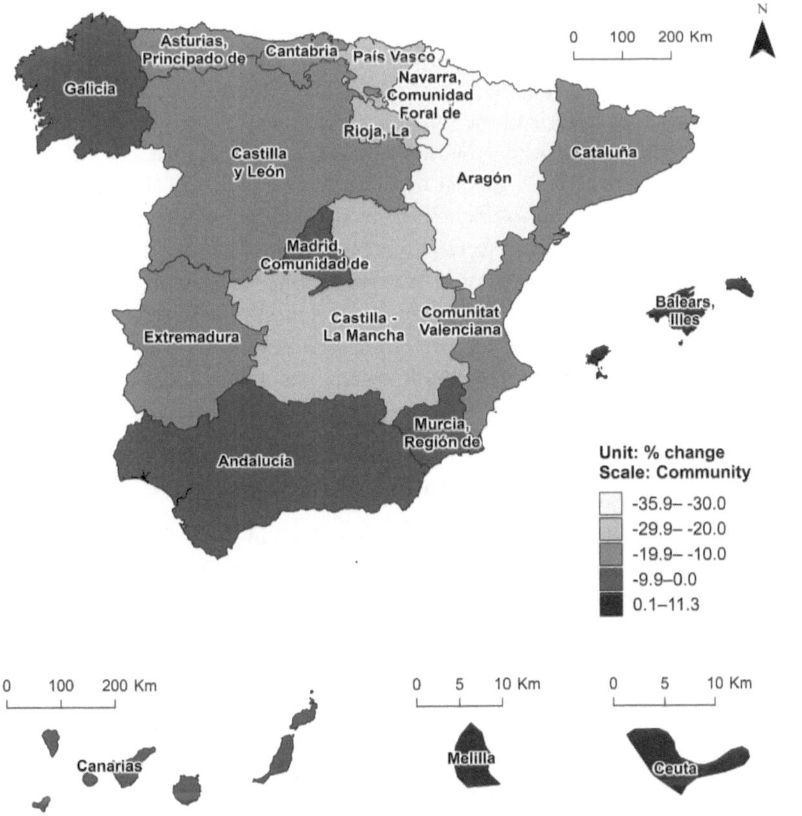

Figure 12.2 House price index (new and second-hand housing) change in Spain (in percent), 2007–2021

Source: National Statistics Institute (2021)

In terms of the Spanish housing market, a special mention should be made about the purchase of homes by foreign citizens. In this sense, and according to the data provided by the Bank of Spain, home acquisitions in Spain by foreign citizens have increased from 4.2% in 2007 to 10.5% in 2019 (Álvarez, Blanco, & García Posada, 2020), although it is foreseeable that after the COVID-19 crisis on the one hand and Brexit on the other this proportion will decrease significantly. Among foreign buyers, those from other European countries stand out, especially those from Northern Europe (in particular, the British), to whom Italy has joined in recent years. It should also be noted that there is also a significant number of Romanian, Moroccan, Chinese, and Russian buyers.

In addition, the real estate boom of the 2000s together with the increase in the standard of living, the rise of international tourism, and the improvement of communications, has led to a high number of second homes in Spain. More than 14% of the Spanish families are holders of a second residence (Montoriol Garriga, 2020, supported by data from the 2011 Housing Census, prepared by the National Statistics Institute), a number that rises to 36% according to other information produced from a survey by the European Central Bank (Díaz Guijarro, 2014), which in any case places Spain among the European countries with the most secondary dwellings.

The average house size in Spain was 144 m^2 in 2015 according to cadastral data, while the size of apartments had decreased to around 97 m^2 during the building boom due to a rise in prices. In the last decade, the size of newly built apartments has increased up to around 110 m^2, as new purchasers replace their previous (smaller) homes (*El País*, 2018, https://bit.ly/3lt1pyt).

The year in which the most homes were built in Spain was 1960, as the country experienced a certain economic stability that allowed a small improvement in the quality of life. This, together with the rural exodus, meant that new buildings were needed. When you add in tourism, these have characterized the Spanish economy until today. According to the cadastre, 17% of the buildings in all the provinces' capitals date from that decade (*El Confidencial*, 2019, https://bit.ly/2OFn22K). A special mention has to be made about the period that has been called 'the building tsunami', which refers to the building-boom years of 1997–2007. For example, in 2005, 800,000 buildings were newly built in Spain (more than in France, Germany, and the United Kingdom combined), which explains why the bursting of the real estate bubble hit Spain so hard (Fernández Durán, 2006). Last, it should be noted that the last 12 months have seen a revolution for housing. The pandemic has slowed down sales and mortgages, which has changed the price trend, boosted technology, and given prominence to new constructions, villas, and outdoor spaces (idealista/news).

2. Regulatory framework

Under Spanish law, the regulatory framework is contained in the Civil Code and in the Mortgage Act (when the purchaser mortgages the property to guarantee the loan). In regard to the building process and the responsibility of the promoter of the building, the Building Planning Act must also be considered, particularly in regard to liability for construction defects.

An important particularity of Spanish law is that the sale contract (or another contract that aims to dispose of the property) is not enough to transfer ownership. According to Art. 609 of the Civil Code, the contract must be followed by the *traditio* (delivery of the good's possession with the intention of transferring the property) for the property to be effectively transferred.

Strictly speaking, no formal requirements are laid down in the Civil Code for the conveyance of property, as the rule is the *freedom of the form* (Art. 1.278 Civil Code). However, if the buyer wants to mortgage the property, it is necessary to register the mortgage contract in the Land Registry, and for this purpose, a public deed granted before a notary public is needed.

However, registration is not necessary to convey the property, although it includes more security; for example, if the owner sells his or her property to buyer B and then to buyer C, the party who first registers his or her title in the Land Registry will be deemed to have the better claim (Art. 1.473 Civil Code). In addition, Art. 34 of the Mortgage Act protects the purchaser, who, in good faith, acquires from the registered owner and registers the acquisition in the Land Registry, even if the transferor's title is later annulled or terminated. In this way, registration in the registry can have a decisive effect, even if it is not compulsory.

3. The conveyance process

In this section, we describe how the conveyance process usually takes place in Spain (see Figure 12.3).

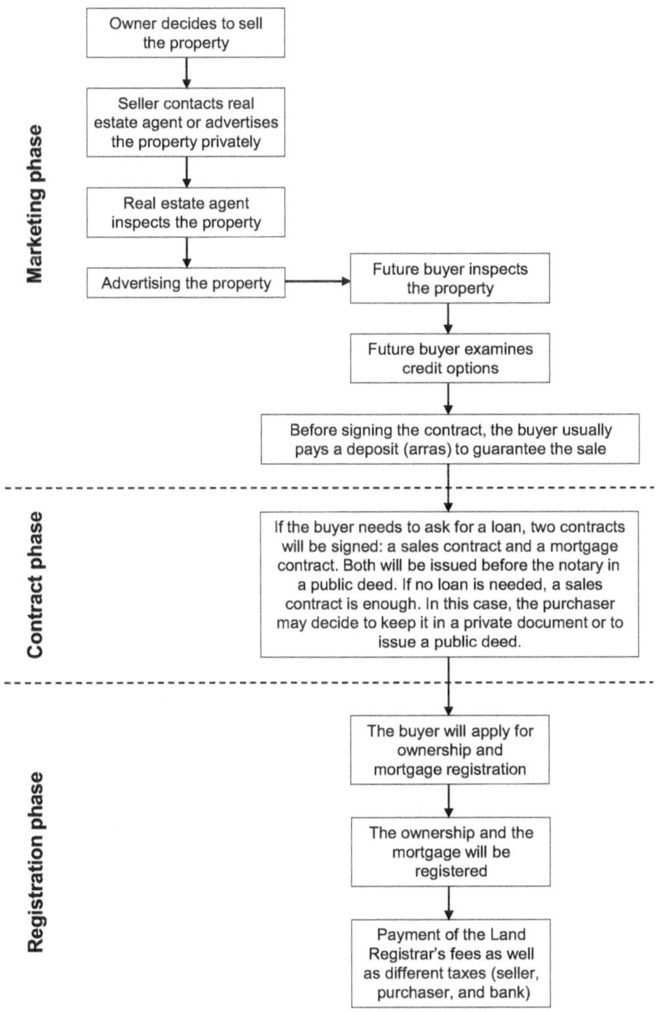

Figure 12.3 The conveyance process for real estate in Spain

Source: Author

Marketing phase

The first phase could be called pre-contract or marketing phase, in which parties exchange information, make an offer and receive an answer or a counter-offer until they reach an agreement. When a real estate agency is involved, this pre-contract phase turns into a marketing phase. However, there is no mandatory intervention of a real estate agent in the conveyance process, and this may explain why only 50% to 60% of the property sales in Spain are carried out through a real estate agent. The activity of estate agents is regulated by Royal Decree 1294/2007 of 28 September, approving the General Statutes of the Official Associations of Real Estate Agents and its General Council. In such regulations, the requisites to become an estate agent in order to belong to an official real estate agency association are established, but it must be taken into account that in Spain, after 2000, the intermediation in conveyance processes can be carried out by anyone, without the need to prove a certain educational qualification or belong to a professional association. The liberalization of the real estate conveyance activity has given rise to many complaints due to abuses such as unreasonable fees, commissions imposed on both contractual parties, and excessive deposit payments. As a result, estate agents are calling for the field to be better regulated.

In Spain, estate agency contracts may be exclusive, or not. The basic duties of the estate agents who belong to an official association are the following: (1) to carry out their activities with efficiency, professional ethics and deontology, reserve and legality, observing due diligence in the exercise of the profession; (2) to respect and ensure full compliance with the legal system; and (3) to observe the legal provisions on consumer protection with regard to conveyance and real estate leasing, insofar as it affects their professional activity (Art. 9 Royal Decree 1294/2007). The estate agent's fee is paid by the principal, which is usually the seller. The fee normally consists of a percentage of the sales price that ranges between 3% and 5% of the sale price. In this marketing phase, the estate agent mainly advertises the property on the real estate agency's website.

The price of the property is fixed by an agreement between the seller and purchaser, following the advice given by the real estate agent. However, before signing a sales contract or delivering any sum, the purchaser should always visit the property to make sure that the offered characteristics meet reality. If the first information regarding the property is contained in brochures or websites, it must be considered that the buyer may demand from the seller the services, facilities, and forms of payment included in the offer, promotion, and publicity of the property, even if they are not included in the contract that will be signed afterwards. The purchaser should also find out the legal status of the property (e.g. who the owners are and whether there are any previous mortgages, liens, or other encumbrances on it, and whether there are any tenants or other stakeholders tied to the building).

Usually, when a deal between the parties is reached, the buyer pays a deposit to guarantee the sale *(arras)* and as a sign of the agreement's conclusion.

In this pre-contract or marketing phase, EU legislation has also been implemented, and since 2013, all homes must have an energy certificate in order to sell or to rent it out. It is the seller's duty to obtain such a certificate before the sales contract is signed.

Contract phase

Due to the *freedom of form*'s principle contained in Art. 1278 of the Civil Code, as soon as the parties agree on the price of the property and the good to be transferred, the agreement is binding and creates obligations between the parties: the main ones are the duty to convey the property (on the seller's side) and the duty to pay the price (on the purchaser's side). The *freedom of form* means that both a verbal sales contract and a sales contract in a private document are valid. Given the importance of the transaction and the relevance of registration, these contracts are usually issued before the notary in a public deed, as only public documents are able to enter into the Land Registry (Art. 3 Mortgage Act). As explained, the contract is not enough to convey the property to the purchaser, as it must be followed by the *traditio* (Art. 609 Civil Code) that can take place by the delivery of the possession over the home (Art. 1.462, paragraph 1, Civil Code) and also by issuing the public deed (Art. 1.462, paragraph 2, Civil Code). If a public deed is issued, a fee based on the property's value has to be paid by the purchaser to the notary.

If the buyer needs to mortgage the property, he or she will need, first, to apply for title registration at the Land Registry, without which no mortgage can be created. Second, he or she will have to negotiate with the bank that granted the loan in order to constitute a mortgage or to subrogate on an existing mortgage. Such an agreement (new mortgage or subrogation on the existing one) has to be necessarily issued as a public deed because it will be valid only if it is registered in the Land Registry (Art. 1.875 Civil Code), and as mentioned, only public documents may enter into the Land Registry, so another fee will have to be paid to the notary. In this case, the fee must be paid by the bank (Art. 14.1.e.ii Act 5/2019 of March 15, regulating real estate credit contracts). The fee is around 0.3% to 0.5% of the secured sum, plus the cost of the copies issued for each of the parties.

When a mortgage is created, the cost of the property valuation (between € 150 and € 400) is paid by the purchaser.

Registration phase

The estate agent or the bank usually assists the buyer with the applications for registration.

The Spanish Land Registry registers the transfer of ownership only when the sales contract has been issued in a public deed and the purchaser requests registration, as it is not compulsory in order to convey the property. However, registration is necessary if the property has to be mortgaged to guarantee a loan. Two fees have to be paid to the Land Registrar: one for the conveyance, based on the sale's amount, and another for the mortgage's registration, which ranges from 0.1% to 0.3% of the guaranteed sum's amount. The conveyance's registration fee has to be paid by the purchaser, while the mortgage's registration fee must be paid by the bank (Art. 14.1.e.iii Act 5/2019).

Post-transaction phase

Certain taxes have to be paid in Spain when selling a home. A tax also has to be paid when a property is mortgaged. Taxes regarding the property's conveyance are imposed over the seller and the buyer. The seller must pay the income tax (on a national level) on any profit from selling the property, and the amount ranges from 19% to 26% of the profit. In addition, a municipal added value tax *(plusvalía)* is charged for the increase in value that a home has experienced during the time it was owned. Its amount is fixed by each city municipality, and it is payable within 30 days of the sale. In the case of a newly built house, the buyer must pay both the VAT (10% of the purchase price) and the documented legal acts tax (0.5% to 1.5% of the price). If it is a second-hand house, they will have to pay the property transfer tax (between 4% and 10% of the price).

With regard to the mortgage, the only tax that has to be paid is the documented legal acts tax, which ranges from 0.5% to 1.5% of the guaranteed sum, and that is paid by the bank.

4. Summarizing the conveyance process and its transaction costs

To summarize, the Spanish conveyance process is quite simple as outlined in Table 12.1. Ownership is transferred with a sales contract followed either by the delivery of the possession of the

Table 12.1 Fees and taxes during the conveyance process in Spain

Phase	*Fees and taxes*	*Seller*	*New*	*Used*
			Buyer	
Marketing	Real estate agent's fee	3–5% of the property's value	-	-
	Energy performance certificate	Around € 1.5 per m²	-	-
Contract	Conveyance deed fee (notary)	-	1–2% of the property's value	
	Mortgage deed fee (notary)	-	0.3–0.5% of the secured sum	
	Property valuation cost	-	€ 150–400	
Registration	Land registration fee (conveyance)	-	0.1–0.3% of the sales amount	
	Land registration fee (mortgage)	-	0.1–0.3% of the sales amount	
	Fees for mortgage registration	-	Approx. € 400	
Post-registration	Income tax	19–26% of the profit	-	-
	Municipal added value tax	Fixed by each city hall	-	-
	VAT	-	10% of the purchase price (if newly built house)	-
	Documented legal acts tax (with regard to the transfer of property)	-	0.5–1.5% of the purchase price (if newly built house)	-
	Property transfer tax	-	-	4–10% of the price (second-hand house)
	Documented legal acts tax (with regard to the mortgage)	-	0.5–1.5% of the secured sum	
	Sales tax	-	-	-
Total (EUR for the model case price)		**€ 3,000–5,000**	**€ 12,250–14,900**	**€ 5,750–13,400**
Total (% share to the model case price)		**3–5%**	**12.3–14.9%**	**5.8–13.4%**
Excluded costs from the model case		Energy performance certificate around € 1.5 per m², 19–26 CGT, municipal added value tax	Mortgage deed fee (notary) 0.3–0.5% of the secured sum, documented legal acts tax (with regard to the mortgage) 0.5–1.5% of the secured sum	

Note. Model case price: € 100,000

Source: Author

home or by issuing a public deed. No registration is needed unless the owner wants to mortgage the property. Nevertheless, most property transfers are registered in the Land Registry, as there are clear benefits to this. In Spain, although only 50% to 60% of the property sales are carried out with the help of a real estate agent, the purchaser receives qualified legal advice both from the notary public and the Land Registrar.

References

Álvarez, L., Blanco, R., & García Posada, A. (2020). *La inversión extranjera en el mercado inmobiliario residencial español entre 2007 y 2019*. Banco de España. Retrieved November 20, 2021, from https://bit.ly/2LQwTB7

Díaz Guijarro, R. (2014). Los españoles son los europeos que tienen más casas de vacaciones. *Cinco Días*. Retrieved November 20, 2021, from https://bit.ly/35Uo1S3

Fernández Durán, R. (2006). El tsunami urbanizador español y mundial. *Boletín CF+S*, 38–39.

Montoriol Garriga, J. (2020). Las segundas residencias en España: ¿mar o montaña? *Caixa Bank Research*. Retrieved November 20, 2021, from https://bit.ly/313h5it

National Statistics Institute (2021) *House price index*. Retrieved November 20, 2021 from https://www.ine.es/en/index.htm

13

UNITED ARAB EMIRATES (UAE)

Walaaeldeen M. Ibrahim

1. The UAE housing market

The Emirati[1] housing market, both in terms of the rental or freehold ownership, is considered a diversative one, as UAE is a federal country constituted of seven units, each called an emirate. There are some differences between these local markets based on differences in the social and economic standards of each emirate. In this context, we present these markets as three cases.

Abu Dhabi market

The rental market: The years 2019 and 2020 witnessed a continuous delivery of new supply of units in the real estate sector that put pressure on the residential rents, with apartments and villa rates declining on average by 2% and 3% compared with the same periods in 2018. However, some mid- to low-quality buildings recorded annual declines of 7% to 10%, and more than 15% in the Q3 of 2020. Regarding that, most experts expect further drops in 2021 (Asteco, 2020, p. 3).

The sales market: As with the rental market, the sales market has suffered from the same issue, namely, that sales transaction volumes and activity remained low in Q1 of 2020. In addition, the average apartment and villa sales prices softened by 3% since Q4 of 2019 and recorded an annual decline of 9% and 7% respectively, which had not changed in the Q3 of 2020 (Asteco, 2020, p. 4).

Dubai market

The rental market: Statistics show in Q3 of 2020 that there was a continuity of decline in the rental rate by 2% and 9% compared with the same period in 2019, which is reflected in the decrease of apartment rents of between 4% and 13%.

The sales market: According to Dubai Land Department (DLD) data, apartment, villa, and office prices declined by 2%, 1%, and 2% over Q3 of 2020 and 10%, 10%, and 8% annually. Despite this situation, there has been a marginal increase in prices for specific unit types in select new developments such as Dubai Hills and West Gate as well as in existing popular communities, but experts believe that this increase will be temporary unless the economic conditions and market sentiment improve significantly in the short term (Marlow, 2012).

DOI: 10.1201/9781003130383-15

Northern Emirates market

The rental market: It was not better than its predecessors in Abu Dhabi and Dubai that the apartment rental rates in the Northern Emirates recorded average quarterly and annual declines of 3% and 10% respectively. The downward pressure on rates is expected to continue, particularly regarding the upper range (Asteco, 2020, p. 20).

The sales market: Similar to the leasing sector, Sharjah sales prices continued their downward trend, declining by 3% over Q3 2020 and by 20% annually due to increased competition with more affordable options (Asteco, 2020, p. 21).

2. Regulatory framework

As stated, UAE is a federal country, and according to its constitution, the real estate organization has been left to local authorities in the constituted Emirates of the union. Therefore, we will present the legislative process in each local market, as mentioned in the first part.

Abu Dhabi market

Freehold ownership: Law No. 19 of 2005 organizes the freehold ownership for the nationals that enables them to own real estate in any area in the emirate and gives the Gulf Cooperation Council member countries (GCCs) nationals and the other nationalities the possibility to obtain real property rights only in the economic zones, which is defined by the government, and not in the remaining areas in the emirate. Foreign nationals are able to own freehold interest/titles to apartments, offices, villas, and other real estate units, but a foreign owner's interest regarding land is limited to a *Musataha* (a concept akin to a long-term development right, for not more than 50 years), a *usufruct* (a long-term use right), and long-term lease, each for a period not exceeding 99 years. The local government has issued Law No. 13 of 2019, which gives foreign nationals more advantages when expanding their right to also own land in the investment areas.

Rental market: Law No. 20 of 2006 organizes the relationship between the landlord and the tenant, which states that tenancy contracts should be drawn up in writing and be authenticated at the relevant municipality in the emirate. The rent price may be increased when renewing the contract, after consent between both parties, but limited to the defined percentage of 5%, as stipulated by law. The law can establish a committee, headed by a judge, to resolve any disputes that may arise from the contract, and its decision is final for the disputes that do not exceed € 11,500, otherwise the decision may be appealed before an appeal committee.

Dubai market

Freehold ownership: Law No. 7 of 2006 organizes the freehold property for the real estate in Dubai, which gives all nationals and GCC nationals the right to own properties in all areas, while foreign nationals can own properties in the economic zones.

Rental market: There are two main laws which organize the matters on Dubai tenancy: Law No. 26 of 2007 and Law No. 33 of 2008. The first law concerns the rights and obligations of both parties, the tenant and the landlord, through a model lease contract, while the second law organizes the mechanism of resolving rental disputes. The rental price is determined according to the Rent Index of the Emirate of Dubai, which is authorized at Real Estate Regulatory Agency (RERA). It may be increased, in the case of renewal, by the consent of the parties and according to a defined percentage that is determined by RERA. The landlord has no right to increase the rent over this limit. This index is subject for review from time to time from RERA, according to the market.

RERA has developed the Rental Disputes Settlement Centre to adjudicate in all the rental disputes that may arise between the parties. The dispute is heard before a tripartite committee headed by a judge and the membership of two experts in the real estate. The decision shall be final if the dispute does not exceed € 23,000; otherwise, any of the parties can appeal the decision before an appellate tripartite committee headed by a judge, with the membership of one judge and a real estate expert.

Northern Emirates market

Five emirates form the Northern Emirates – Sharjah, Fujairah, Ajman, Umm Al Quwain, and Ras Al Khaimah – seem to use the same methods in terms of real estate property. Their local laws allow nationals the possibility to own real estate, while foreign nationals, despite what their nationality may be, can enjoy extended rent for 99 years. The rental market is the same as other emirates in that it depends on a contract between the tenant and the landlord, subject to registration in the relevant municipality, and the tenant price may be increased by a certain percentage (Place Community Managers, n.d.). This percentage is subject to review by the local government in the emirate and according to the economic situation and the market. Most local laws in the Northern Emirates stipulate that the landlord cannot increase the rental value within the three years of the lease contract.[2] Any disputes that may arise between the parties can be heard before a committee, subject to the entitled municipality in the emirate arriving at a decision that resolves the dispute when the limit is not more than 100,000; otherwise, the dispute can be referred to the competent court. The first level committee is a tripartite committee headed by a judge and the membership of two real estate experts, while the appellate committee is composed entirely of judges.

3. The conveyance process

In the UAE conveyance process, real estate management companies play a leading role in the marketing, contract, and registration phases. Most landlords assign one of these companies to work on their behalf when exercising their roles, for various reasons. The most common reason is because these companies have experience and the facilities to effectively carry out such tasks. Moreover, most of proprietors are businessmen that need to save money and time.

Marketing phase

In this stage, management companies use all its facilities and abilities to increase the demand for the real estate units, relying on planning and promotions to attract buyers. Here, the companies rely on real estate brokers to expand their market.

Pre-contract and contract phase

In this stage, the management companies work on behalf of the owner in signing the contracts (the lease or the sale and purchase contracts). These companies often have a legal department that exercise this task, or they may make agreements with legal firms to exercise and supervise this step. Some owners prefer to keep this process with them to be secure, but this is rare in practice (www.tamimi.com).

Table 13.1 Fees and taxes during the conveyance process in UAE

		Abu Dhabi		*Dubai*		*Northern Emirates Market*	
Phase	*Fees and taxes*	*Seller*	*Buyer*	*Seller*	*Buyer*	*Seller*	*Buyer*
Marketing	Real estate brokerage (where applicable)	2% + 5% VAT	-	2% + 5% VAT	-	2% + 5% VAT	-
	Real estate financing for the bank fees (where applicable)	-	Often 1% of the financed amount + 5% VAT	-	-	-	-
	Real estate financing registering fees (where applicable)	-	0.1%	-	-	-	-
	Registering of usufruct contract	-	-		2% of the real estate value	-	-
Contract	Registering real estate mortgage	-	-	-	-	-	0.05% of the mortgage value + € 115 issue the mortgage pact
	Ownership pact cost	-	€ 230	-	€ 58	-	-
	Real estate assessment fees	-	€ 575–690 + 5% VAT	-	€ 920–3,450 + 5% VAT	-	-
	Authentication for initial sale and purchase a real estate contract	-	-	-	-	-	230 + 5% VAT
Registration	Municipality fees	-	2%	-	-	-	-

(*Continued*)

Table 13.1 (Continued)

		Abu Dhabi		*Dubai*		*Northern Emirates Market*	
Phase	*Fees and taxes*	*Seller*	*Buyer*	*Seller*	*Buyer*	*Seller*	*Buyer*
	Registry fees	-	-	-	4% of the real estate value	2% of the real estate value + € 115 issue the ownership pact	-
	Registering lease contract ended in ownership of the real estate unit	-	-	4.25% is distributed as follows: 4% of the real estate value, 0.25% of the total value of the lease contract	-	€ 460 + 5% VAT	-
Post-registration	Delay fine for the delay of registering real estate transactions	-	-		-	-	€ 920–1,840
	Sales tax	-	-	-	-	-	-
Total (EUR for the model case price)		**2,100**	**€ 2,834–2,955**	**€ 6,100**	**€ 7,024–9,681**	**4,698**	**€ 1,277–2,197**
Total (% share to the model case price)		**2.1%**	**2.8–3.0%**	**6.1%**	**7.0–9.7%**	**4.7%**	**1.3–2.2%**
Excluded costs from the model case		-	Often 1% of the financed amount + 5% value added tax, real estate financing registering fees (where applicable) 0.1%	Registering lease contract ended in ownership of the real estate unit: 0.25% of the total value of the lease contract	-	-	Registering real estate mortgage: 0.05% of the mortgage value

Note. Model case price: € 100,000

Source: Author

Registration phase

Regarding the rental market, according to the rental local laws in the seven emirates in UAE, there is a consensus that all the rent contracts must be registered in the municipality against 5% fees of the rental value which is included in the contract (Executive Council Decree, 2013). The process of registering the rent contracts is linked with the public services to the leased unit. Consequently, unless the lease contract is registered, no public services (e.g. electricity, water, parking) are available.

In terms of the ownership market, we can see in the Civil Transactions Law provided in Art. 1277 that ownership of immovable property and of other real rights are not transferred, either between parties or with third parties, except by registration in accordance with the relative law provisions. This means that the ownership of any real estate unit cannot be conveyed or transferred legally to the buyer unless the transaction be registered at the entitled authority. There is not a federal registry in the country, but rather the local authority in each emirate manages this process, given some fees, which may differ between the three markets in UAE.

Table 13.1 outlines the fees and charges applicable to the sale of residential property in each of the states Abu Dhabi, Dubai, and Northern Emirates Market.

Notes

1 1 UAE Dirham = 0.2300 Euro (as of 1 October 2021).
2 See Art. 13, Law 2 of 2007, for organizing the relationship between the landlord and the tenant in Sharjah Emirate.

References

Asteco. (2020). *UAE real estate report – Q3 2020*. Asteco. Retrieved from https://www.asteco.com/links//reports/2020q3/Asteco_UAE_Q32020_Real_Estate_Report.pdf

Executive Council Decree. (2013). *No. 30*. Retrieved from https://www.tlg.ae/source/uploads/ck_files/1559023342.pdf

Marlow, T. (2012). Real estate within the UAE: A summary of legislative development. *Tamimi.com*. Retrieved from www.tamimi.com/law-update-articles/real-estate-within-the-uae-a-summary-of-legislative-development/

Place Community Managers. (n.d.). *Main page*. Retrieved November 20, 2021, from www.place.ae/

14
CROATIA

Peter Gladoić Håkansson and Jasenka Kranjčević

1. The Croatian housing market

According to the Central Bureau of Statistics in 2011 85.1% (1,912,901) of a total of 2,246,910 dwellings were suitable for permanent housing in Croatia[1] (Tkalec, Vizek, & Žilić, 2018). The total area of all the apartments was 168,651,195 m^2, which means that the average area of an apartment was 75 m^2, while in multi-apartment buildings, the area of the average apartment was about 60 m^2 (Rašić, Slijepčević, Stojčić, & Vizek, 2019, p. 12). When the residential real estate market is monitored, it is divided into apartments and family houses.

According to the Central Bureau of Statistics, the number of occasionally used dwellings (holiday dwellings and dwellings used during seasonal agricultural work) was 262,769, with a total area of 15,116,934 m^2 and an average of 59 m^2. Unfortunately, data on the number of other types of buildings in Croatia are not available. The percentage of construction in Croatia is 6.5% (Rašić Bakarić et al., 2020). According to the data from the Central Bureau of Statistics, a total of 4,933 buildings were built in 2018, of which 3,824 were residential and 1,109 were non-residential (Central Bureau of Statistics, 2019). When these data are compared with the figures from 2017, in 2018, we see a decrease in the total number of constructed buildings by 0.1%, an increase in the number of constructed residential buildings by 3.4%, and a decrease in the construction of non-residential buildings by 10.6%.

From 2012 to 2017, around 540,000 transactions were realized on the real estate market in Croatia, with the sale of real estate as the most common transaction from 2012 to 2017. The number of leases or rents is neglected and often left unrecorded to evade taxes. Most of the transactions relate to the sale of agricultural land (220,259), followed by apartments (103,193), and construction land (74,378) (Tkalec et al., 2018, pp. 13–14).

In terms of apartment prices, Croatia is divided into three geographic areas: the Adriatic, the continental part, and the city of Zagreb (Figure 14.1). The locations with the highest price on the Adriatic were in Dubrovnik and its surroundings, Split and part of the central Dalmatian islands, Opatija and Krk, and the western part of Istria. The high prices in these locations can partly be explained by the interest of foreign investors. Foreigners often buy real estate in Istria or Dalmatia, and data from 2020 show that Germans, Austrians, and Swiss invested the most in real estate on the Adriatic.

The cheapest locations on the Adriatic coast were around Opuzen, the hinterland of Zadar and Karlobag and Senj. As for the continental part, the most expensive locations were in larger

DOI: 10.1201/9781003130383-16

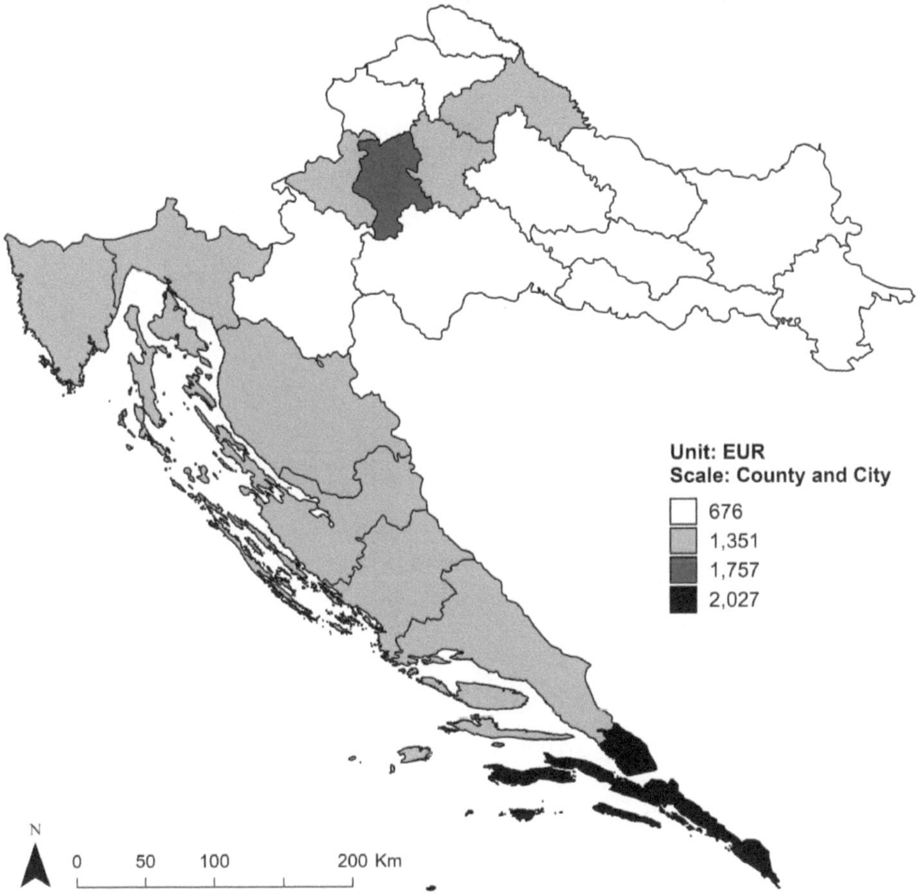

Figure 14.1 Map of distribution of house prices in Croatia

Source: https://dnevnik.hr/galerija/vijesti/hrvatska/nova-analiza-trzista-nekretnina-provjerili-smo-razlike-u-cijenama-609960.html/61903737/609960

city centres (Osijek, Slavonski Brod, Varaždin, Bjelovar, Koprivnica), then in the vicinity of Zagreb and in the interior of Istria, while cheaper locations were in predominantly rural municipalities. The city of Zagreb should be viewed as the capital of Croatia, which generates more expensive real estate prices.

In the east of Croatia (Slavonia), residential real estate continues to fall in price (particularly in rural areas due to strong depopulation), but in the city of Osijek, a small increase was recorded. However, in general, prices of real estate rose sharply from 2002 to 2008. If we compare the average prices in 2002 and 2008, we can see that real estate prices in Croatia nearly doubled in that period, with the largest increase in prices being in the Adriatic (+125%). In the city of Zagreb, growth was somewhat more moderate in that period (+93%), while in the other regions, the increase was +68%. Real estate prices peaked in 2008; however, a decline began in parallel with the slowdown in economic activity that hit Croatia along with the rest of the world due to the 2008 global financial crisis. The total decline in real estate prices from 2008 to 2015 was 20%, but if we look at the period (2002–2015), we can see that real estate prices in Croatia

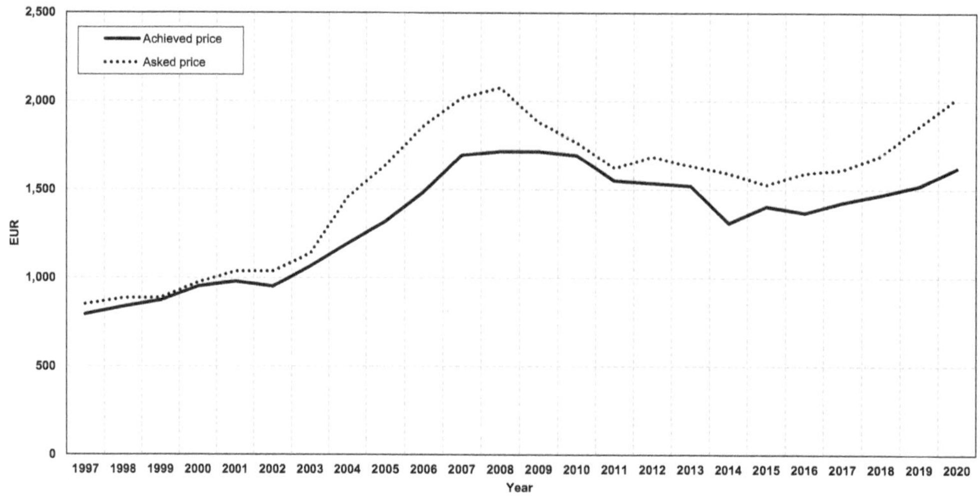

Figure 14.2 Change in house prices in Euros charting both the asking price and the achieved price, 1997–2020

Source: Burza Nekretnina (2021) www.burza-nekretnina.com/statistike

increased by 61% (Kunovac & Kotarac, 2019). Further, as Figure 14.2 shows, after 2015, prices again started to increase.

From 2012 to 2017, between 80,000 and 100,000 purchase and sale transactions were recorded annually. These transactions recorded an annual increase of approximately 13.1%. The most significant share of the value of transactions in 2017 on the real estate market was for apartments, which account for 35.4% of the value of contracted transactions on the real estate market, followed by various real estate (construction land and business premises) (Tkalec et al., 2018, p. 14).

In the real estate market in 2019, around 112,000 transactions were realized, which, when compared with 2018, represents a growth of 6.5%. Most of the transactions were sales of agricultural land, in total, 38,315, which amounts to 34.3% of the total number of transactions. This is followed by sales of flats, which accounted for 26,480, as in, 23.7% of the total number, and transactions of construction land, which accounted for 17,100 (15.3% of total transactions). Purchase transactions of family houses were recorded as 14,006, which accounted for 12.5% of the total number of transactions in the real estate market in 2019 (Rašić Bakarić, Slijepčević, Stojčić, & Vizek, 2020).

Many real estate properties in Croatia do not yet have proper ownership documentation, which is a requisite for the purchase, especially in conjunction with a bank loan. According to Bejaković and Mrnjavac (2020), an estimate of just under 50% of all real estate in Zagreb was not properly registered in official registries, which, of course, makes transaction more difficult. However, in 2012, the government initialized a legalization process for illegally constructed buildings (Official Gazette, 2019a). The scope of illegal construction is sufficiently indicated by the data of the Ministry of Construction on legalized square footage in the period from January 2012 to June 2019. A total of 884,608 illegally constructed projects in Croatia were legalized in that period. According to the register of issued decisions on the derived condition, almost 1 billion m^2 (more precisely 974,260,497 m^2) have been legalized (mgipu.gov.hr). Even according to the cheapest estimate, assuming that the value per square metre is € 100 (not only legalized buildings, family houses, and apartments but also garages, renovated parts of buildings, balconies, and ancillary buildings and facilities), € 100 billion in newly created real estate value appeared on the domestic market in just seven years.

Another problem that hinders transactions is that real estate agencies charge commissions to both parties, although this is not determined by the Real Estate Brokerage Act. Further, the COVID-19 crisis and devastating earthquakes in 2020 in Zagreb and Sisak-Moslavina County will certainly change some current trends in real estate prices and sales.

2. Regulatory framework

The real estate market in Croatia is regulated by four basic laws and two ordinances. The Law on Ownership and Other Real Rights (which defines real estate, among other things. Official Gazette, 2017), the Real Estate Valuation Act (which enables real estate valuation, Official Gazette, 2015a), the Real Estate Brokerage Act, the Real Estate Sales Tax Act (Official Gazette, 2019b) and two ordinances, one on real estate valuation methods (Official Gazette, 2015b) as well as the ordinance on the real estate market information system (Official Gazette, 2015c).

With the adoption of the Real Estate Valuation Act in 2015, the real estate market information system was established in Croatia using eNekretnina [trans. *e-RealEstate*], of which a collection of sales contracts is an integral part. The establishment of the information system provides insight into basic data on real estate (appraiser market, appraisal commissions, real estate agents, etc.). The Ministry of Physical Planning, Construction and State Assets publishes the *Real Estate Market Overview* (mgipu.gov.hr) in Croatia once a year.

3. The conveyance process

In this section, we outline the conveyance process in Croatia.

Marketing phase

Following the model of a conveyance process outlined by Palm, Jingryd, and Kordić (2020), the first activity is the marketing phase. In Croatia, conveyances on the property market are usually brokered by estate agents who are required to be registered with the Croatian Chamber of Commerce (Official Gazette, 2007). In the process of a real estate conveyance, the real estate agent deals with the relationship between the principal (often the seller) and a third party (often the buyer) and negotiates and prepares for legal transactions that are the subject of a particular property. The principal is a natural or legal person who signs a written brokerage contract with a real estate agency (Palm, Jingryd, & Kordić, 2020).

The agent connects the principal with the third party, who then enter into a sales agreement. The agent must examine the legal status of the property in the Land Register and gather the requested legal documents required for concluding a sales pre-contract/contract. The agent advertises the property on the relevant portals and property search engines as well as on the agency's website.

Once an interested buyer is found, the agent engages a solicitor to prepare a preliminary contract, a sales contract, and a real estate tax form and to carry out the ownership registration at the Land Registry. The agency is obliged to pay the solicitor and land registry fee; however, all these costs are included in the commission, which is calculated in accordance with the Real Estate Act (Official Gazette, 2007). The principal pays the commission to the agency if a sales agreement is reached. If the agent represents both the seller and the buyer (in which case they are both principals), the buyer is charged with an additional fee, but the total amount of the commission must not exceed the limit defined by the Real Estate Law. The highest total commission fee is 6% of the total achieved amount of the selling price. If the commission turns out to be 4%, the seller and the buyer normally pay 2% each. Further, the seller must provide all data and documents regarding the ownership of the property and inform the agent of any changes related to the property. The brokerage agreement is concluded in writing. Also, unless otherwise agreed, the brokerage agreement is deemed to be concluded after a period of 12 months but may be extended several times by the agreement of the parties.

When acquiring a property, the buyer should examine the property, including legal and other cadastral data such as the Land Registry, the building permit, use permit, and energy certificate. Although estate agents are obligated to check all the documentation, they are not responsible for any irregularities. The first step is to check the ownership registered in the Land Registry of the municipal court in the area where the property is located. The Land Registry entry consists of three parts: (1) a census sheet listing the property (i.e. its main features, including the number of Land Registry parcel, surface, and culture); (2) an ownership list, in which the owner of the property is registered; and (3) cargo tickets, which reveal whether there are third-party burdens (such as mortgages, real-offs, reopen, rent, redemption, custody, or seizure). If the purchase is to be financed via bank loan, it is the buyer's responsibility to investigate this possibility. The requested loan is usually 10% or 25% lower than the value of the property. In some cases, the bank approves housing loans in the ratio of 1:1 but with increased security instruments such as private participation or a guarantor.

Pre-contract phase

When an interested buyer is found and the price is agreed, the buyer and the seller sign a pre-contract. The pre-contract obliges the parties to conclude the main contract but does not convey ownership of the property. The pre-contract has the same form as the main contract: it must identify the property and the price in writing. As previously mentioned, it is the agent's responsibility to engage and pay the solicitor for preparing a pre-contract. However, it is common for a buyer to hire a lawyer to review pre-contractual and contractual agreements.

The price of the real estate is negotiated between the buyer and the seller, whereas the commercial value of a property is defined by the tax authority. However, in order to allocate credit from the bank, a licensed appraiser must establish an estimate of the real estate's value. When signing the pre-contract, the buyer gives the seller a deposit (down payment) as a sign of concluding a contract and as a warranty that the contractual obligations will be performed. The deposit is a certain amount of money or exchangeable assets, and in most cases, the deposit is 10% of the total amount of the contractual price in line with the legal maximum. If the contractual obligations are met, the price is reduced by the size of the deposit or returned to the buyer. This means that the buyer bears the risk of not obtaining financing.

Contract phase

The sales contract is the final purchase document. It obliges the seller to transfer the property to the buyer who becomes the owner of the purchased property after the stated amount is paid. This phase takes place at the buyer's bank and in a licensed notary office. If the purchase has been financed by a bank loan, the contract includes mortgage activities. The required documentation when applying for a housing loan includes standard documentation to prove the creditworthiness of the borrower, additional documentation confirming the purpose of the loan (pre-contract), proof that the seller is the owner of the property, and other basic documents of the seller. In the next step, the bank appraises the property. The future borrower pays the cost of the appraisal. The estimated appraisal cost for a flat of 50 m^2 is approximately € 176. After the processing and approval of the loan, the public notary formalizes the loan contract. The borrower (the real estate buyer) pays the cost, and this cost depends on the loan amount, the number of participants in the loan, and collateral. On average, this cost ranges from € 338 to € 541. Further, the borrower is also obliged to pay a fee for processing the loan. On average, the bank mortgage fee ranges from € 135 to € 406.

Registration phase

The seller is required to both certify their signature on the sales contract by a licensed notary office and have the sales contract verified by the Land Registry, which is operated by the municipal courts. Namely, in order for real estate buyers to become owners, it is necessary to register the ownership in the Land Register based on a sales contract and tabular statements (real estate ownership is not acquired automatically). These documents represent a certified statement that allows the seller to transfer ownership to the buyer. The buyer submits this statement with other documents to the Land Registry, after which, the ownership is registered. It is also possible to submit an electronic proposal for registration in the Land Register, regardless of which municipality the property is located, through all public notaries and through attorneys who have requested such an opportunity. Title registration is constitutive; namely, it is the registration that makes the sale binding towards third parties.

The buyer is responsible for paying the sales tax, which is 4% of the market price for the property purchased. The buyer is also required to report the amount of the sales contract to the tax authority within 30 days and must pay the tax upon receipt of the official real estate tax document. The sales tax is municipal, which means that the revenue belongs to the municipality in the area where the real estate is located.

The conveyance process is shown in Figure 14.3.

4. Summarizing the conveyance process and its transaction costs

To summarize, one could say that the Croatian conveyance process in many ways follows the continental European model. According to Palm et al. (2020), the Croatian model is in most respects a typical continental European civil law regime, with the same benefits and shortcomings as in other civil law countries. Table 14.1 gives a summary of the transaction costs that may appear in the transaction process.

Croatia faces a number of challenges when it comes to the real estate market. For example, COVID-19 and the subsequent downturn in economic activity will probably influence the real estate market; however, in what way and how much is difficult to say as of spring 2021. Also, the earthquakes that hit Croatia in the spring and winter of 2020 will affect transactions and prices in these areas. Given these challenges, certain reforms are important. For property to become

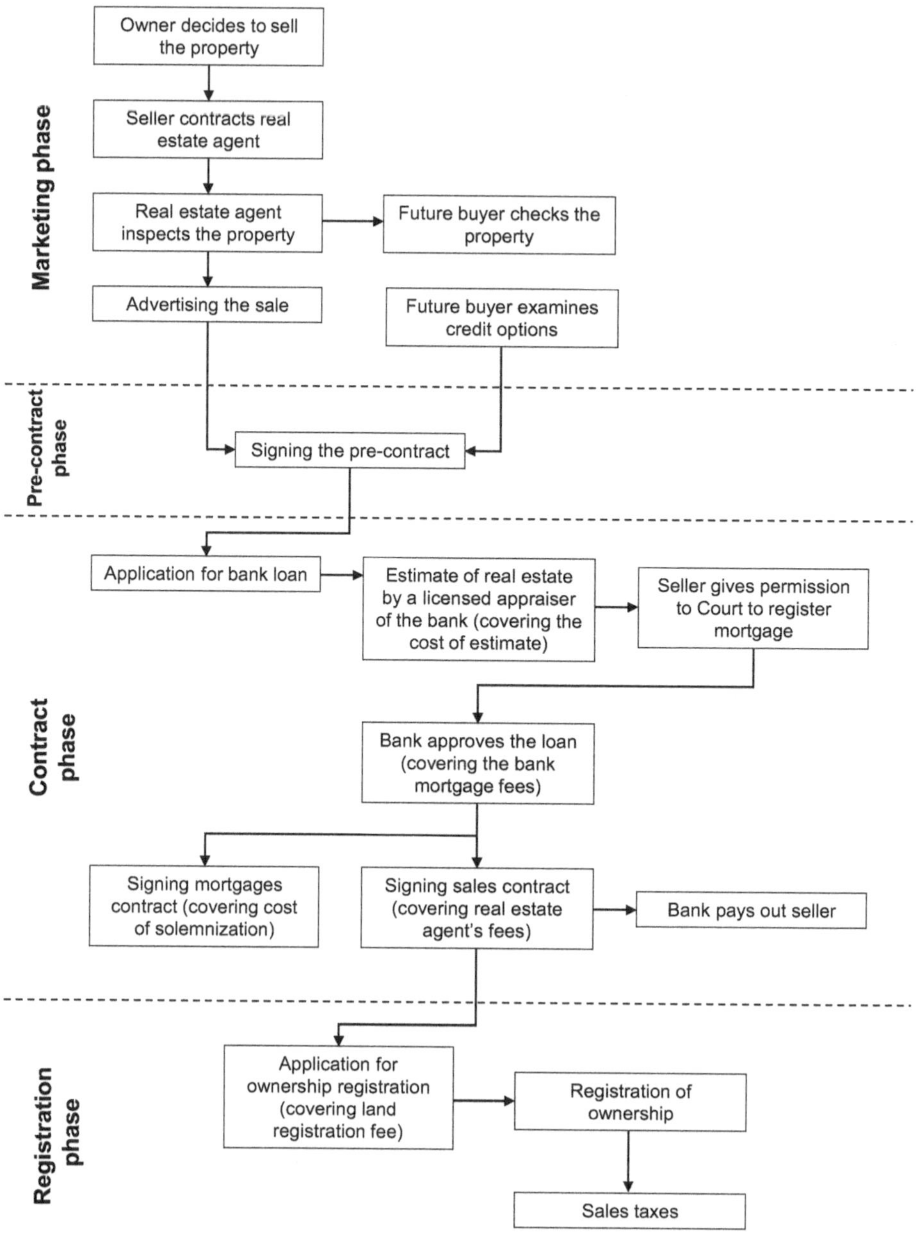

Figure 14.3 The conveyance process for real estate in Croatia

Source: Palm, Jingryd, & Kordić (2020)

capital, it is particularly important that the legalization reform continues. Without legal documents, it is difficult to sell property and even more difficult to get a loan. Transaction costs are essential, and these costs affect the real estate market and slow down mobility, which, in turn, affect the efficiency of the economy as a whole.

Table 14.1 Fees and taxes during the conveyance process in Croatia[2]

Phase	*Fees and taxes*	*Seller*	*Buyer*
Marketing	Real estate agent's fees	2–3%	2–3% (occasionally [1]
Pre-contract & contract	Legal fees	-	345–413
	Appraiser fee	-	€ 176–500 [2]
	Bank mortgages fees	-	€ 135–406 [3]
Registration	Land registration fee	-	34
Post-registration	Sales tax	-	3%
Total (EUR for the model case price)		**€ 2,000–3,000**	**€ 5,690–7,353**
Total (% share to the model case price)		**2.0–3.0%**	**5.7–7.4%**
Excluded costs from the model case		-	-

Note. Model case price: € 100,000

Source: Author

Notes

1 1 Croatian Kuna = 0.1331 Euro (as of 1 October 2021).

2 The table shows which party is obliged to pay the different fees and taxes. However, it is debatable whether that party is also the one to ultimately bear the cost.

References

Bejakovic´, P., & Mrnjavac, Ž. (2020). Housing policy and labour market in Croatia. In P. Gladoić Håkansson & H. Bohman (Eds.), *Investigating spatial inequality: Mobility, housing and employment in Scandinavia and South East Europe*. Bingley: Emerald Publishing Limited.

Burza Nekretnina (2021) www.burza-nekretnina.com/statistike

Central Bureau of Statistics (Državni zavod za statistiku). (2019). Završene zgrade i stanovi u 2018. Priopćenje br. 3.1.3. Retrieved from https://www.dzs.hr/

Kunovac, D., & Kotarac, K. (2019). *Cijene stambenih nekretnina u Hrvatskoj, Pregledi – 41, Hrvatska narodna banka*. Retrieved from www.hnb.hr/documents/20182/2626448/p-041.pdf/a46c4569-30fc-4bb9-80e5-4f5953762d25

mgipu.gov.hr, Ministry of Physical Planning, Construction and State Assets. Retrieved from https://mgipu.gov.hr/istaknute-teme/legalizacija-nezakonito-izgradjenih-zgrada/8314; Retrieved March 30, 2021, from https://mgipu.gov.hr/UserDocsImages/86

Official Gazette. (2007). *Real estate law, OG 107/2007*. Retrieved from https://eur-lex.europa.eu/legal-content/EN/TXT/PDF/?uri=CELEX:52019XC1018(02)&from=EN

Official Gazette. (2015a). Zakon o procjeni vrijednosti nekretnina NN 78/2015.

Official Gazette. (2015b) Pravilnik o metodama procjene vrijednosti nekretnina (NN 115/15).

Official Gazette.(2015c). Pravilnik o informacijskom sustavu tržišta nekretnina (NN 114/15, 122/15).

Official Gazette. (2017). Zakon o vlasništvu i drugim stvarnim pravima NN br. 91/96., 68/98., 137/99., 22/00., 73/00., 114/01., 79/06., 141/06., 146/08., 38/09., 153/09., 143/12, 152/14, 81/15, 94/17).

Official Gazette. (2018). Zakon o porezu na promet nekretnina NN 115/16, 106/18.

Official Gazette.(2019a). Zakon o postupanju s nezakonito izgrađenim zgradama [Law on the Treatment of Illegally Constructed Buildings] (OG 90/11 86/12, 143/13, 65/17, 14/19).

Official Gazette. (2019b). Zakon o posredovanju u prometu nekretnina NN 107/07, 143/12, 144/12, 14/14, 32/19.

Palm, P., Jingryd, O., & Kordić, L. (2020). Housing and transaction costs. In P. Gladoić Håkansson & H. Bohman (Eds.), *Investigating spatial inequality: Mobility, housing and employment in Scandinavia and South East Europe*. Bingley: Emerald Publishing Limited.

Rašić Bakarić, I., Slijepčević, S., Stojčić, N., & Vizek, M. (2020). *Pregled tržišta nekretnina Republike Hrvatske: 2019*. Retrieved from https://repozitorij.eizg.hr/islandora/object/eizg%3A335

Rašić, I., Slijepčević, S., Stojčić, N., & Vizek, M. (2019). *Regled tržišta nekretnina Republike Hrvatske: 2018*. Retrieved from www.eizg.hr/userdocsimages/publikacije/serijske-publikacije/pregled%20tr%C5%BEi%C5%A1ta%20nekretnina/Nekretnine-2018-finalno-web.pdf

Tkalec, M., Vizek, M., & Žilić, I. (2018). *Pregled tržišta nekretnina Republike Hrvatske 2012.-2107*. Zagreb: Ministarstvo graditeljstva i prostornog uređenja, Ekonomski institut Zagreb.

15
GERMANY

Christine Helbach, Sandra Schilling, Fabian Thiel

1. The German housing market

The German residential market is constituted of two main forms: tenancy and ownership. The legislation differs between freehold, condominium, and leasehold, and there is a difference between owning a house as freehold or as condominium. For condominiums, the building with the apartment and the ground it is built on as well as all technical infrastructure is the common property of all the owners. This requires certain restrictions and is regulated in the declaration of partition *(Teilungserklärung)*. A leasehold is a particular form of ownership. A notarized leasehold contract defines the right to construct a building on land owned by another person for a defined period of time. This specific right can be applied on any form of real estate. The leaseholder is granted the right to use the land and has to pay a ground rent. This places the leaseholder in a position equivalent to an owner. The proportion of leasehold properties in Germany is estimated rather low, at around 5%, and it is mostly found in larger cities such as Hamburg or Frankfurt am Main (JLL, 2019). This form of ownership has its implications on the conveyance process, and the differences will be pointed out throughout the chapter.

In contrast to the rather steady increase in the German consumer price index, real estate prices have increased considerably faster in Germany, as shown in Figure 15.1a. Over the past decade, residential property prices increased by ~60% on average compared with a ~15% increase in the German consumer price index (The Bank for International Settlements (BIS), n.d.; Statistisches Bundesamt (Destatis), 2021c). Rising real estate prices and continuous growth in rents has led to a higher political interest in the housing market. This rise in prices is caused by the gap between the demand and supply of adequate housing particularly observed in major cities (e.g. the Big 7 cities: Berlin, Dusseldorf, Frankfurt am Main, Hamburg, Cologne, Munich, and Stuttgart) and metropolitan areas within the whole country. As a result, increased regulatory interventions on the housing market have been enacted in recent years, such as the rent cap. The relevance of leaseholds also seems to intensify, as it enables municipalities to allow and boost the construction of housing yet still own the land in a long-term perspective (JLL, 2019).

In contrast to the competitive real estate markets in most urban surroundings, rural areas can experience a shrinking population, higher unemployment rates, and struggling real estate markets. However, the average price for one- and two-family homes increased by more than 30% over the last five years in all areas of Germany (see Figure 15.1b).

DOI: 10.1201/9781003130383-17

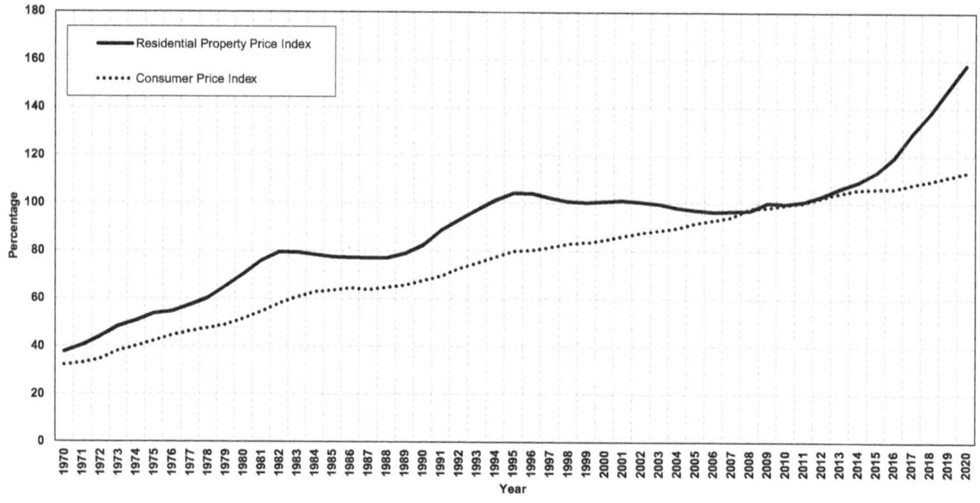

Figure 15.1a Consumer price index and residential property price index (nominal) in Germany (in percent).

Source: National sources, BIS residential property price database, Statistisches Bundesamt (Destatis) (2021b)

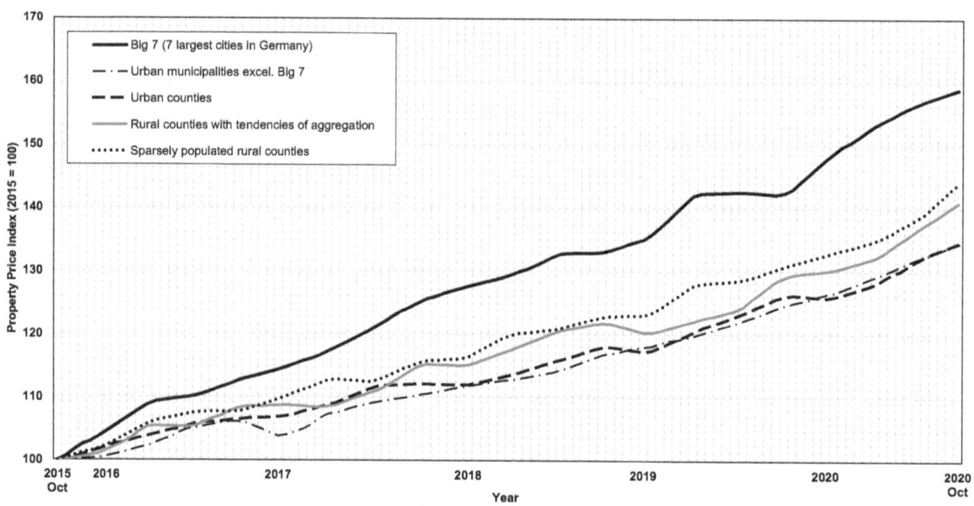

Figure 15.1b Property price index for one- and two-family homes in Germany

Source: Destatis Preisindizes für Wohnimmobilien

Annually, 318,000 condominiums are transacted in Germany, corresponding to about 1.5% of the total stock. In addition, 277,000 single-family homes are sold yearly, corresponding to approximately 2% of the total stock (Immobilienmarktbericht Deutschland, 2019; LBS, 2018). The price levels vary greatly within Germany, with an average price for newly built condominiums of € 2,980/m^2, and for existing houses, € 1,740/m^2 (Immobilienmarktbericht Deutschland, 2019). For the distribution within Germany, see Figure 15.2 focusing on detached one- and two-family homes.

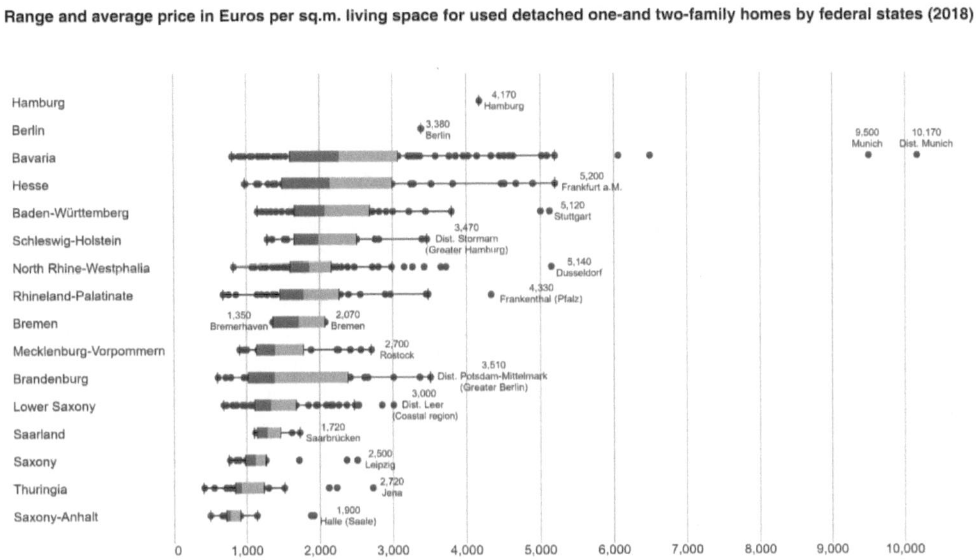

Figure 15.2 House prices in Germany per square metre for used detached one- and two-family homes, 2018

Source: Adapted illustration based on AK OGA, Immobilienmarktbericht Deutschland (2019)

The average price for a condominium in one of the seven largest cities is about € 5,800/m^2, and in smaller cities, it is € 3,600/m^2. Surprisingly, the most expensive houses and condominiums are not found in the German capital Berlin (average price of € 5,000/m^2) but rather in Munich (the capital of the state of Bavaria), with € 9,200/m^2 (Immobilienmarktbericht Deutschland, 2019).

In Germany, the tenancy rate is much higher than the ownership rate and is one of the highest in comparison with other members of the Organisation for Economic Co-operation and Development (OECD) (Deutsche Bundesbank, 2020). In 2018, of a total of about 41 million households, 56% rented and only 44% owned their property (Destatis, 2018). The highest renting ratio in Germany is found in Berlin, with about 83%, and the lowest ratio with 35% is found in the Federal State of Saarland. There are about 37 million units in residential buildings, of which 40% are located in single-family homes (~15 million). Out of 22 million apartments in multi-dwelling houses, about 78% are let and only 22% are inhabited by their owners (Destatis, 2018).

According to Destatis (2018), the average size of a single-family home is 129 m^2 and the average apartment size is 97 m^2, ranging from 75 m^2 in Berlin to 112 m^2 in northern Bavaria. In general, condominiums are about 25% larger than rental units. The majority of the units (42%) were built between 1949 and 1978, and another 25% date back before 1948. However, about 27% of units in single-family homes are newer than 1990, but only 15% of the units in multi-dwelling houses were built after 1990 (Destatis, 2019).

Between 2009 and 2018, an average of 220,000 residential units per year have been built (Statistisches Bundesamt (Destatis), 2021a), although research has found that more than 340,000 newly built units per year are needed to meet the demand (Institut der deutschen Wirtschaft Köln e.V., 2019).

2. Regulatory framework

In Germany, all plots of land are surveyed, and the details are entered in cadastral maps *(Liegenschaftskataster)*. Land generally is divided into cadastral units and cadastral plots *(Flurstücke)* and registered under a plot number in the land survey register kept by the cadastral offices *(Katasteramt)*. It is important to ensure that the information on the cadastral maps is coherent with the land register *(Grundbuch)*. The land register, which is located at the civil district court *(Amtsgericht)*, serves as the legal basis for transactions regarding plots of land and rights to them. Legal relations that are significant with a plot of land are evident from the land register.

Under German civil law, agreements related to land are effective only if they have been executed by a notarial deed (Section 311b German Civil Code *[Bürgerliches Gesetzbuch]*). Integral parts of the property that cannot be separated from the property without destroying the property (immovable fixtures) are not subject to the independent property right (Section 93 German Civil Code).

The most common form of (private) title in land is absolute ownership *(Eigentum*, Section 903 German Civil Code; Article 14 para 1 sentence 1 German Basic Law – *Grundgesetz)*. The so-called absolute owner, however, is subject to general restrictions and obligations imposed by law ('property entails obligations' according to Article 14 para 2 German Basic Law – '*Eigentum verpflichtet*') or any rights of third parties. The absolute owner has control of the land and any upstanding buildings on it.

The German Leasehold Law *(Erbbaurechtsgesetz)* creates a theoretical construct that enables the separate sale of the plot and the building right or building. In the land registry, there are two separate indices – one for the plot and one for the leasehold.

The creation of condominium property is based on the German Condominium Act *(Wohnungseigentumsgesetz)*, which includes the provisions on the creation of residential property as well as its management and the division into individual property and joint property. In the second section, it regulates the rights and obligations of the homeowners' association, including the allocation of usage and costs.

If the real estate purchase involves the financing via a mortgage, the offer for that credit is based on the German Pfandbrief Act *(Pfandbriefgesetz)*, the German Ordinance on the Mortgage Lending Value *(Beleihungswertverordnung)*, and the European Capital Requirements Regulation (CRR).

The purchase price is expressed in Euros. However, a purchase price in foreign currencies may also be agreed.

3. The conveyance process

This section gives an outline of how the transaction of real estate in Germany proceeds. There is no difference between the selling of a freehold property or a condominium; however, the sale of a leasehold needs the additional agreement of the landowner. Figure 15.3 shows the conveyance process.

The law on real estate matters is based on Section 925 ff. of the German Civil Code. Private real estate sales typically do not have a pre-contract phase. A binding conveyance process must include a contract drawn up before a (German) civil law notary; otherwise, there is no transfer of ownership and no entry in the land register.

Marketing phase

The sale of private real estate property involving an estate agent has no strong tradition in Germany. However, in the previous one or two decades, this has changed due to online platforms

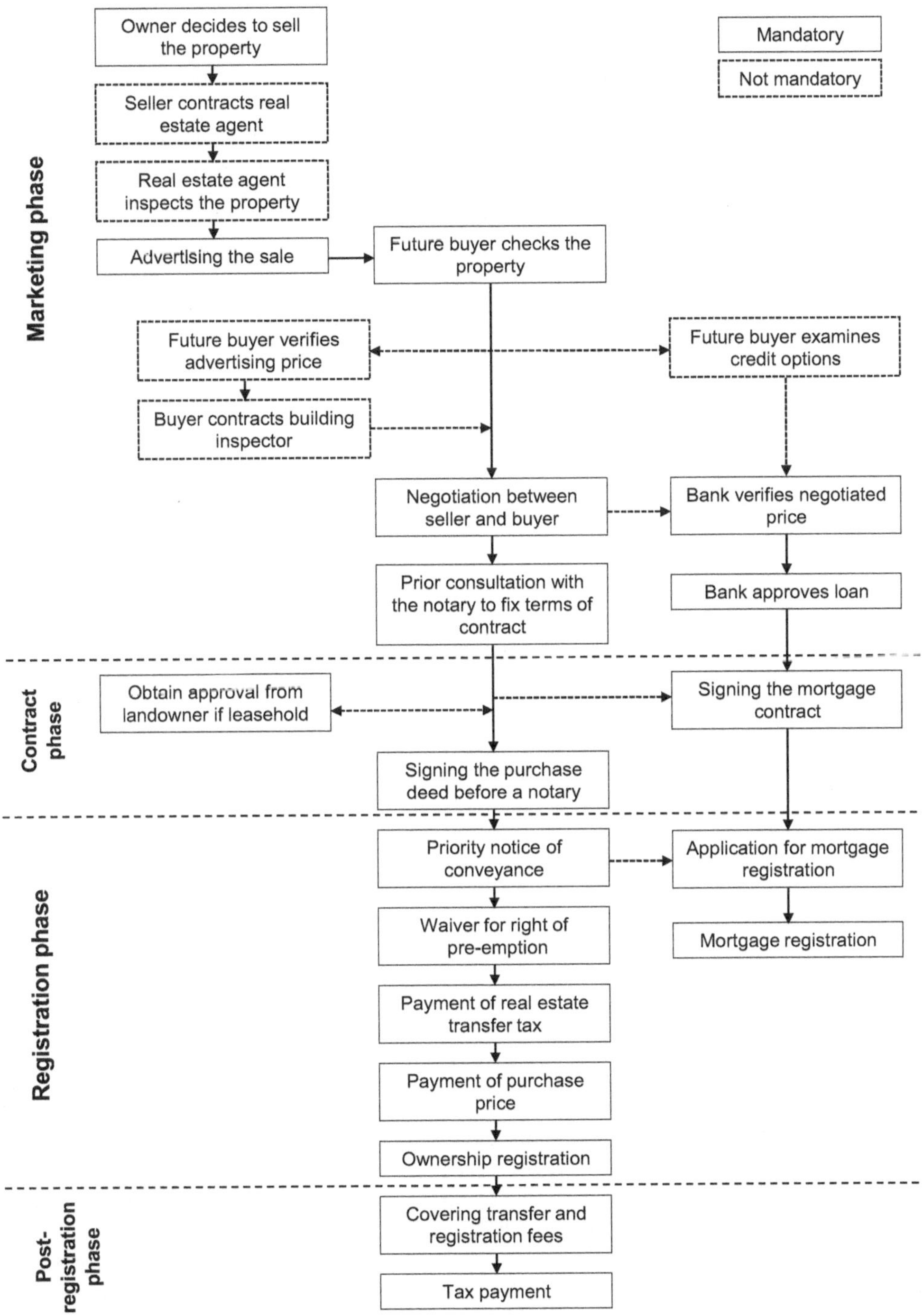

Figure 15.3 The conveyance process for real estate, respectively condominium ownership, in Germany

Source: Author

for advertising real estate sales. The price of such an online advertisement depends on its size, duration, and prominence of display. In urban areas, it is more likely that a property is sold with the involvement of an estate agent (>50%) than in more rural parts of Germany (Institut der deutschen Wirtschaft Köln e.V., 2017). The current law states that the estate agent's commission has to be paid either by the client (mainly the seller) in total or split in equal parts. In Germany, the involvement of an estate agent in a real estate sale is not mandatory, and the real estate profession is not licensed. However, organizations like the IVD *(Immobilienverband Deutschland)* ensure that their members meet certain quality standards.

In Germany, there is the freedom of contract within the regulations of the German Civil Code. With that in mind, every estate agent has to sign a contract with one's clients based on Section 656a ff. of the German Civil Code. Consequently, the estate agent is bound to the client, but the agent is not allowed to lie or conceal any vital information concerning the marketed real estate property.

Estate agents are usually commissioned by the seller, but potential buyers can commission the search for a suitable property. If a real estate property is put on the market via an agent, the agreement contains the marketing price and the lowest sale price. It also stipulates the obligations and rights of both parties. The contract is open ended and is renewed for another month if it is not recalled. When there is a signed contract, the estate agent receives a commission *(courtage)*. In general, the commission ranges between 3% and 7% of the purchase price including value added tax. An estate agency markets the property not only on its own website but also often on larger online platforms. The largest platform in Germany belongs to an international group of financial investors who also may own similar platforms in other European countries. Depending on the local real estate market, the purchase price may vary from the asking price.

Potential buyers should enquire with their bank about the parameters for property credit before starting the process. If the buyer is to accept a concrete proposal, the bank will conduct the mortgage offer for credit based on the relevant legislations. The lending value of smaller objects will be determined by reference prices, and for larger objects, a real estate appraiser will ascertain the lending value which is on the long-term below the market value. Depending on the buyer's financial situation, the credit limit is about 60% to 90% of the lending value. The gap between the credit and the purchase price has to come from the buyer's private equity.

Due to EU regulation, all homes must have an energy performance certificate. The energy efficiency of the property must also be stated in the real estate advertisement.

Contract phase

The modalities of the real estate sale are fixed within a contract of purchase drawn up by a German civil law notary. Such a contract is mandatory for all real estate sales. The purpose is that it should work as a safeguard for the parties involved.

Only with such a notary contract is the sale legal and able to be registered within the land register. The German land registry is a public register in which all real estate properties, their ownership structure, and their liabilities are recorded. This means every real estate property is individually registered. Having a property listed in the register is the only way owners can prove that they legally own the property and can mortgage it.

In general, the buyer chooses the notary, but the notary has to maintain a strictly neutral position. In the contract draft, all details such as the purchase price, payment, covering for the mortgage, transfer date, and new ownership structure are outlined, and all parties receive a copy.

In a formal appointment, the final contract is read out by the notary and finally signed by all parties involved. If the ownership consists of a leasehold, the leasehold grantor has to agree to the sale and there could possibly be changes (runtime, lease) to the leasehold contract. If the property transaction contract has not been notarized it is void, and neither the seller nor the buyer has any claim based on the contract.

There is the possibility that if one or all parties are not able to attend this formal appointment then a third party can sign the contract with a notarized letter of attorney concerning the sale. All parameters to be considered, for example, tax and foreign laws have to be clarified beforehand by the concerned party, as this is not part of the notary process. Once the contract is signed, the parties can withdraw from the sale only under certain circumstances that are regulated by the contract.

Registration phase

After signing the contract, the notary will procure a priority notice of conveyance with the land registry for the property concerned *(Auflassung)*. This is to prevent unapproved changes in the register entry (e.g. a second sale by the same seller).

In Germany, the local municipality has a specific right of pre-emption for which the notary has to obtain a waiver. Subsequently, the notary monitors that all necessary official permits and waivers of statutory pre-emption rights have been obtained, and finally, that documents 'freeing' the property from any encumbrances to be cancelled have been submitted. Thereafter, the notary will send the buyer a demand of payment for the purchase price. The buyer is now required to pay the cost as stated in the sales contract. In addition, the buyer will receive a demand of payment from the local tax office to pay the real estate transfer tax *(Grunderwerbsteuer)*. Depending on the federal state the property is located in, the amount ranges between 3.5% and 6.5% of the sales price.

After the notary obtains the information that the sales price and the tax is paid fully, the notary will ensure the complete ownership transfer of the sold property in the land registry.

The time frame for the whole process is up to eight weeks after signing the contract. The cost for the notary and the ownership transfer at the land registry is about 1.5% of the purchase price. The fees to be paid to the notary and the land register are fixed on a scale laid down by statute, and therefore, are not subject to negotiation.

The costs of the agreement, its implementation, and the fees payable to the notary and the land registry are borne by the buyer. In practice, the question of who is responsible for payment should be regulated in the contract. Regardless of any contractual agreement, the contracting parties remain jointly and liable to the notary, the land registry, and the payment of the real estate transfer tax to the (local) taxation department.

The risk of the accidental destruction or obsolescence of the property passes to the buyer when the property is transferred to the buyer by the contract. At this point (the transfer date or 'closing'), the buyer becomes entitled to the benefits of the property and bears all burdens attached to it (Section 446 German Civil Code). Although legal ownership is transferred to the buyer only afterwards, it is important to note that the property is insured in favour of the buyer from the transfer date to the extent desired by the buyer.

Post-transaction phase

With the transfer of the ownership, the sale process itself is completed.

In the case that the real estate is resold within ten years and has been let during this time, there could be a tax on speculative gains for the seller. This tax depends on the profit and the individual tax rate (maximum tax rate currently 42%).

All property owners have to pay an annual local tax *(Grundsteuer)*. The tax is determined based on the size of the building plot, the living space, and a local multiplier. In the 100 largest German cities, this multiplier varies between 381% and 910% (av. 560%). For a standard single-family home, this means an annual tax of € 323 in Gütersloh, and in the neighbouring city of Bielefeld, it would mean a tax of € 559, while in Frankfurt am Main, this tax would be € 424 per year, and in the nearby city of Offenbach am Main, the tax would add up to € 758 per year (IW Consult GmbH, 2021).

4. Summarizing the conveyance process and its transaction costs

The German conveyance process is straightforward. The notary guides the transfer of the ownership, and signing the contract with a notary is mandatory. If a real estate agency is involved in the sale, this affects only the marketing side, as real estate agents are not allowed to give legal advice. Any legal or tax advice has to be sorted out with a specialist.

Table 15.1 summarizes the transaction costs in Germany for real estate. The real estate transfer tax and the real estate agent fee differs between federal states.

The high amount of real estate transfer tax that has to be paid by the buyer for every real estate sale is one point that contributes to the low resell frequency of private real estate properties and the comparatively high tenancy rate. Due to this, most people tend to rent real estate or own the property they live in for a long period – often for a lifetime – rather than sell it and buy a new one.

Table 15.1 Fees and taxes during the conveyance process in Germany

Phase	*Fees and taxes*	*All real estate*	
		Seller	*Buyer*
Marketing	Real estate agent's fees	~3.5% (~7%)	~3.5% (0%)
	Energy performance certificate	~ € 350	-
Contract	Bank mortgage fees (Disagio)	-	(max 5%)
	Notary fees	-	~1.5%
Registration	Real estate transfer tax	-	3.5–6.5%
	Land registry	-	~0.5%
Post-registration	Capital gains tax (ownership < 10 yrs)	Personal tax rate (e.g. 40%) of sales profit	-
	Real Estate Ownership Association annual fee (not mandatory)	-	(~ € 50–100)
	Sales tax	-	-
Total (EUR for the model case price)		**€ 3,850–7,350**	**€ 5,550–12,100**
Total (% share to the model case price)		**3.8–7.3%**	**5.55–12.1%**
Excluded costs from the model case		40% CGT	5% bank mortgage fee

Note. Model case price: € 100,000

Source: Author

References

AK OGA. (2019). *Immobilienmarktbericht Deutschland 2019 – Arbeitskreis der Oberen Gutachterausschüsse, Zentralen Geschäftsstellen und Gutachterausschüsse in der Bundesrepublik Deutschland*. Retrieved January 1, 2021, from https://redaktion-akoga.niedersachsen.de/immobilienmarktbericht_deutschland/download/downloads-183285.html

The Bank for International Settlements (BIS). (n.d.). *Residential property price database*. Retrieved January 1, 2021, from www.bis.org/statistics/pp.htm

Deutsche Bundesbank. (2020). *Research brief – Gründe für die niedrige Wohneigentumsquote in Deutschland 30. Ausgabe*. Retrieved January 18, 2021, www.bundesbank.de/de/publikationen/forschung/research-brief/2020-30-wohneigentumsquote-822090

Jones Lang Lasalle (JLL). (2019, March). *Leaseholds in major German cities*. Retrieved March, 9, 2021, from www.jll.de/en/trends-and-insights/research/leaseholds-in-major-german-cities

Institut der deutschen Wirtschaft. (2017). *Kurzbericht – Das Bestellerprinzip für Immobilienkäufe*. Retrieved March 21, 2021, from www.iwkoeln.de/studien/iw-kurzberichte/beitrag/alexandra-toschka-michael-voigtlaender-das-bestellerprinzip-fuer-immobilienkaeufe-321820.html

Institut der deutschen Wirtschaft. (2019). *Pressemitteilung – Deutschland muss mehr bauen*. Retrieved February 9, 2021, from www.iwkoeln.de/presse/pressemitteilungen/beitrag/ralph-henger-michael-voigtlaender-deutschland-muss-mehr-bauen.html

IW Consult GmbH. (2021). *Grundsteuerranking 2021 – Haus & Grund Deutschland in collaboration with IW Consult GmbH*. Retrieved May 15, 2021, from www.iwconsult.de/aktuelles/projekte/grundsteuerranking-2021

LBS. (2018). *Wohnungsbestand*. Retrieved February 9, 2021, from www.lbs-markt-fuer-wohnimmobilien.de/inhalt/bestandszahlen/

Statistisches Bundesamt (Destatis). (2018). *Wohnverhältnisse privater Haushalte – Fachserie 15 Sonderheft 1–2018*. Retrieved January 1, 2021, from www.destatis.de/DE/Themen/Gesellschaft-Umwelt/Wohnen/Publikationen/Downloads-Wohnen/evs-wohnverhaeltnis-haushalte-2152591189004.pdf?__blob=publicationFile

Statistisches Bundesamt (Destatis). (2021a). *Bautätigkeit und Wohnungen – Fachserie 5 Reihe 3–2019*. Retrieved January 1, 2021, from www.destatis.de/DE/Service/Bibliothek/_publikationen-fachserienliste-5.html

Statistisches Bundesamt (Destatis). (2021b). *Verbraucherpreisindex für Deutschland – Lange Reihen ab 1948 – Oktober 2021*. Retrieved January 1, 2021, from www.destatis.de/DE/Themen/Wirtschaft/Preise/Verbraucherpreisindex/Publikationen/Downloads-Verbraucherpreise/verbraucherpreisindex-lange-reihen-pdf-5611103.html

Statistisches Bundesamt (Destatis). (2021c). *Preisindizes für Wohnimmobilien nach Kreistypen, Statistisches Bundesamt*. Retrieved April 1, 2021, from www.destatis.de/DE/Themen/Wirtschaft/Preise/Baupreise-Immobilienpreisindex/Tabellen/haeuserpreisindex-kreistypen.html

16

JAPAN

Kimitaka Aoki and Eigo Tateishi

1. The Japanese housing market

Residential real estate in Japan[1] can be broadly classified into three types: residential land, detached houses, and condominiums (including tenement houses). Detached houses and condominiums are further classified into new and used. Apart from the ownership of the house, the residential land on which the house is built is generally subject to ownership or leasehold *(shyakuchi-ken)*. Also, condominiums and tenement houses are generally subject to unit ownership *(kubun shoyū-ken)* for both the land and the building. The transactions (selling and buying, as this chapter does not elaborate on renting) of these five types of real estate are based on a common legal framework called the Real Estate Brokerage Act, yet various transaction costs such as brokerage fees, registration fees, and taxes differ across the real estate types. In general, real estate transactions in Japan involve three main actors: the seller, the real estate broker, and the buyer. As this chapter explains, the combination of these five real estate types and three major actors complicates residential real estate transactions in Japan, especially its overall picture of transaction costs.

As Figure 16.1 shows, the change in the real estate price index (where the average price in 2010 is indexed as 100) between 2008 and 2020 is about a 1.5-fold increase for condominiums, while detached houses and residential land remain almost unchanged (MLIT,[2] 2020a). However, such changes in real estate prices are not evenly distributed spatially in Japan, where the population is declining nationwide. Figure 16.2 shows a map of price changes (€/m^2) for residential land between 2009 and 2020 at the municipal level. It is clear from the map that land prices are rising mainly in the major metropolitan areas, while prices continue to decline in other areas. This spatially heterogeneous distribution of price changes is reflected in the sale prices of residential land. According to data from MLIT (2020b), the national average price (per m^2) of residential land as of 2020 is € 559. On the other hand, in Tokyo, the average land price is € 2,874. In the central part of Tokyo, the average transaction price of used condominiums exceeds € 0.01 million (per m^2) (Tokyo Kantei, 2021).

The average size of a detached house in Japan is about 127 m^2, and that of a condominium is about 51 m^2 (SBJ,[3] 2018). The total stock of housing in 2018 was about 52 million units (not including vacant houses and houses with unidentified ownership).

In Japan, the average age of a typical house before demolition is about 30 years (MLIT, 2018). Together with the incentive mechanism in public loan and tax scheme for new home

DOI: 10.1201/9781003130383-18

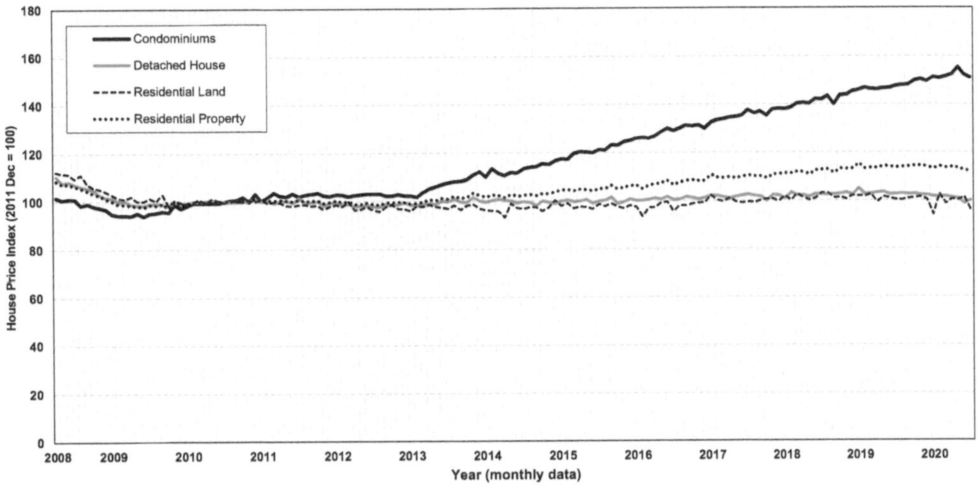

Figure 16.1 Real estate price index for condominiums, detached houses, residential land, and all types, 2008–2020

Note. The average price in 2010 is indexed as 100

Source: MLIT (2020a) 'Real Estate Price Index (Residential)'

purchases and the large transaction costs of used houses, the transactions of used houses are very limited and new houses account for nearly 80% of the total volume of residential real estate transactions (Sunahara, 2018, pp. 36–46). As of 2018, the annual transaction volume of new houses in Japan is about 950,000 units. Of these, about 110,000 are new condominiums for sale, 430,000 are new detached houses,[4] and 390,000 are new residential real estate for rent. About 60% of these new houses are concentrated in the three major metropolitan areas of Tokyo, Nagoya, and Osaka, with the Tokyo metropolitan area alone accounting for 50% of the supply of new condominiums (MLIT, 2018).

Although there is such a traditional preference for new housing construction, the Basic Act for Housing introduced in 2016 may bring a major change in the Japanese real estate market (MLIT, 2016). By promoting the market distribution of existing houses and the active renovation of vacant houses, the act explicitly sets the goal of transforming the real estate market so that the existing housing stock will be more actively utilized.

2. Regulatory framework

The Real Estate Brokerage Act (Law No. 176 of 1952, hereafter, 'the Act') is a main legal framework for real estate transactions in Japan. The Act aims to protect buyers and sellers and renters in real estate transactions and to facilitate the smooth distribution of the real estate. To achieve these aims, the Act regulates real estate companies that engage (1) in their own transactions (sale, purchase, and exchange) of residential lands and buildings, and (2) in agency, mediation, and brokerage of clients' transactions (sale, purchase, exchange, and rent) of residential lands and buildings. Specifically, the Act stipulates various regulations and obligations for the real estate business. For example, the Act (1) regulates how the real estate is to be advertised, (2) obliges real estate companies to inform buyers and renters about the state of real estate to be transacted or rented and the conditions of the transaction), (3) obliges real estate companies

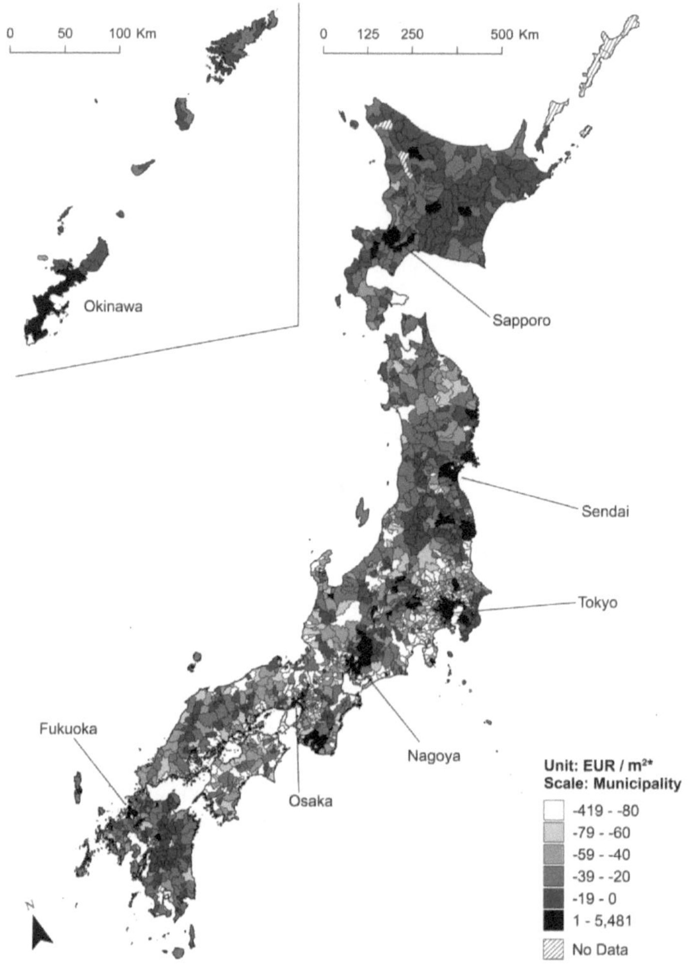

Figure 16.2 Change in the average residential land price (€/m^2) at the municipal level, 2009–2020

Note: Due to data limitations, prices for detached houses and condominiums are not available, so residential land prices are used as a proxy indicator. *1 Euro = 130 Japanese Yen (as of 2021 September).

Source: MLIT (2020b) 'Time series trend table of price change rate and average price.' The authors added spatial information to the data.

to issue documents describing the important matters of the transaction, and (4) regulates the contents of contracts when acting as a seller themselves.

In Japan, all companies operating real estate business should be authorized by the national or local government[5] based on the Act, and these authorized companies are commonly called authorized building lots and buildings traders (hereafter, 'real estate traders'). For real estate traders to act as agents or mediators for the sale, purchase, or exchange of residential lands or buildings, they need to hire a certain number[6] of real estate transaction agents *(takuchi tatemono torihiki shi)* who have a national license certified by the Act. The authorization allows real estate traders to (1) inform buyers about important matters of the transaction, (2) issue a disclosure statement (stating the important matters), and (3) certify that the transaction contract has no mistakes.

3. The conveyance process

The following section provides an overview of the four phases of brokerage transactions for residential land, used houses, and used condominiums in Japan. Although the details of the transactions and various transaction costs differ depending on the type of real estate, this chapter summarizes the basic transaction processes common to each type (Figure 16.3). The detailed differences in transaction costs are summarized in Table 16.1. The sales transaction of new detached houses and new condos are the same as the brokerage transaction flow that is described

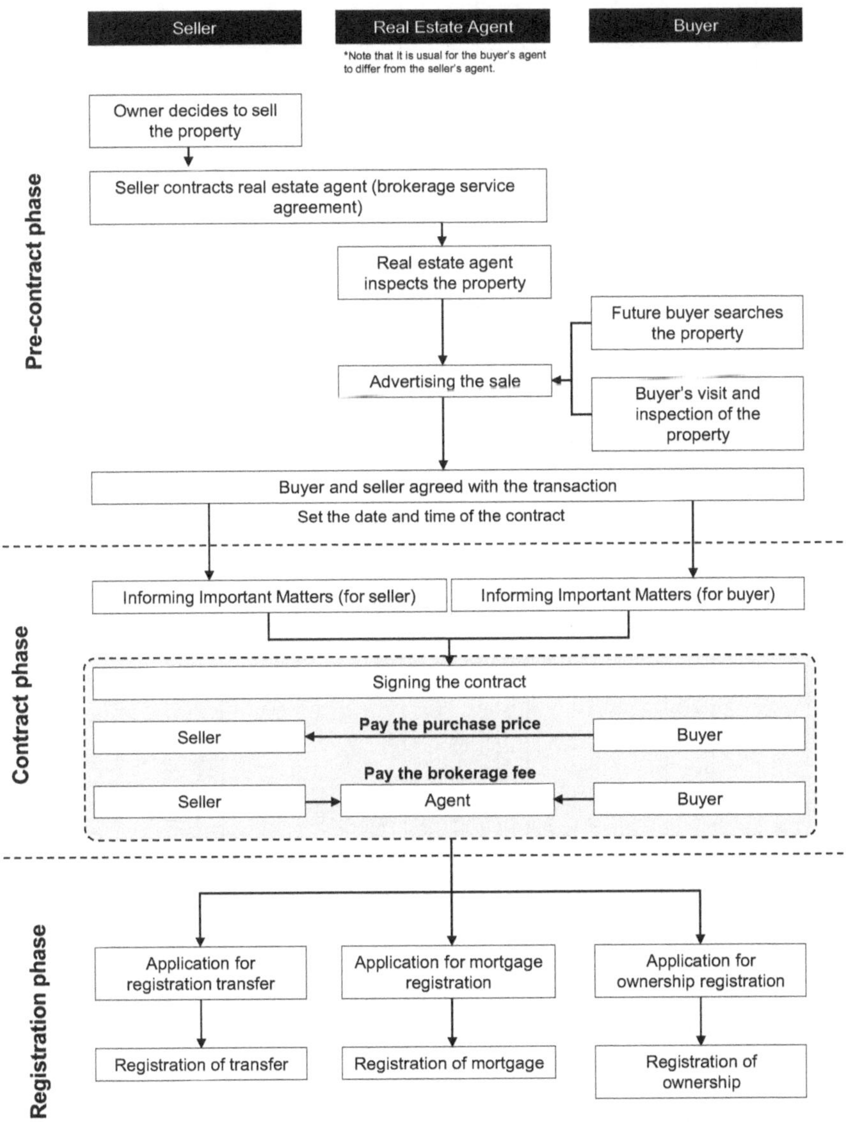

Figure 16.3 The conveyance process for a typical real estate brokerage transaction in Japan

Source: Author

Table 16.1 Fees and taxes during the conveyance process in Japan

			Land		*Property*	*Detached house/condo*	
						New	*Used*
Phase	*Fees and taxes*		*Seller*	*Buyer*	*Seller*	*Buyer*	
Marketing	Real estate agent's fees		-	-	-	-	-
	Marketing fees		-	-	-	-	-
Contract	Consumption tax		-	-	-	10%	-
	Stamp tax		€ 76				
	Bank	financing charge	-	about € 760	-	about € 760	
		Stamp tax	-	€ 152	-	€ 152	
		Loan guarantee fee	-	0.15–0.45%	-	0.15–0.45%	
		Fire and earthquake insurance	-	-	-	Varies depending on warranty (€ 1,140–3,040)	
	Real estate agent	Brokerage fee (Sales price € 30,400 or more)	3% + € 456 + consumption tax			-	3% + € 456 + consumption tax
Registration	Stamp tax		-	Loan agreement € 152	-	Loan agreement € 152	
	Registration and license fees (% of real estate's assessed value for property tax)		-	2%	-	0.4%	2%
	Registration and license tax for mortgage registration		-	0.4%	-	0.4%	
	Total amount of expert's request and administrative fees		-	about € 1,520	-	about € 1,520	
	Housing Loan Repayment and Mortgage Cancellation Fees		about € 760	-	about € 760	-	-

Post-registration	Transfer of income tax (After the filing is confirmed)	Income tax (short-term) = Transferred income × 30.63% Income tax (long-term) = Transferred income × 15.315%	-	Income tax (short-term) = Transferred income × 30.63% Income tax (long-term)= Transferred income × 15.315%	-	-
	Resident tax (After the filing is confirmed)	Resident tax (short-term) = Transfer income × 9% Resident tax (long-term) = Transfer income × 5%	-	Residential tax (short-term) = Transfer income × 9% Residential tax (long-term) = Transfer income × 5%	-	-
	Real estate acquisition tax (% of the value of the property acquired)	-	4%	-	4%	
	Sales tax	-	-	-	-	
Total (EUR for the model case price)		**€ 4,186**	**€ 10,005**	**€ 4,186**	**€ 17,800–19,700**	**€ 11,145–13,045**
Total (% share to the model case price)		**4.1%**	**10.0%**	**4.1%**	**17.8–19.7%**	**11.1–13.0%**
Excluded costs from the model case		15–36% CGT, 5–9% residential tax	0.15–0.45% loan guarantee fee, registration & license fees 2% assessed value, registration & license tax for mortgage registration 0.4%	15–36% CGT, 5–9% residential tax	0.15–0.45% loan guarantee fee, registration & license fees 2.0% of assessed value, registration & license tax for mortgage registration 0.4%	

Note. Model case price: € 100,000. The figures in the table exclude conditional relief measures.

Source: Author

in this chapter, except that the real estate trader (the developer) of the house or condo in the sale functions as the seller, and there is no pre-contract phase.

Marketing phase

The seller of real estate first needs to find a real estate trader to act as a broker (hereafter, the 'broker'). Given that there are many real estate brokers in Japan, it is the seller's choice which broker should handle the transaction. Once the seller decided on a broker to work with, both parties will conclude a brokerage service agreement. The content of the brokerage service agreement is specified by Articles 34 and 35 of the Act. This agreement functions in two ways: (1) it clarifies the operational content of the broker and the brokerage fee, and (2) it legally certifies that the broker is requested by the seller to sell the real estate.

When a brokerage service agreement is concluded, the real estate in the sale will be registered in REINS (Real Estate Information Network System), which was created to ensure the transparency of the real estate transaction market and to facilitate market distribution of the real estate. This system can be accessed by different authorized brokers, and the entire real estate industry can work together to match potential buyers for real estate.

On the other hand, the buyer of real estate usually gathers the sales information of the real estate from real estate advertisements. Although there are various types of mediums for real estate ads, such as newspapers, magazines, and postings in stores, nowadays the most common way to obtain information about real estate is on the internet. The content of these real estate ads is regulated by the guidelines voluntarily established by the real estate industry, which is also certified by the Fair Trade Commission. Once the buyer has selected a real estate to purchase, they will contact the broker to inquire about the conditions for purchase and go to a private viewing (note that it is usual for the buyer's broker to differ from the seller's broker). During this process, the broker will advertise the real estate and start negotiating the price.

Contract phase

The Japanese Civil Code considers that the contract for a real estate transaction is finalized when both the buyer and seller agree with the transaction. Therefore, at the preliminary stage of signing a contract, the buyer will prepare a letter of intent for buying and the seller will prepare a letter of intent for selling. Once both parties agree with the transaction, they also confirm the content of the transaction contract and the date and time of actual signing of a buyer-seller contract. Although the agreement is legal enough to make a transaction, it is customary to draft a buyer-seller contract to clarify and document matters to be agreed upon at this time. When a broker is involved in the sale and purchase of the clients, the broker is obligated to prepare a buyer-seller contract (Article 37 of the Act). The sales price, terms of payment, ownership rights and collateral, terms and conditions for the cancellation of the contract, and compensation for any damages that need to be repaired are all agreed upon at the time of the signing the contract. In addition, a licensed real estate transaction agent of the trading company must inform the buyer and seller about important matters of the transaction (such as relevant urban planning regulations, the condition of the land or building, the condition of the contract) as a written document before the contract is made (Article 35).

Once the contract is signed by both parties, the seller is obligated to convey the registration of ownership to the buyer, and the buyer is obligated to pay the purchase price. The buyer usually pays a deposit first rather than paying the full amount of the purchase price. Also, the buyer will often postpone the payment of the remaining balance because it usually takes time for the seller to complete the obligations stated by the contract, including the cancellation of the real estate ownership on the registry. The remaining balance will be settled once a *shiho-shoshi* laywer (Japanese solicitor) confirms that both the buyer and seller have completed all the documents required to convey the ownership from the seller to the buyer. When the buyer takes out a housing loan and/or the seller wants to cancel out the mortgage by spending the selling profit, the settlement of the purchase will usually be made with the bank or mortgage firm involved in the loan and/or mortgage.

When both the seller and the buyer sign a buyer-seller contract and the transaction is concluded, they both need to pay brokerage fees to the broker as compensation for handling the negotiation and contract paperwork. In general, the seller and buyer need to pay brokerage fees when a used house, a new house, or residential land is sold and purchased.[7] However, when real estate is purchased directly from the seller, no brokerage fee is charged. Also, when a brokerage transaction is not concluded, no brokerage fee is charged. The maximum amount of the brokerage fee must be calculated based on the formula defined in Article 46 of the Act. This formula varies depending on the sales price of the real estate, but in general, for properties of € 0.03 million or more, it can be calculated as (the property price × 3% + € 456) × 1.1 (10% consumption tax) (e.g. the brokerage fee for a € 0.38 million property is € 13,042). In this example, if the same broker mediates for both the seller and the buyer, the brokerage fee for this broker is € 26,083, which is the sum of € 13,042 from the seller and € 13,042 from the buyer. If the broker of the seller and the buyer are different, € 13,042 will be paid to each broker.

Registration phase

After signing a buyer-seller contract and paying the full purchase price, it is necessary to register the real estate. The purpose of real estate registration *(fudousan touki)* is to clarify and record the physical condition (address, area size, etc.) and rights (ownership, mortgage, etc.) of the real estate. Given that real estate registration requires legally complex paperwork, it is generally handled by *shiho-shoshi* lawyers. The *shiho-shoshi* lawyer collects a certified copy of the real estate register, prepares the necessary documents such as the signed buyer-seller contract, and attends the actual handover of the real estate between the seller and the buyer.

After seeing that the transaction has been completed correctly, the *shiho-shoshi* lawyer takes care of the registration of transfer of ownership *(shoyū-ken iten touki)* at one of the local Legal Affairs Bureaus *(houmu kyoku)* on behalf of the seller and the buyer. However, in recent years, this procedure has been increasingly being carried out online. Once the registration of transfer of ownership is completed, the *shiho-shoshi* lawyer mails the buyer a certificate of completion of the registration and a certified copy of the real estate register; then, the registration process is complete. The registration process incurs transaction costs, such as fees for experts including *shiho-shoshi* lawyers and various administrative fees. These transaction costs vary depending on the type of real estate and the professional's fees, but they are usually about € 760 for sellers and about € 1,520 for buyers.

Post-registration phase

Once the registration is completed, the seller and buyer will each need to pay the relevant taxes. The seller needs to pay the tax on the capital gain as well as resident tax determined by the volume of capital gain. The capital gain from the real estate transaction is calculated by (the selling price of the real estate) minus (acquisition costs) minus (transaction costs). The amount of these two taxes will vary depending on how long the seller has resided in the real estate (in general, there are 5-year and 10-year thresholds). On the other hand, the buyer will need to pay real estate acquisition tax that is a one-time taxation on the acquisition of real estate. Also, the buyer (i.e. the new owner of the real estate) needs to pay property tax as well as city planning tax annually.

4. Summarizing the conveyance process and its transaction costs

This chapter outlined the real estate market and the basic process of real estate brokerage transactions in Japan. In the Japanese real estate market, the share of new real estate is high, and the number of transactions of used real estate is low. In Japan's real estate brokerage transactions, the efficient matching of buyers and sellers is realized by REINS, a nationwide information system for the brokers. Laws, administrative procedures, transaction costs, taxes, and so on for sales transactions vary depending on the type of real estate, and thus their details are rather complex. In particular, the complexity of legal procedures and costs during the registration phase, and the various types of taxes after registration, could be an interesting point of comparison with the situation in other countries. Recently, the number of used real estate transactions has been gradually increasing in the market. In particular, since 2016, the number of transactions of used condos became higher than that of new condos. Together with the policy change in the national government, we can view this trend that Japan is entering into as an era to actively utilize existing housing stock under nationwide depopulation. Therefore, we would like to draw the reader's attention to the possibility that this trend of utilizing existing housing stock will change the landscape of the Japanese real estate market and its transaction processes elaborated in this chapter.

Notes

1 1 Japanese Yen = 0.0076 Euro (as of 1 October 2021).
2 Ministry of Land, Infrastructure, Transport, and Tourism.
3 Statistics Bureau of Japan.
4 Including (1) the houses built by large developers for sale and (2) the custom-built houses built by building contractors.
5 More specifically, MLIT or the Prefectural Governor will authorize the trader regardless of individual trader or corporate trader.
6 Each trader office must have one full-time real estate transaction agent for every five employees in the office.
7 Brokerage fees are also charged for the brokerage of rental houses/apartments.

Bibliography

Morishima, Y. (2011). *実践不動産学教科書 (Practical real estate science textbook)*. Tokyo: Toyo Keizai Shinbun.

Maruyama, E., Tanaka, K., & Ishihara, S. (1987). *不動産学概論* (*Introduction to real estate science*). Tokyo: Recruit Publishing.

MLIT. (2021). *日本における不動産取引に関連する法律* (*Laws related to real estate transactions in Japan*). Retrieved September 20, 2021, from www.mlit.go.jp/common/001121685.pdfReferences

MLIT. (2016). *住生活基本計画* (*The basic act for housing*). Retrieved September 20, 2021, from www.mlit.go.jp/jutakukentiku/house/jutakukentiku_house_tk2_000032.html

MLIT. (2018). *我が国の住宅ストックをめぐる状況について（補足資料） (The situation surrounding Japan's housing stock (supplementary material))*. Retrieved September 20, 2021, from www.mlit.go.jp/policy/shingikai/content/001323215.pdf

MLIT. (2020a). *不動産価格指数（住宅） (Real estate price index (residential))*. Retrieved September 20, 2021, from www.mlit.go.jp/totikensangyo/totikensangyo_tk5_000085.html

MLIT. (2020b). *変動率及び平均価格の時系列推移表 (Time-series trend table of price change rate and average price)*. Retrieved September 20, 2021, from www.mlit.go.jp/totikensangyo/totikensangyo_fr4_000043.html

SBJ. (2018). *平成30年度住宅土地調査 (H30 house and land survey)*. Retrieved September 20, 2021, from www.e-stat.go.jp/stat-search/files?page=1&layout=datalist&lid=000001248658

Sunahara, Y. (2018). *新築がお好きですか? 日本における住宅と政治 (Do you like a new house? Housing and politics in Japan)*. Kyoto: Minerva Shobo.

The Real Estate Transaction Promotion Center Foundation. (n.d.). *Fudosan Japan*. Retrieved from www.fudousan.or.jp/

Tokyo Kantei. (2021). *中古マンション70㎡価格月別推移 (Used condominium 70m2 prices by month)*. Retrieved September 20, 2021, from www.kantei.ne.jp/report/70m2/866

17
POLAND

Sylwia Lindqvist

1. The Polish housing market

In Poland,[1] there are three types of real estate: real estate premises (including residential and non-residential premises), built-up land properties (including those with residential buildings, agricultural, and others), and undeveloped land properties (including agricultural, built environment, and forest properties).

According to Statistics Poland (GUS, 2019), a total of 468,300 transactions were recorded in the Register of Real Estates Prices and Values (RREPV) in 2019. Those transactions constitute 83.3% of the number of notarial deeds related to real estate sales recorded by the Ministry of Justice, of which most of the transactions took place in urban areas (61.1% in cities). In addition, most of the transactions registered at RREPV were related to premises, at 49.4%, and 16.9% involved built-up land properties. On the Polish market of residential premises, sales on the secondary market dominated (61.8% of the number), and the most frequently sold premises were those with an area ranging from 40.1 m^2 to 60 m^2.

According to the notary report conducted by the Ministry of Justice, 562,400 notarial deeds concerning the sale of real estate were signed in 2019, which is an increase of 7.1% from the previous year. Also, the sale of plots with a residential building increased by 14.7%. The sale of premises constituted 38.9% of all notarial real estate deeds, while the share of plots with a residential building was 11% and the sale of co-operative ownership rights to premises was 9% (GUS, 2020).

In Poland, there are two forms of residential property ownership: direct ownership, in the form of owner-occupied housing (both one- or two-family homes and owner-occupied flats), and indirect ownership, in form of housing co-operatives. The legislation differs between direct and indirect ownership. Owner-occupied flats are regulated by the Act of 24 June 1994 on premises ownership, and an owner's co-operative right to an apartment is regulated by the Act of 15 December 2000 on Housing Co-operatives and the Act of 16 September 1982 Co-operative Law.

According to Eurostat, approximately 84% of the population has direct and indirect ownership, while the remaining 16% live in rented flats. Most homeowners (approximately 73%) do not have mortgage loans on their properties. In 2018, Poland and Denmark were the only EU member states in which a majority of the population lived in a detached house (Eurostat, 2020; Ministerstwo rozwoju, 2020; OF, 2019).

 DOI: 10.1201/9781003130383-19

It is worth mentioning that membership in housing co-operatives in Poland is not mandatory. A limited property right (tenant-ownership) can also be transformed into a separate property. With Polish tenant-ownership, the consent of the housing co-operative is required when the tenant-owner performs certain activities, but not all. For example, the consent of the housing co-operative is not required to rent out the apartment or transfer it to a family member for free use. The right to a dwelling is also transferrable, can be passed to heirs, and is subject to execution. Transformation into a property in itself is related to wider rights related to the possibility of disponing the premises.

The owner of the flat has a share in the land, common areas used by all residents of the block, and fixtures located in the apartment itself (heaters, installations, windows, etc.). In the case of an owner-occupied flat, registering with the Land and Mortgage Register is mandatory, while establishing the Land and Mortgage Register in case of tenant-ownership is not mandatory. However, in this case, it is sometimes recommended for those who intend to mortgage the apartment. Tenant-owned apartments may also be easier to sell when it is registered.

Property prices

In 2019, there was a price rise of residential premises by 8.7% compared with 2018, and in 2018, the prices rose by 6.5%. The growth in prices was observed on both the primary and secondary markets as indicated in Figure 17.1.

The average price of residential premises was € 71,852 on the primary market and € 56,672 on the secondary market. The average price per 1 m² of usable floor area of residential premises sold on the primary market amounted to € 1,265 (median € 1,187) and € 1,065 (median € 933) on the secondary market. The average price of properties built up with residential buildings amounted to € 72,842 (GUS, 2019).

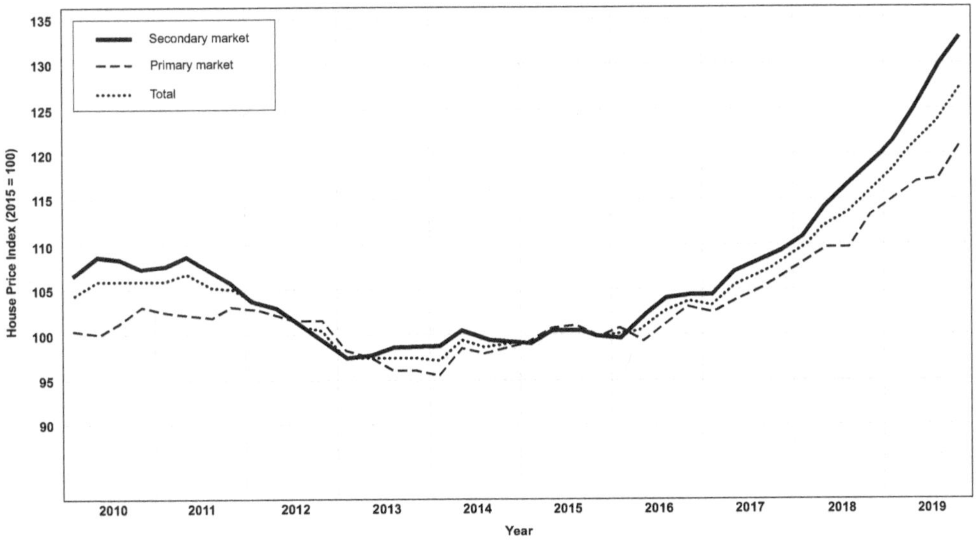

Figure 17.1 Quarterly price indices of residential premises for Poland

Note. 2015 = 100

Source: GUS (2019)

Figure 17.2 Change in house prices (PLN/m^2) in the primary market (in percent), 2006–2021

Source: Narodowy Bank Polski (NBP)

Figure 17.2 illustrates the change of housing for the primary market in all major cities in Poland over a 15-year period.

2. Regulatory framework

The important regulatory framework in Poland consists of the Act of 23 April 1964 Civil Code (CC), the Act of 21 August 1997 Real Estate Management Act (REMA), and the Act of 14 February 1991 Law on Notaries.

Conveyance of real estate and co-operative ownership rights for apartments

The only mandatory professional in an estate transaction process in Poland is a notary. According to Article 27 of the REMA and Article 158 of the CC, the sale of real estate requires the conclusion of the contract in the form of a notarial deed. Transferring the right by contract requires in turn an entry in the Land and Mortgage Register. A notary thus prepares a notary deed and the transfer of property ownership. According to the Law on Notaries, a notary is a lawyer and a person of public trust appointed by the Ministry of Justice and who is authorized to prepare notarial acts and perform notarial activities. In Poland, even mortgage deeds must be notarized.

A notary acts on behalf of both parties in a transaction and must not favour any party. Another regulatory framework relevant to a notary is the Code of Ethics issued by the National Chamber of Notaries. In addition to drawing up a notary deed, the role of the notary is to verify important transaction issues, such as examining the subject of the contract (i.e. the property) in terms of debt, property rights, value, and so forth. According to Article 19a, the notary is subject to compulsory third-party liability insurance for damage caused in the course of performing the notarial activities.

In Poland, the notarial deed is also required when selling co-operative ownership rights to the premises; however, this does not require registration in the Land and Mortgage Register. The extract of this deed must be immediately sent to the co-operative. Housing co-operatives are required to keep a register of premises for which separate Land and Mortgage Registers have been established (Chapter 2, Article 17 of the Act on Housing Co-operatives).

According to the Law on Notaries, Chapter 3, Articles 91–92, a notary draws up a notarial deed if required by law or if it is the will of the parties.

A notarial deed should contain the following elements:

- The day, month, and year of the deed, and if necessary or at the request of the party, the hour and minute of the commencement and signing of the deed
- The place where the act was drawn up
- The name, surname, and seat of the notary's office – also, if the deed was prepared by a person designated to replace the notary or authorized to perform notarial activities, then the name and surname of that person
- The names, surnames, parents' names, and place of residence of natural persons, name and seat of legal persons or other entities participating in the deed, first names, surnames, and place of residence of persons acting on behalf of legal persons, their representatives or proxies, and other persons present at the preparation of the act
- Declarations of the parties, referring, if necessary, to the documents presented with the act
- A statement, at the request of the parties, of the facts and essential circumstances that occurred during the drawing up of the act
- A statement that the act has been read, accepted, and signed
- Signatures of those taking part in the act and those present at the preparation of the act
- The signature of a notary

If a notarial deed concerns a legal act, the act should contain content relevant to this act. In addition, the notarial deed should contain all other statements that are required in the deed as resulting from this act, from specific provisions or parties request.

Immediately after the deed is signed, a notary submits an application for entry in the Land and Mortgage Register via the ICT system handling/supporting court proceedings. The Central Repository of Electronic Extracts of Notarial Deeds is administrated by the National Council of Notaries, according to Article 92a § 1. The main legislation concerning the register is the Act of 6 July 1982 in the Land and Mortgage Register.

Estate agents

In Poland, the involvement of real estate agents in property transactions is not mandatory. According to Article 179a of the REMA, a real estate agent is an entrepreneur who conducts business in the field of real estate brokerage involving the paid performing of activities aimed at the conclusion of a contract by other persons, mainly a purchase or sale of the right to real estate, the purchase or sale of a co-operative ownership right to an apartment (premises), or the rental or lease of real estate or part thereof. According to Article 180, the scope of real estate brokerage activities is specified in the agency agreement. The contract must be entered into in writing or electronically to be valid and binding.

According to Article 181, a real estate broker is subject to compulsory third-party liability insurance for damage caused in connection with the real estate brokerage. If the real estate broker performs activities with the help of other persons acting under their supervision, that person(s) is also subject to third-party liability insurance for damages caused by the actions of these persons.

As of 1 January 2014, there has been a liquidation of state real estate agent licences granted under REMA, and as a consequence, the requirements involving education, qualifications, and criminal records have changed. However, the Polish organization PFRN has launched the PFRN Real Estate Broker licence, with the aim of regulating the real estate market and providing clients with reliable information about brokerage services (see pfrn.pl, n.d.). Many are opposed to deregulation, as the lack of clear legal regulations creates room for unfair practices and negatively impacts the image of the entire industry. Companies and brokers who provide services without a licence and without proper preparation can operate on the market, but the quality of their services may be questionable.

3. The conveyance process

It is difficult to define whether a typical transaction process involves an estate agent, as there are no clear figures on the involvement of real estate agents. It is appreciated that the involvement is between 25% and 50% of residential transactions in urban areas (ZERP, 2007). The transaction process can be divided into several phases, such as marketing, preliminary contract and final contract, and the registration and post-transaction phase (see Figure 17.3).

Marketing phase

If a real estate agent is involved, the agreement between the agent and the principal is compulsory and must be in writing (or carried out electronically). The exclusive agreement is considered the more common one.

According to Article 181a, a real estate agent performing the activities concluded in a brokerage agreement has the right to inspect the following items:

- Land and mortgage registers
- Real estate cadastres
- Records of utility infrastructure, taxation tables, and maps
- Local plans
- Studies of the conditions and directions of spatial development of the city council
- City council decisions on building conditions and land development
- Registers of real estate prices

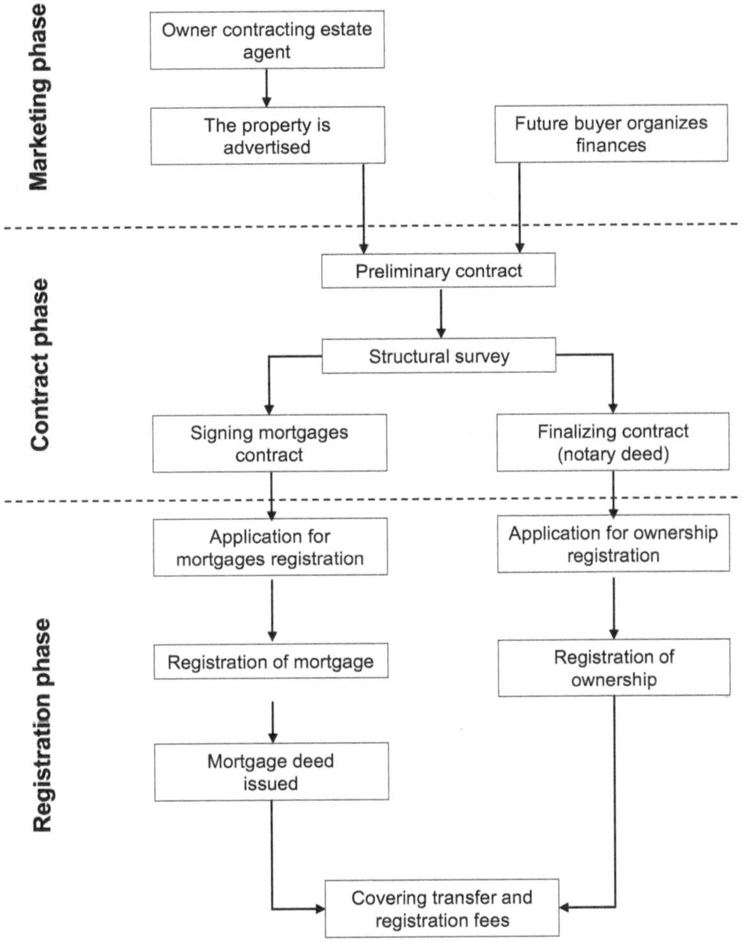

Figure 17.3 The conveyance process in Poland

Source: Author

- Registers of persons who are entitled to the rights specified in Article 179b point 2 and the rights of separate ownership of the premises
- Population records in the scope of the registration obligation
- Energy performance certificates

In addition, the real estate agents are entitled to collect the necessary copies/extracts from these inspections. The agent's fee is usually between 2.5% and 4%; however, 5% is also common.

A variety of marketing systems can be found in Poland. These range from large websites aggregating advertisement, including those publishing only developer advertisements, such as RynekPierwotny.pl or Obido.pl, to websites for both primary and secondary markets, such as GetHome.pl or OtoDom.pl. The price of advertising online varies according to how many ads are placed per month and what extra services are required.

The bidding procedure concerning the definition of an offer and time limit for the expected answer is regulated in Articles 66–72 of the CC. According to Article 28 of the REMA, the sale of properties can take place through a tender or without a tender procedure. In the case of a tender, the conditions of the sale shall be given in the announcement of the tender. In case of a non-tender procedure, the conditions of sale shall be determined between the seller and buyer.

As a result of the Act on the energy performance of buildings, which has been in place since 9 March 2015, all homes placed on the market for sale must have a building energy performance certificate (Kalkulatory Budowlan, 2021). If the seller fails to comply with this obligation, the buyer may request in writing for the seller to fulfil this obligation no later than two months from the date of delivery of the request. The buyer has 14 days from the conclusion of the contract to make such a request. If the seller still fails to provide the energy certificate, the buyer may have the energy performance certificate drawn up at the expense of the seller (within 12 months).

The price of the energy performance certificate depends on certain factors, such as the size of the house and the amount of time the specialist spent on the energy audit. According to KB.pl, prices of certificates range from € 110 to € 220 and up. For example, in Warsaw (and cities such as Kraków, Wrocław, and Poznań), the expense should be assumed to be closer to € 176–220. However, in smaller cities, the prices will be in the range of € 110–€ 132. Some other sources list the lowest prices for residential premises as starting from € 88.

Contract phase

After the negotiations are completed and the parties agree upon the conditions of the sale, a preliminary contract is drafted, which gives the parties time to prepare for the final contract. A preliminary contract does not require the involvement of a notary; however, if made in the form of notarial deed, the agreement has more binding power. If the contract is not drafted by a notary, the parties commit themselves to either a down payment, a penalty sum, or an agreement to annul the contract without any penalty. However, if one of the parties fails to meet the contract provisions, then the preliminary contract can be enforced only through a court procedure if it was drafted by a notary. When all the conditions are fulfilled, the final sale contract can be drafted.

Before the final contract is drawn up, the buyers and sellers choose a date convenient for both parties and for preparing the appropriate documents. These may be an identity card or passport, an excerpt from the Land and Mortgage Register, a preliminary contract, and a bank certificate confirming the granting of a loan (if applicable), which the notary usually informs accordingly in advance. Thereafter, based on the documents and a contract prepared in advance between the buyer and the seller, the notary office prepares a notary deed. The notary is obliged to control certain important aspects of the transaction. At the conclusion of the contract, ownership is transferred unconditionally to the new owner following payment of the purchase sum, and the contract is signed by both parties and the notary.

Registration phase

The registration application is made by the notary on behalf of the parties; however, the registration is merely declaratory. According to Chapter 5, Article 10 of Act of 9 September 2000 on tax on civil law transactions, a notary is required to bear the tax on civil law transactions in the form of a notary deed, in which the notary is a so-called tax carrier. A notary is obliged to make the carrying out of a civil law transaction conditional upon the prior payment of tax. They are also obliged to keep a tax register and pay the collected tax on behalf of taxpayer no later than the seventh day of the month following the month in which the tax was collected. The tax rate on the contract of a real estate sale or co-operative right to a dwelling is 2% of the market value of the property, and the cost of obtaining a mortgage is 0.1% of the amount of the secured debt.

According to Chapter 3 of the Act of 28 July 2005 on court costs in civil cases (Article 42), there is also a fixed fee in the amount of € 44 for an application for an entry into the Land and Mortgage Register. For establishing a Land and Mortgage Register, according to Article 44, a fixed fee of € 22 is collected, irrespective of the fee for the entry of ownership or co-operative ownership right to the premises. The registration cost is paid by the buyer to the notary in connection with the preparation of the notary deed.

The notary's fee is paid by the party to the notarial act. In the case of real estate purchase, it is customary for the buyer to pay the notary fee, but this is not explicitly stated in the regulations. According to Article 89 § 1 of the Law on Notaries, the parties on notarial actions are jointly and equally liable for the remuneration due to the notary. The amount of the remuneration depends on the value of the subject of the notarial activity.

The maximum rates of the notary, according to § 3 of the Regulation of the Ministry of Justice of 28 June 2004 are as follows:

- Up to € 660: € 22
- Over € 660 to € 2,200: € 22 + 3% of the surplus above € 660
- Over € 2,200 to € 6,600: € 68 + 2% of the surplus above € 2,200
- Over € 6,600 to € 13,200: € 156 + 1% of the surplus above € 6,600
- Over € 13,200 to € 220,000: €2 22 + 0.4% of the surplus over € 13,200
- Over € 220,000 to € 440,000: € 1,049 + 0.2% of the surplus above € 220,000
- Over € 440,000 to: € 1,489 + 0.25% of the surplus over € 440,000, but not more than € 2,200

For particular notarial activities, the fee may constitute a specific part of this amount. For example, the maximum notary fee is half the rate for the notarial deed concerning the sale of a co-operative right of the premises or preliminary contract. For drawing up a notarial deed documenting the establishment of a mortgage to secure a bank loan granted for housing purchase, the maximum notary fee is a quarter of the rate. In addition, a VAT charge of 23% should be added to the notary fee.

Post-transaction phase

In Poland, a property sales tax is also charged on capital gains when a taxpayer decides to sell their property within five tax years from the date of its acquisition. Generally, however, income from the sale of a home five years after its purchase is exempt from the tax. The tax is 19% on the profit from the sale. According to the law, it is possible to take advantage of a special exemption if the profit is used towards another residence. From January 2019, the length of the term has been extended to three years (see the Act of 26 July 1991 r. on personal income tax).

4. Summarizing the conveyance process and its transaction costs

To summarize, the Polish conveyance process for both real estate and tenant-ownership involves a notary system, where the notary plays an important and necessary role in the transaction process. Also, the real estate agent profession has been deregulated, and as a consequence, the requirements to become an agent have been significantly reduced. When the agent is involved

Table 17.1 Fees and taxes during the conveyance process in Poland

Phase	*Fees and taxes*	*Real estate*		*Tenant-ownership*	
		Seller	*Buyer*	*Seller*	*Buyer*
Marketing	Real estate agent's fees	2.5–5%	-	2.5–4%	-
	Marketing fees	Depending on tools	-	Depending on tools	-
	Energy performance certificate	€ 110 –	-	€ 88 –	-
Contract (VAT on notary activities 23%)	Notary deed	-	€ 22–2,200 (€ 347 for the model price)	-	1/2 Notary rate
	Preliminary contract with notary	-	1/2 Notary rate	-	1/2 Notary rate
	Notary mortgage deed	-	1/4 Notary rate	-	1/4 Notary rate
Registration	Registration tax	-	2%	-	2%
	Mortgage tax	-	0.10%	-	0.10%
	Registration fee	-	€ 44	-	€ 44
	Mortgages fees	-	€ 22	-	€ 22
Post-transaction	Capital gains tax	19% (of profit)	-	19% (of profit)	-
	Sales tax	-	-	-	-
Total (EUR for the model case price)		**€ 2,610–5,110**	**€ 2,673**	**€ 2,588–4,088**	
Total (% share to the model case price)		**2.6–5.1%**	**2.7%**	**2.6–4.1%**	**2.5%**
Excluded costs from the model case		19% CGT	0.1% mortgage tax	19% GCT	0.1% mortgage tax

Note. Model case price: € 100,000

Source: Author

in a transaction, they help with the preliminary contract, while the final contract is drawn up by a notary as a notarial deed.

Table 17.1 summarizes the transaction costs in Poland both for real estate and tenant-ownership dwellings. However, it is difficult to calculate exact costs, as costs may vary depending on different factors and the complexity of the transaction.

Note

1 1 Polish Złoty = 0.22 Euro (as of 1 October 2021).

References

Act of 23 April 1964 Civile Code.
Act of 6 July 1982 on Land and Mortgage Register.

Act of 16 September 1982 Co-operative Law.
Act of 14 February 1991 Law on Notaries.
Act of 26 July 1991 r. on personal income tax.
Act of 24 June 1994 on premises ownership.
Act of 21 August 1997 on Real Estate Management.
Act of 9 September 2000 on tax on civil law transactions.
Act of 15 December 2000 on Housing Co-operatives.
Act of 28 July 2005 on court costs in civil cases.
Act of 29 August 2014 on the energy performance of buildings.
Eurostat. (2020). *Housing statistics*. Retrieved November 20, 2021, from https://ec.europa.eu/eurostat/statistics-explained/index.php/Housing_statistics#Type_of_dwelling
Kalkulatory Budowlane (KB.pl). (2021). *Świadectwo energetyczne budynku – cena, kto i dlaczego musi uzyskać certyfikat energetyczny?* Retrieved November 20, 2021, from https://kb.pl/porady/swiadectwo-energetyczne-budynku-jak-i-po-co-je-uzyskac/
Ministerstwo rozwoju. (2020). *Raport: Stan mieszkalnictwa w Polsce*. Retrieved from https://www.gov.pl/web/rozwoj-technologia/raport-o-stanie-mieszkalnictwa
Observator Finansowy.pl (OF). (2019, March 6). *73 proc. Polaków mieszka we własnych mieszkaniach, bez kredytu*. Retrieved November 20, 2021, from www.obserwatorfinansowy.pl/tematyka/makroekonomia/wskazniki-ekonomiczne/73-proc-polakow-mieszka-we-wlasnych-mieszkaniach-bez-kredytu/
Pfrn.pl, Polska Federacja Rynku Nieruchomości (PFRN). (n.d.). *Main page*. Retrieved November 20, 2021, from www.pfrn.pl
Regulation of the Ministry of Justice of 28 June 2004 on the maximum rates of notary fees (Journal of Laws of 2020, item 1473).
Statistics Poland (GUS). (2019). *Municipal infrastructure*. Retrieved November 20, 2021, from https://stat.gov.pl/en/topics/municipal-infrastructure/municipal-infrastructure/
Statistics Poland (GUS). (2020). *Statistical analyses – real estate sales in 2019*. Retrieved from https://stat.gov.pl/en/topics/municipal-infrastructure/municipal-infrastructure/real-estate-sales-in-2019,2,12.html
ZERP, Centre of European Law and Politics, University of Bremen. (2007, December). *Study COMP/2006/D3/003 conveyancing services market – country fiches*. Retrieved from https://ec.europa.eu/competition/sectors/professional_services/studies/csm_study_complete.pdf

18 RUSSIAN FEDERATION

Nikolay Volovich and Evgenia Afanaseva

1. Analysis of housing stock

According to official statistics of the Russian Federation[1] for the beginning of 2020, the total area of premises was 3,857 million m^2, including 2,780 million m^2 in the cities (with 282 million m^2 in Moscow). Including apartment buildings, this area was 62.5% or 2,409 million m^2, including 161.8 million m^2 in houses of blocked building. The area of premises in individual houses (18.163 million units) was 1.255 million m^2 (37.5%) (The Federal Service for State Registration, Cadastre and Cartography (Rosreestr), n.d.). This statistic does not include rooms in student and other residences and also other rooms suitable for accommodation. The population at the beginning of 2019 was 146.8 million people, among which 75% live in cities (in Moscow, as of 2020, the population was 12.7 million people). The average availability of housing in the country is 25.8 m^2 per person, including 25.4 m^2 in the cities (for Moscow, this indicator is lower at about 20 m^2). The map in Figure 18.1 illustrates the spatial structure of apartment prices in the Russian Federation.

One feature of the housing stock in Russia is the high availability of private property for citizens (89% for the beginning of 2019). In fact, the individual housing stock is now completely the private property of citizens. In apartment buildings (47.8 million units of housing), citizens own about 2.084 million m^2, legal entities own approximately 80 million m^2, and municipalities and the state own about 127 million m^2 of the total area of premises (Ministry of Construction, Housing and Communal Services of the Russian Federation, n.d.).

As a result, the percentage of rental housing in apartment buildings, according to estimates by the Federal Institute of Development (dom.rf), is 13.2%, including social leasing from the state and municipalities (4.8%), most of which, according to the law, can be privatized free of charge. A total of 97% of the lessors at market (commercial) conditions are individuals, but about a third of them do not register lease agreements officially.

2. Regulatory framework

The situation with housing stock and also the geographical redistribution of the places of residence, namely, the accelerated migration from remote areas to Moscow, the cities with over a million people, and the southern regions, caused the change of state housing policies in the last 20 years.

DOI: 10.1201/9781003130383-20

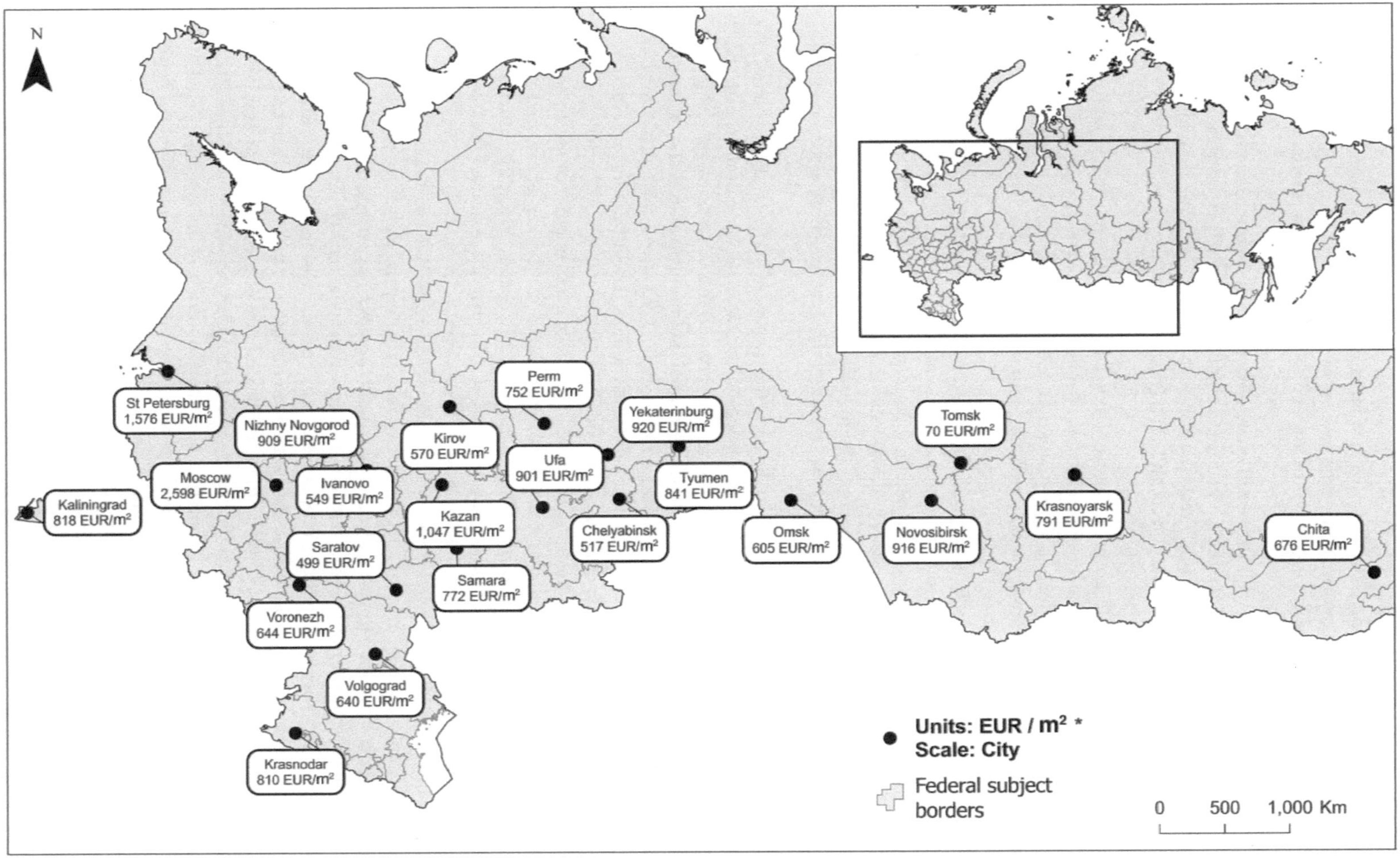

Figure 18.1 Average cost of 1 m^2 in housing units within apartment buildings in some Russian Federation cities with a population of more than 1 million

Note. *1 Euro = 81.21 Russian Ruble (as of 2021 September)

Source: **Статистика и показатели**

During the Soviet era, all premises in apartment buildings were the property of the state, which maintained them. The reformation of the residential real estate market happened in the early 1990s after the public housing stock was transferred to the residents as their private property free of charge.

In 1998, the general database with all the real estate objects and the rights to them was established (General State Real Estate Register, organized and maintained by Rosreestr). It includes information about all land plots and other real estate objects, including apartment buildings and certain apartments. However, not all objects and the rights to them were registered, but this generally concerns individual apartments.

Since 1 September 2006, the government has carried out an amnesty policy, and it has already allowed for the registration of the rights for 14 million real estate objects in this simplified order: land plots, individual apartments, dachas, garden houses, and so forth. Today, the State Duma adopted the law that will prolong 'country' amnesty for five years, until 1 March 2026 and also expands it.

The regulation of the market had originally been planned since the beginning of the 21st century by only indirect methods, through the taxation system and introduction of a system of town-planning zoning. Now each city has public town-planning regulations. It is not obligatory to register land plots under apartment buildings, and if apartment owners' associations want to do this, they are met with nonchalance from the municipalities, which are not interested in losing their monopoly on allocating land for new buildings. This causes an increase in conflicts between citizens and the municipalities.

With all the land plots, the owners pay land tax, and the owners of houses and apartments in apartment buildings pay property tax. The tax is based on the cadastral value of real estate and determined by regional institutions and revised once every two to five years. Citizens' real estate taxes are stated by regulatory legal acts of representative bodies of municipal units and by laws of the federal cities and cannot exceed 0.3% of the taxation value. Tax rates can be reduced to zero by regulatory legal acts of representative bodies of municipal units (laws of the federal cities).

At the same time, the Tax Code of the Russian Federation states considerable privileges. For example, pensioners are exempted from a tax on one house or one apartment. For most people, the tax base for apartments is defined as its cadastral cost reduced for the size of cadastral value of 20 m^2 of the total area of the apartment. For owners of individual houses, this deduction is calculated as the amount of the cadastral value of 50 m^2 of total area of the house (Tax Code of the Russian Federation Housing code of the Russian Federation).

After the beginning of the accelerated free privatization of apartments in apartment buildings, the authorities removed their duty to allocate money for capital repairs and other costs on the maintenance of these houses. However, due to poverty and the lack of experience of how to collectively manage apartment houses, many citizens do not make capital repairs. Therefore, a considerable part of the Soviet housing stock during these years became worthless. In 2007, the State Fund to reform housing and public utilities had as its main objective to help citizens and regions with the capital repairs of housing. However, the main expenditure of the fund goes towards the financial support of municipalities to resettle citizens from emergency housing stock. Since 2013, the fund also gives funding for the modernization of municipal infrastructure systems.

A considerable share of the apartment buildings requires capital repairs, and about 0.7% (in Moscow 0.05%) are officially recognized as in a state of needing emergency repairs. According to the law, municipalities are obliged either to pay owners market value (including compensation if they do not make capital repairs) or to move their inhabitants to other apartments. Only 83% of housing stock are provided with the central water supply system and 78% of housing stock are provided with the sewerage (water disposal). Today, about 5% of apartment buildings each year receive capital repairs with the financial resources of the state and municipalities. Growth of the payment for utilities led to a reduction of irrational resource consumption by the population. In particular, from 2010 to 2018, water consumption per inhabitant was reduced by 25%, although only about 50% of apartment buildings are equipped with metering devices.

In 2015, work began on the law obliging the owners of apartments to pay for capital repairs by charging a tariff depending on the size of the apartment. The tariffs are established by the regional government, and payments go into a regional fund. The amounts are not enough, and therefore, the state co-finances these actions. At the same time, certain apartments owners are not required to pay this tariff; for example, senior citizens over 80 years and owners of apartments in new buildings (up to five years).

However, it is clear that capital repairs to a considerable part of apartment buildings from the Soviet period are economically unfeasible because of new engineering and ecological requirements. As a result, their destruction and the construction of new multi-floor houses is the only alternative. Therefore, the state policy is directed towards the construction of new housing. Since 2019, the national project, 'Housing and Urban Environment'. For this purpose, at the Ministry of Construction, a special institute for development was created – Dom.rf, which develops state programmes in this sphere. This institute will take on the roles as developer, among others. The purpose is to provide middle-income families with affordable housing. This purpose has led to problems in terms of a rapid increase in volume of new housing from 80 million m^2 in 2018 up to 120 million m^2 in 2024. At the same time, financing sources have changed. In 2018, only 16.3 million m^2 of apartment housing were financed by mortgage lending. The goal for 2024 is 53 million m^2.

At the same time, since 2018, the transition has begun to finance housing construction by accumulating funds from investors on escrow accounts in specially authorized banks (as a rule, state banks) and also the necessity of joint financing the social and engineering infrastructure by developers. Figure 18.2 shows the price levels by using the average cost of 1 m^2 in the primary market had to grow from € 816 to € 1,056 in 2024. Due to inflation in 2020, this average cost

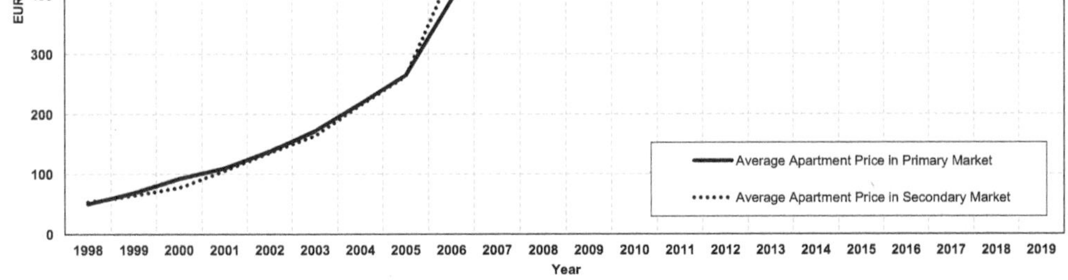

Figure 18.2 Average price of typical medium-quality apartments in primary and secondary markets in Russia, 1998–2019

Note. *1 Euro = 81.21 Russian Ruble (as of 2021 September)

Source: The Russian Federal State Statistics Service (2021)

was reached by 2021. At the end of 2020, the average cost of a square metre in projects of new buildings of a mass segment in Moscow reached € 2,436.

Other offered mechanisms of influence on the housing market are renovation programmes in the cities. The pioneer was Moscow, where, within renovation, people move to new apartments with the increased area. The project started in Moscow in 2017 and will provide resettlement into more than 350,000 apartments by 2032. Within this programme, 30 million m^2 will be constructed – more than twice the number will be constructed than destroyed. The total area of each new apartment transferred to the owners of apartments in the demolished houses at consent is nearly 20% to 30% more than apartments they lived in before (the demolished houses). Inclusion in this programme requires the consent of two-thirds of the apartment owners.

In other cities in the country, the potential for this does not exist. However, the federal government has promised financial support to regions to implement renovation programmes with the consent of most of the inhabitants of the houses set to be demolished.

3. The conveyance process

The following stages are included in the conveyance process in Russia:

Marketing phase

If the buyer has available the full amount for the purchase, then it is possible to prepare a sale agreement. However, the buyer is often not able to pay the full amount all at once. They may need to receive money from the sale of real estate or receive credit from a bank. In these cases, it is better to make and sign the provisional agreement. For it to be valid, an advance payment is required. The seller is to provide documents regarding title of the real estate and documents with technical characteristics of the object (e.g. the cadastral passport) and also the information about all persons registered on this address. Execution of the transaction with the help of the real estate agent means additional expenses for the seller, approximately 1.5% to 3% of the amount of the transaction.

The buyer checks the reliability of the title document given by the seller. To ensure the safest deal, one should check whether the apartment has any encumbrance or arrests. For example, one should check the rights of the citizens living there before privatization. The Federal Registration Service offers a service for checking the legality of the offered transaction.

Contract phase

The transaction can be ensured by a notary or the agreement can be made in a simple written form. The use of notarization of the sale-purchase, is not required by law, but is common as it is the equivalent of buying title insurance. The cost ranges between 0.7% and 1.5%. The seller and the buyer agree on the price of the transaction, which is then registered in the agreement. The amount of income tax for the seller is calculated from this amount. Therefore, the seller often insists that the price specified in the agreement should be lower than the real price. However, since 2021, banks are obliged to check the compliance of both the real and the contractual amounts. The payment can be made in three different ways.

The most popular, for today, is through a safe in a bank or settlement centre. Approximately 70% of all transactions take place according to this scheme. The buyer places the stipulated amount in the safe, and the seller gets access to it only after the conditions defined by the agreement have been satisfactorily performed. In case of purchase and sale of the apartment this condition is obtaining the property right by the buyer.

Payment by credit has become more and more popular. In this case, the buyer opens a bank account, and the seller has access to it only after they provide the bank with the documents confirming that the terms of the contract have been fulfilled.

And the last method is the transfer of all amounts to the bank account of the seller. The payment conditions are registered in the agreement. As a rule, it is specified therein that the full amount is transferred after the acceptance of the apartment in which all registered persons were removed.

Registration phase

State registration serves as the main proof of the property right for any real estate object. When giving a document package in registration authorities (through the multipurpose centre of any area or directly in person at the Federal Registration Service of the area where the real estate is located or in electronic form on the internet), it is necessary to check all filed documents. The slightest mistake will cause the documents to be returned, and then one must submit the papers anew.

The Federal Registration Service maintains the General State Real Estate Register (GSRER), which keeps all data on houses, apartments, and land plots (e.g. the cadastral number, area, and the name of the owner). It is necessary to register the transition of the property rights from the old owner to the new owner.

During electronic registration, participants of the transaction send all necessary documents to the Federal Registration Service online independently on the website of the Federal Registration Service or by means of the intermediary. The notary, real estate agency, firm builder, or bank can act as the intermediary. During the independent filing of the application by citizens on registration at the multipurpose centre or on the website of the Federal Registration Service, it is necessary to pay only the state fee (€ 24). Notaries have the right to carry out transactions remotely, irrespective of the location of the subject of the transaction.

Federal Registration Service can refuse to register the property rights if a suspension was made and its causes were not corrected within three months. If the registration is denied after this time, the fee is not returned. The law specifies several reasons for granting the registrar the right to suspend or deny the registration of the transaction. In the case of the suspension of the process of registration of the transaction, notices with the detailed reasons for this decision are sent to the stakeholders.

On average, the Federal Registration Service carries out the transaction within two to seven days. The Federal Registration Service requests some data from state agencies independently. If the registrar can consider the paper application up to ten days, the electronic execution of the transaction takes place twice as quick. Registration takes place even faster if the transaction was certified by the notary. Notaries send documents to the Federal Registration Service free of charge, so it is necessary to pay for only the execution of the transaction.

In 2020, the Federal Registration Service finished the transition to a general federal system of accounting of the rights for real estate and data of the State Real Estate Property Cadastre. It allows for the storing and processing of information centrally, to make both the registration of transactions and the statement of objects for the cadastral registration available in one system. It is supposed that, by 2024, all accounting and registration actions will be carried out within a day, with most of the transactions being registered online.

In general, in Russia, about 100 million residential and non-residential buildings, constructions, and rooms are registered, in addition to 60 million land plots, a quarter of which do not have clear boundaries. Also, there is no exact information about the owners of 46 million real estate objects (about a third of all objects).

Post-transactions phase

Owners of apartments or other residential real estate who have owned it longer than the minimum term are exempted from a tax in the case of its sale, and there is no need to submit a tax declaration for these. At the same time, the minimum term of ownership can be three years (if housing was purchased in any manner since 1 January 2016 or in other particular cases) or five years in other cases. Particular cases are inheritance or a gift from a family member or a close relative (parents, children, grandfathers/grandmothers, grandsons, full- and half-brothers and half-sisters). Also, no tax is due if the housing was privatized or received as a result of the cession of property under the agreement of lifetime maintenance with dependence. In these cases, the minimum term of ownership for tax-exempt sale is also three years.

In cases where the taxpayer is not exempted from this tax, the tax base is defined according to the price of the sale of residential property, but it cannot be lower than 70% of its approved cadastral value. The income tax rate is 13% for residents and 30% for non-residents of the Russian Federation. But this tax is paid from the amount of the transaction, less a non-taxable minimum of € 12,000, or less expenses upon the acquisition of this property.

4. Conclusions

To conclude, this chapter portrays the Russian property market of residential buildings as outlined in Table 18.1. The Russian market is characterized with a high degree of owning

Table 18.1 Fees and taxes during the conveyance process in Russia

Fees and taxes	*Real estate*	
	Seller	*Buyer*
Real estate agent's fees	1.5%–3%	-
Marketing fees	€ 50–170	-
Pre-contract related fee	-	€ 20
Legal fees	-	€ 300 + 0.1% of sum, which exceeds € 115,000
Bank mortgages fees	-	€ 250–850
Real estate registration fee	-	€ 23 for individuals € 250 for organizations
Capital gains tax	13% for profit for individuals (Owners of apartments or other residential real estate, who own it longer than the minimum term, are exempted from a tax) 20% for profit for organizations	-
Sales tax	-	-
Total (EUR for the model case price)	**€ 1,550–3,170**	**€ 578–1,178**
Total (% share to the model case price)	**1.6–3.2%**	**0.6–1.2%**
Excluded costs from the model case	13% CGT for individuals, 20% GCT for organizations	Real estate registration fee: Calculated as individuals

Note. Model case price: € 100,000

Source: Author

and a low degree of renting, mostly due to the free privatization of housing in the post-Soviet period.

In addition, this chapter considered the development of the state system of regulating the housing market, including the taxation system, mortgage lending, housing renovation programmes, support of capital repairs of apartment buildings, and the creation of modern infrastructure. The chapter concluded with what is involved in the process of registration of transactions in terms of residential real estate, the development of the federal state system of registering real estate objects, and also the transition of the rights to them. On the basis of modern digital technologies, the newly created Federal Registration Service system allows for the increased transparency and the market efficiency of the real estate.

Note

1 1 Russian Ruble = 0.012 Euro (as of 1 October 2021).

References

The Federal Service for State Registration, Cadastre and Cartography (Rosreestr). (n.d.). *Main page*. Retrieved November 20, 2021, from https://rosreestr.gov.ru/site/

Ministry of Construction, Housing and Communal Services of the Russian Federation. (n.d.). *Main page*. Retrieved November 20, 2021, from https://minstroyrf.gov.ru/

Tax Code of the Russian Federation Housing code of the Russian Federation. Retrieved from https://data.nalog.ru/html/sites/www.eng.nalog.ru/Tax%20Code%20Part%20Two.pdf

Nordic law countries

Sweden
Finland
Nordic law countries

19

FINLAND

Saija Toivonen and Kirsikka Riekkinen

1. The Finnish housing market

In the Finnish real estate market, dwellings are mainly owner-occupied or rented. The ownership of dwellings can be organized either directly, in the case of real properties, or by owning shares in an apartment house company. The legislation in both cases differs in terms of transactions, for example (as is described later).

Finland is a sparsely populated country, which also has an impact on its real estate market conditions. In 2020, the average population density in Finland was circa 18 persons/land area km^2, whereas in the capital city of Helsinki circa 3,000 persons/land area km^2 (Statistics Finland, 2020a). Since 2015, the yearly transaction volume has been approximately 3,000 apartments (2/3 located in apartment houses and 1/3 in attached houses) representing 3% of the total stock of apartments. In turn, the average transaction volume for detached houses has been 15,000 dwellings, which is 1% of the total stock of detached houses (Statistics Finland, 2020b, 2020c). The impacts of urbanization can be seen in the real estate market development. Price levels vary significantly within the country as illustrated in Figure 19.1 with higher prices in the southern more densely urbanized parts of the country.

Nominal price indices for detached houses and dwellings in apartment house companies have grown faster in the metropolitan area than the rest of the country (see Figure 19.2 and Figure 19.3). In 2020, the average price in the metropolitan area for dwellings in apartment houses was € 4,200/m^2 and € 3,400/m^2 for attached houses and detached houses. The average price outside the metropolitan area was € 1,600/m^2 for dwellings located in apartment houses or in attached houses and € 1,500/m^2 for detached houses.

In Finland, 63% of households live in owner-occupied dwellings, 34% rent, and 2% live in right-of-occupancy dwellings. However, renting is much more popular in the growth centres than outside them. A total of 33% of the rented dwellings are built on state-subsidized housing loans. Approximately 89% of the dwellings are occupied by permanent residents (Statistics Finland, 2020b). In 2020, there was, in addition, circa 500,000 free-time residences (including low-equipped summer cottages) which represents approximately 25% of the total number of buildings in Finland (Statistics Finland, 2020c).

DOI: 10.1201/9781003130383-22

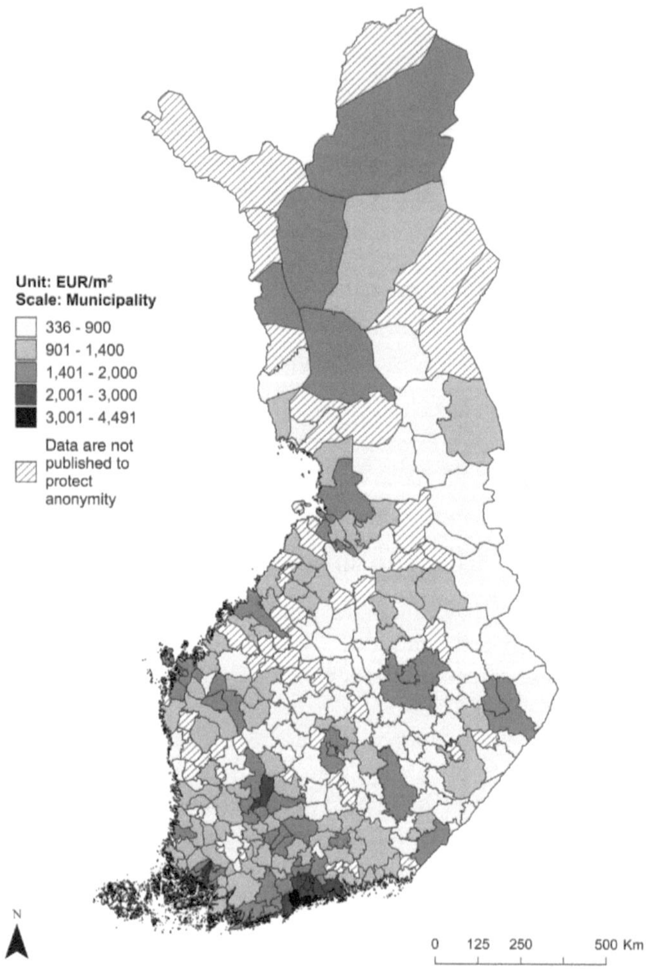

Figure 19.1 Average prices of old dwellings in housing companies – all building types, 2020

Source: Statistics Finland (2021)

The distribution of household dwelling units in Finland (Statistics Finland, 2020b) is as follows:

- 39% live in detached houses (95% owner-occupied and 3% rented)
- 14% live in attached houses (63% owner-occupied, 18% rented, 14% state-subsidized rented, and 3% right of occupancy)
- 46% live in apartment houses (37% owner-occupied, 39% rented, 19% state-subsidized rented, and 3% right of occupancy)
- 1% other

In 2019, the average floor space per person was approximately 41 m^2 in Finland, and 35 m^2 in the three biggest cities of Helsinki, Espoo, and Vantaa. The average dwelling size was 81 m^2. The Finnish population structure is aging, which is also reflected in the increasing number of

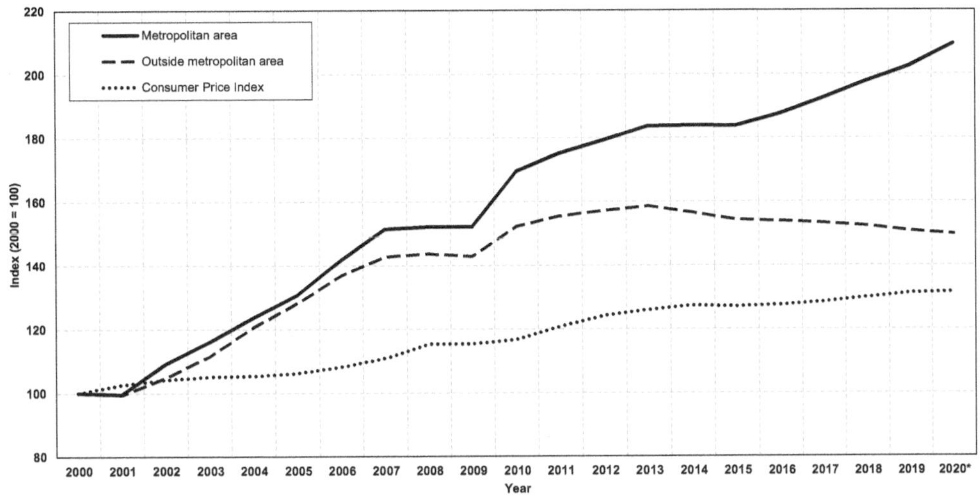

Figure 19.2 The nominal price indices of dwellings in apartment house companies compared with the consumer price index (Statistics Finland, 2020d, 2020f).

Note. Prices are from secondary market transactions

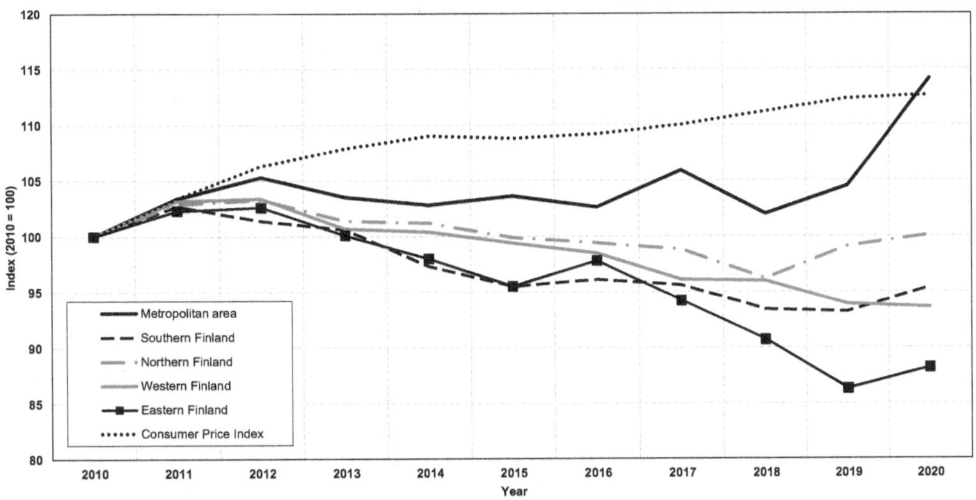

Figure 19.3 The nominal price indices of detached houses and the consumer price index

Note. Prices are from secondary market transactions

Source: (Statistics Finland, 2020e, 2020f)

single households (45% in 2019) (Statistics Finland, 2020b). A total of 40% of the apartment houses were built between 1960 and 1979, and 14% have been built since 2010, whereas 55% of the attached houses were built between 1970 and 1989, and 9% have been built since 2010. Approximately half of the detached houses have been built since 1980, and 8% have been built since 2010. On average, approximately 36,000 new dwellings have been built yearly between 2015 and 2020 (Statistics Finland, 2020g).

2. Regulatory framework

The Finnish conveyance of a real property can be done by writing and signing the contract on paper or in an electronic format. The Code of Real Estate (540/1995) requires that the conveyance be legally binding; therefore, it has to be done either in written format or by using electronic conveyance system. The contract follows the stipulations of the Code of Real Estate, where the clauses of contract need to include at least the following: (1) purpose of conveyance (sale, gift, inheritance, etc.), (2) the object of conveyance, (3) the seller's and buyer's names, and (4) the price and/or other compensation. If a pre-contract is written, the same stipulations apply. In addition to the clauses of contract, the Code of Real Estate stipulates what happens in the case of errors and breach of contract. The seller of the real property is responsible for defects occurring in the real property for five years after the signing of the contract.

The conveyance of an apartment is actually a transaction of shares of an apartment house company, and the stipulating legislation is the Housing Transactions Act (843/1994). These shares entitle their owner to use and possess a certain apartment within the company as well as use shared spaces. The apartment building is governed by the owner's association, and the ownership of shares provides the right to participate in the decision-making of the association. The owner of the shares is responsible for participating in the maintenance and repair costs of the company. Given that the shares of apartment house companies are regarded as movable property, the freedom of contract is applied in the conveyancing process and clauses of contract. When the conveyancing is between two private owners, it can be done by oral contract. The Housing Transactions Act stipulates errors and breach of contract, which differ from the stipulations of the Code of Real Estate. The seller of the apartment is responsible for defects occurring in the apartment for up to two years after signing the contract.

The public purchase witness is needed only in real property (not apartment) conveyances and only in cases of sale agreement written on paper. The qualifications for public purchase witnesses are stipulated in the Act on public purchase witness (573/2009). Public purchase witnesses are qualified either due to being an authority official or a person authorized by the National Land Survey of Finland (often real estate agents or bank officials).

3. The conveyance process

Next, we go through the conveyance process of a real property in two cases: the signing a written contract on paper or using the electronic conveyancing, and the conveyance of an apartment in an apartment house company. In all cases, the key parties are the seller and the buyer. A real estate agent or legal assistance can be used but is not required. Figure 19.4 shows the basic steps of a conveyance process for a real property and apartment.

Marketing phase

The marketing of real property is typically done on websites, of which, two are the most prominent (Oikotie.fi and Etuovi.com). A seller may use a real estate agent to assist with the marketing and sale, and they are used in most of the real property conveyances when the sale is between two private persons.

The National Land Survey collects and maintains a sale price data register on real properties in Finland. This information can be used when deciding on the asking price of the object of conveyance. This register is accessible to the public for a fee. The Housing Finance and

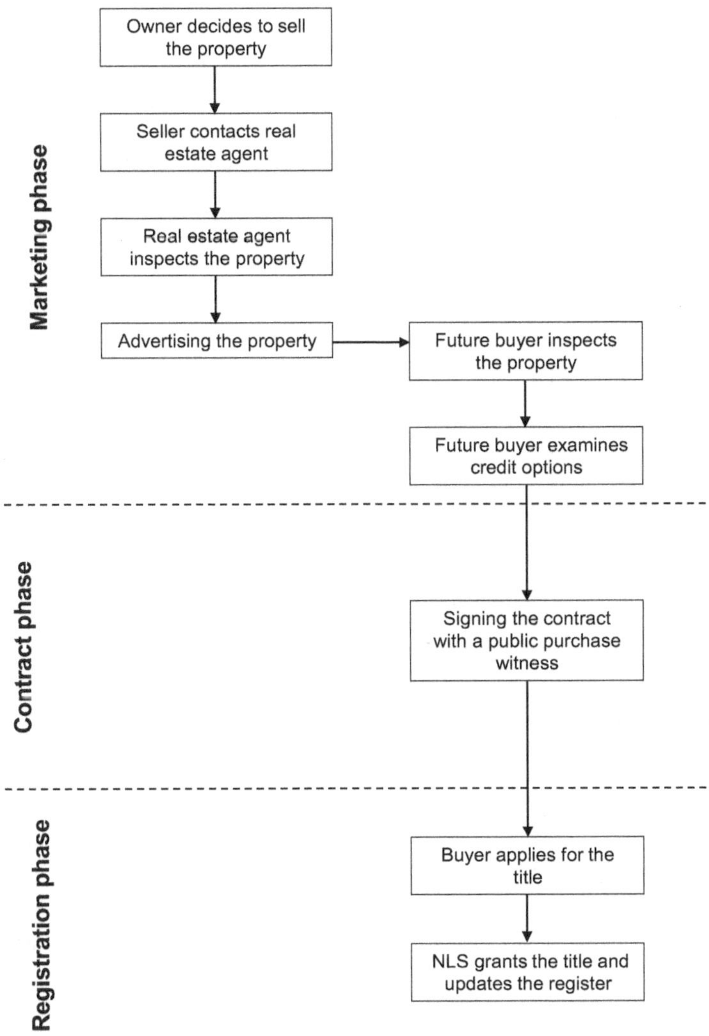

Figure 19.4 The conveyance process in Finland

Source: Author

Development Centre of Finland ARA and the Ministry of the Environment publish sale price data on apartment transactions in Finland, which is collected from the biggest real estate agents and the Central Federation of Finnish Real Estate Agencies (ARA, 2021). However, these data are not comprehensive because they are collected from real estate agents and other commercial actors, not from private sellers. Also, they are available to the public for only one year, and the information required to identify a single sale is difficult due to lacking addresses or other factors affecting the identification of the sale. Finally, the data are available only when there has been a minimum number of sales in one area, practically meaning that the service is of use only in bigger cities. The data depend on the participating agencies' willingness to have their business data seen by the public.

Contract phase

The contract and conveyancing of a real property need to follow the stipulations of the Code of Real Estate (554/1995). If the contract is signed on paper, the seller, the buyer, and a public purchase witness are required to be present. The public purchase witness is required to verify the identities of the seller and buyer and to check that they are capable and allowed to do the transaction (e.g. seller is the actual and only owner of the object of the transaction). They are also required to check that the contract follows the stipulations of the Code of Real Estate. The public purchase witness will attach a certificate on each copy of the contract and inform the municipality about the transaction. The buyer pays the purchase price to the seller and pays the transfer tax to tax authority. If the conveyance is done in the e-conveyance system by the National Land Survey, there is no need for public purchase witness to verify the transaction, as this is done by the system. The identities of the buyer and the seller are also checked with strong a verification tool, and therefore, they do not need to be present at the same place.

The contract of an apartment transaction is within the freedom of contract. The contract is typically written and signed by the buyer and seller, but in theory, an oral contract would be sufficient.

Registration phase

After signing the contract of real property transaction, the buyer must apply for the title from the National Land Survey within six months. The buyer may also certify their bank to apply for the title. This is common if the bank is otherwise involved in the process, for example, if the property is mortgaged for a loan.

In case of apartment conveyance, a new law (Act on the Residential and Property Information System, 1328/2018) that came into force in 2019 changed the registration process of ownerships so that housing management companies are no longer responsible for keeping a register (digital or paper-based) of the owners of apartments and transactions. However, during the transition time until 2029, the housing manager can continue to handle the registration of ownerships. Starting from 2019, the buyer should apply for registration of ownership to the Residential and Commercial Property Information System from the National Land Survey within two months of the signing of the contract. The registration of ownership is possible only if the apartment house company is also in the register. All apartments that are constructed and sold after 2019 are automatically in the digital system, but old apartment house companies need to be added by the companies themselves. Transactions of old apartments as well as the nullification of paper share (apartment) certificates and their transfer into digital ones is the responsibility of the buyer in conveyance or else the owner's responsibility if there is no conveyance before the transition time ends.

Transaction costs

Transaction costs differ in the cases of real properties and apartments (Table 19.1). All costs presented here are based on year 2021 price levels and taxation. In the case of real properties, transaction costs depend on whether the conveyance is done on paper with a public purchase witness or using the e-conveyance system. In both cases, the transfer tax is 4% of the purchase price, and it is to be paid by the buyer. The public purchase witness cost is € 120 per conveyance, excluding possible travel costs, and the registration fee for the title is € 119. If the conveyance is done electronically, the fee for the system is € 185, including the title fee. In addition,

Table 19.1 Mandatory fees (€) and other voluntary costs (in grey) related to real property and apartment conveyance calculated for € 100,000 transaction, based on 2021 price levels, including VAT

Phase	*Fees and taxes*	*Real property (transaction on paper)*		*Real property (transaction online)*		*Apartment*	
		Buyer	*Seller*	*Buyer*	*Seller*	*Buyer*	*Seller*
Marketing	Energy/other mandatory certificates	-	€ 250	-	€ 250	-	€ 150
	Marketing fee	-	€ 150	-	€ 150	-	€ 150
	Real estate agent	-	2–5% of sale price	-	2–5% of sale price	-	1–4% of sale price
	Building inspection	€ 1,500	-	€ 1,500	-	€ 1,200	-
Contract	Public purchase witness	€ 120	-	-	-	-	-
Registration	Transfer tax	4%	-	4%	-	2%	-
	Ownership registration	€ 119	-	€ 185	-	€ 60	-
	Capital gain tax (If owner-occupied for over two years, then no requirement to pay the tax.)	-	30% of the capital gain up to 30,000, 34% of more than 30,000	-	30% of the capital gain up to 30,000, 34% of more than 30,000	-	30% of the capital gain up to 30,000, 34% of more than 30,000
Post-registration	Sales tax	-	-	-	-	-	-
Total (EUR for the model case price)		**€ 5,739**	**€ 2,400–5,400**	**€ 5,685**	**€ 2,400–5,400**	**€ 3,260**	**€ 1,300–4,300**
Total (% share to the model case price)		**5.7%**	**2.4–5.4%**	**5.7%**	**2.4–5.4%**	**3.3%**	**1.3–4.3%**
Excluded costs from the model case		-	30–34% CGT	-	30–34% CGT	-	30–34% CGT

Source: Author

the energy certificate must be available (around € 250), and some real estate agent fees may have to be paid, typically varying from 2% to 5% of the selling price. These fees are usually paid by the seller, but Finnish real estate agents take the role of mediator, representing both the seller's and the buyer's interest (KKV, 2021). A buyer or seller may use any kinds of building inspector, humidity concentration measurements, lawyers, and so forth. Just as with the agent's fees, these are all voluntary costs. Typically, the buyer also needs to acquire a bank loan to buy the real property, which also incurs costs.

In apartment transactions, the costs are lower than in real property transactions. The transfer tax is 2% of the purchase price. Additionally, there used to be no other mandatory costs; however, due to the new apartment and apartment owner system, the registration fee of ownership is now € 60 to be paid by the buyer to the National Land Survey. Some real estate agent fees may also have to be paid, typically varying from 1.5% to 4% of the selling price, paid by the seller when the transaction is realized. A certificate from the apartment house manager, including the energy certificate, should be available, costing around € 150 for the seller. In both real property and apartment transactions, if the seller decides not to use a real estate agent, then they typically have to pay the marketing fee of online platforms, and this cost is around € 150.

A seller is obliged to pay a capital gains tax of 30% (maximum € 30,000 gain) or 34% (gain above € 30,000). However, the tax does not have to be paid if the seller has owned the dwelling and lived there for a minimum of two years before the sale (Capital Gains Tax Act, 1535/1992). The buyer is responsible for a transfer tax, which is 4% for real properties and 2% for apartments. The government is supportive of those buying their first apartment or house by exempting them from paying the transfer tax (Transfer Tax Act, 931/1996).

References

Finish Competition and Consumer Authority (KKV). (2021). *Main page*. Retrieved March 31, 2021, from www.kkv.fi/en/

The Housing Finance and Development Centre of Finland (ARA). (2021). *Asuntojen.hintatiedot.fi. Data on completed home sales during the past 12 months*. Retrieved March 3, 2021, from https://asuntojen.hintatiedot.fi/haku/

Statistics Finland. (2020a). *Official statistics of Finland, population structure*. Helsinki: Statistics Finland. ISSN:1797-5395. Retrieved March 31, 2021, from www.stat.fi/til/vaerak/index_en.html

Statistics Finland. (2020b). *Official statistics of Finland, dwellings and housing conditions*. Helsinki: Statistics Finland. ISSN:1798-6761. Retrieved March 31, 2021, from www.stat.fi/til/asas/index_en.html

Statistics Finland. (2020c). *Official statistics of Finland, buildings and free-time residences*. Helsinki: Statistics Finland. ISSN:1798-6796. Retrieved March 31, 2021, from www.stat.fi/til/rakke/index_en.html

Statistics Finland. (2020d). *Official statistics of Finland, price of dwellings in housing companies*. Helsinki: Statistics Finland. ISSN:2323-8801. Retrieved March 31, 2021, from www.stat.fi/til/ashi/index_en.html

Statistics Finland. (2020e). *Official statistics of Finland, real estate prices*. Helsinki: Statistics Finland. ISSN:2342-8902. Retrieved March 31, 2021, from www.stat.fi/til/kihi/index_en.html

Statistics Finland. (2020f). *Official statistics of Finland, consumer price index*. Helsinki: Statistics Finland. ISSN:1799-0254. Retrieved March 31, 2021, from www.stat.fi/til/kihi/index_en.html

Statistics Finland. (2020g). *Official statistics of Finland, building and dwelling production*. Helsinki: Statistics Finland. ISSN:1798-9590. Retrieved March 31, 2021, from www.stat.fi/til/ras/index_en.html

Legislation

Act on Public Purchase Witness. (573/2009). *Laki kaupanvahvistajasta (573/2009)*. Retrieved from www.finlex.fi/fi/laki/ajantasa/2009/20090573

Act on the Residential and Property Information System. (1328/2018). *Unofficial translation*. Retrieved from www.finlex.fi/fi/laki/kaannokset/2018/en20181328

Capital Gains Tax Act. (1535/1992). *Tuloverolaki (1535/1992)*. Retrieved from https://finlex.fi/fi/laki/ajantasa/1992/19921535

Code of Real Estate. (540/1995). *Unofficial translation*. Retrieved from https://finlex.fi/fi/laki/kaannokset/1995/en19950540

Housing Transactions Act. (843/1994). *Unofficial translation*. Retrieved from https://finlex.fi/fi/laki/kaannokset/1994/en19940843

Transfer Tax Act. (931/1996). *Varainsiirtoverolaki*. Retrieved from https://finlex.fi/fi/laki/ajantasa/1996/19960931

20

SWEDEN

Ola Jingryd and Peter Palm

1. The Swedish housing market

The Swedish[1] real estate market for homes is constituted of three main forms: rental, ownership, and tenant-ownership. The legislation differs between ownership and tenant-ownership, with tenant-ownership regarded as personal property and not real property. This has implications on both the conveyance process and transaction costs, as we explain in this chapter.

The Swedish real estate market, and its housing market in particular, has a vast number of regulations and policies. As Figure 20.1 shows, over the past decade, real estate prices have increased rapidly in Sweden; for instance, house prices have increased by 67% (Boverket, 2019). Consequently, the government has introduced several restrictions on the financial market, including two constraints on mortgage loans in 2016 – one in relation to the loan-to-value

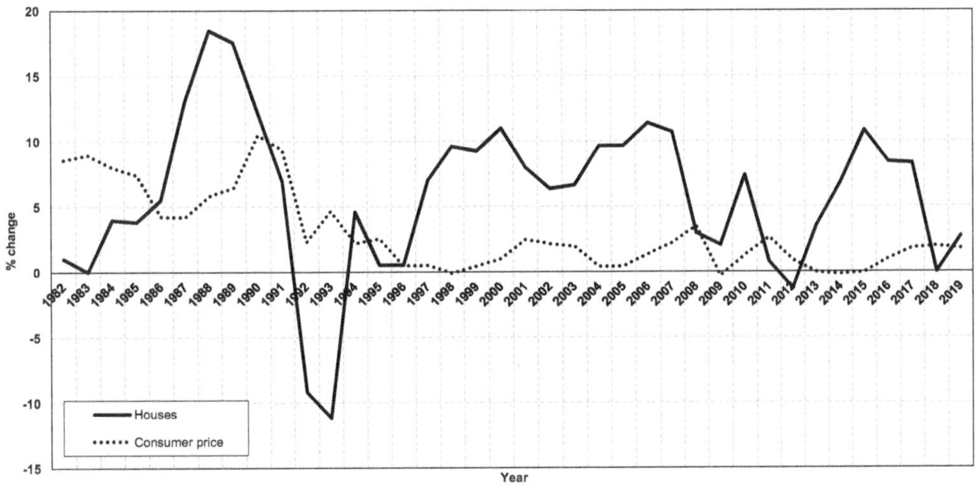

Figure 20.1 Yearly change in consumer price and house prices in Sweden (in percent)

Source: Statistiska centralbyrån (2020)

DOI: 10.1201/9781003130383-23

ratio and one in relation to income. Additionally, two regulations concerning amortization have been introduced during 2018 – one loan-to-value ratio and one loan-to-income ratio. These regulations have caused the housing market to cool down some, with prices no longer increasing at the same rapid pace as in previous years. In 2018 some locations even decreased, but the market had a positive increase once again in 2019.

The yearly transaction volume in Sweden is about 110,000 tenant-owned apartments, corresponding to approximately 10% of the stock, and 55,000 houses, responding to approximately 3% of the stock (Mäklarstatistik, 2020; Lantmäteriet, 2019). The price levels vary greatly within the country, with an average price for tenant-ownership apartments of € 4,100 per m^2 and for houses € 2,500 per m^2 (Mäklarstatistik, 2020). Not surprisingly, the most expensive tenant-owned apartments are found in the capital Stockholm, with an average price of € 9,500 per m^2.

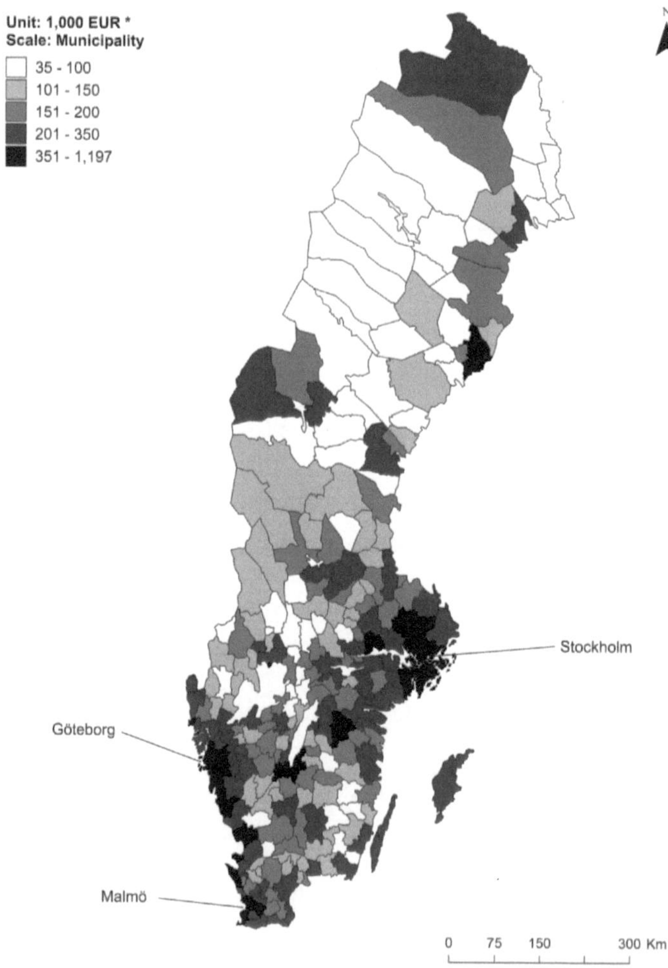

Figure 20.2 Average purchase prices of one- and two-dwelling buildings (no site leasehold right), 2020

Note. *1 Euro = 10.2 Swedish Krona (as of 2021 September)

Source: Statistiska centralbyrån (2020)

The distribution of homes in Sweden is (SCB, 2019a) as follows:

- 42% detached houses (91% ownership, 5% tenant ownership, 4% rental)
- 30% rental apartments (47% municipal housing companies, 36% private companies, 17% private owners)
- 21% tenant-ownership apartments
- 5% special housing (e.g. student housing)
- 2% other

The average house size in Sweden is 122 m^2, and the average apartment is 68 m^2. Most apartments (51%) were built during 1951–1980, while houses have a more diverse building period. A total of 20% of the houses were built before 1930, and 45% were built between 1961 and 1980. However, rapidly increasing prices on the real estate market have resulted in a relatively large increase in new construction, with over 50,000 units built yearly over the last five years (Boverket, 2019) – a growth in construction that Sweden has not seen since 1992 (SCB, 2019b). The rate of new construction is abating but is nevertheless predicted to continue at a higher level than during 1992–2013.

2. Regulatory framework

The Swedish conveyancing process is simple insofar as it does not require the intervention of any professional or government body apart from the Land Registry *(Lantmäteriet)*. The formal requirements to accomplish a legally binding conveyance are laid down in Chapter 4 of the Land Code (SFS, 1970, p. 994; hereinafter referred to as LC). Under 4:1, the binding agreement requires a written contract, signed by the buyer and seller, specifying the sales price and a declaration that the property is conveyed to the buyer. The sales price specified in the sales contract is binding; any side agreement on another sales price is deemed void.

The specific terms of the contract do not, as a rule, require any specific form to be binding. However, 4:3 LC specifies three categories of clauses which must be included in the sales contract in order to be binding: (1) contingency clauses, (2) clauses releasing the seller from title-related liability, and (3) limitations to the buyer's right to convey or encumber the property. While the sales contract *(köpekontrakt)* is the only document prescribed by law to accomplish a binding sale, it is customary to seal the transaction with a sales deed *(köpebrev)*. Typically, the sales deed is signed on the possession date once the sale price has been paid in full. Therefore, in virtually all conveyances, the sales deed is the confirmation that the buyer is the new proprietor. The sales deed must be signed by two witnesses for the title registration to be granted (20:7 LC). Also, the buyer is required under 20:1–2 LC to apply for title registration within three months of the purchase. The function of title registration is declaratory, not constitutive. However, while title registration does not confer ownership per se, it can be decisive in case of a title dispute. For instance, if A sells their property to B and then to C, the party – either B or C – who first applies for title registration will be deemed to have the better title (17:1 LC), unless it is established that, at the time of purchase, C was aware, or had reason to believe, that the property had previously been sold (17:2 LC). For more instances where title registration may affect the outcome of title disputes see 16 & 18 LC.

3. The conveyance process

In this section, we outline the conveyance process for both real estate and tenant ownership apartments in Sweden. There are differences in the process, although we would claim that most private persons do not fully recognize these differences and their implications on their purchase or ownership of the home. Figure 20.3a and Figure 20.3b show the conveyance process.

The Swedish conveyance process differs from most others, as it does not include any pre-contract phase; however, the LC demands a written contract to have a binding sale. In light of the foregoing, the typical conveyancing process takes the following form.

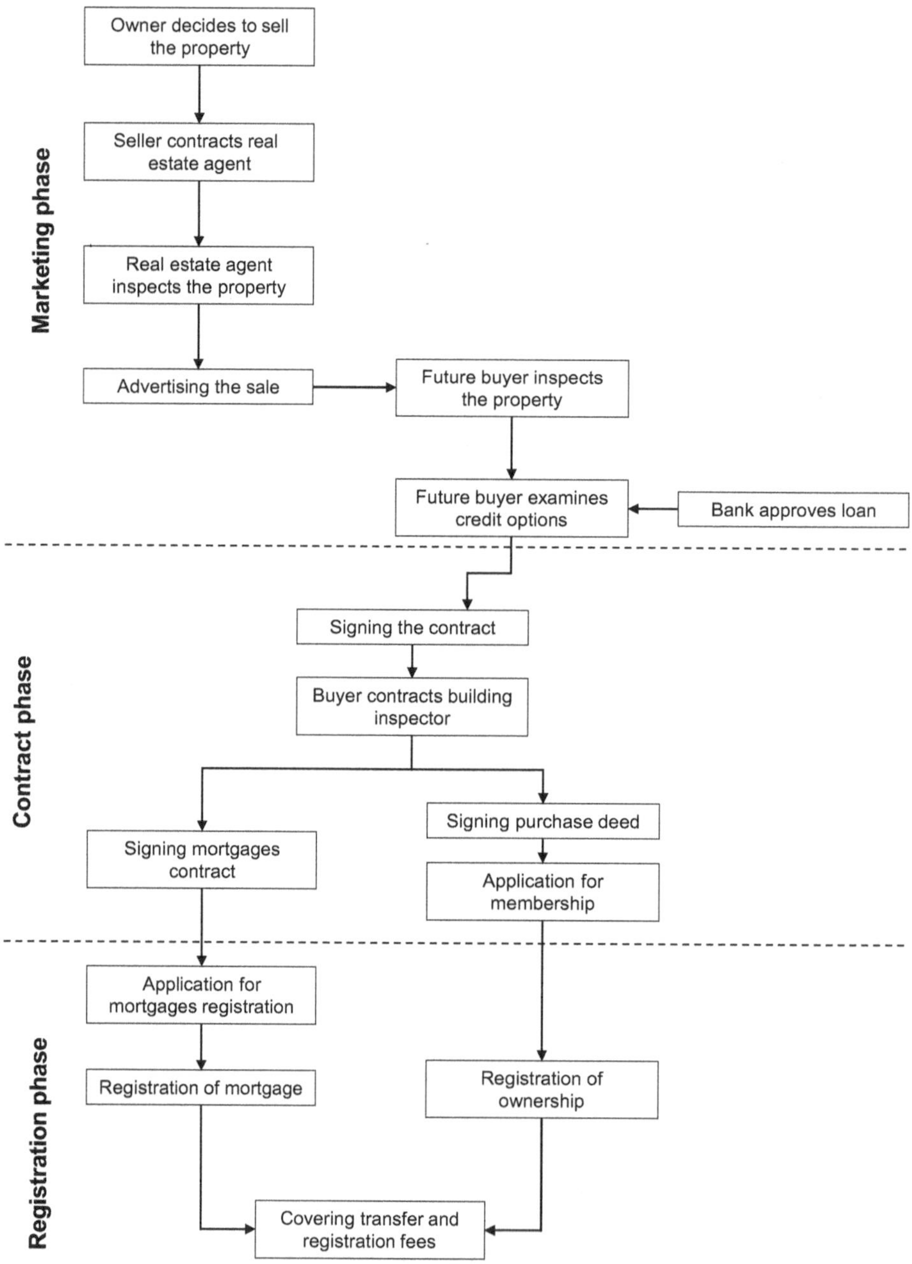

Figure 20.3a The conveyance process for real estate, respectively tenant ownership context, in Sweden
Source: Author

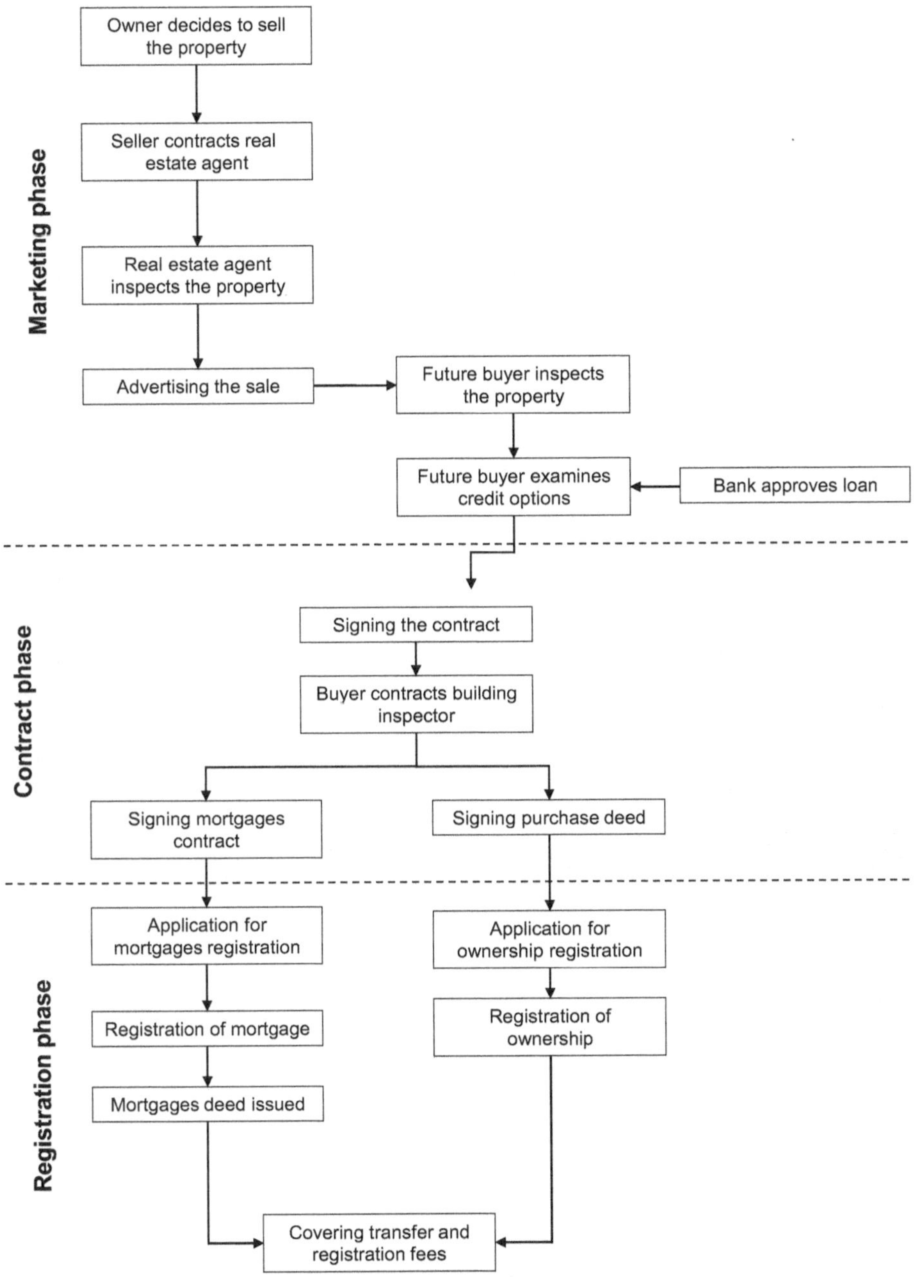

Figure 20.3b The conveyance process for real estate, respectively real estate context, in Sweden

Source: Author

Marketing phase

This first phase of the conveyance process most often involves an estate agent. While, as stated, there is no mandatory intervention in the conveyancing process, as a practical matter, more than

90% of all conveyances are in fact accomplished through estate agents. The activity of estate agents is regulated by the Estate Agents Act (SFS 2011, p. 666; hereinafter referred to as EAA), and estate agents must have a licence from the Estate Agents Inspectorate *(Fastighetsmäklarinspektionen)* to be allowed to practice the profession.

In Sweden, estate agency contracts are usually exclusive; however, 9 § EAA sets a three-month cap for the period of exclusivity. This period cannot be renewed until at least two months of the three-month period have passed. Also, estate agents are held to a duty of impartiality and must safeguard the interests of buyer and seller alike (8 §), with the exception that, to the extent possible without infringing upon the rules laid down in the EAA, the estate agent must take particular heed to the economic interests of their client. As a practical matter, while there is no legal impediment to an estate agent being hired by the buyer, the hiring party is virtually always the seller. However, due to the various obligations and prohibitions laid down in the EAA, the agent's chances of favouring the seller in order to maximize the sales price should not be overestimated. One practical consequence of the duty of impartiality is that the agent is not allowed to sign the sales contract on behalf of buyer or seller (15 § EAA).

Despite acting as an intermediary, the agent's fee is paid by the principal (23 § EAA), which is most often the seller. The agent is usually remunerated by means of a percentage of the sales price. Percentages range between 1% and 5%, with lower percentages in metropolitan areas where property prices are higher. The agent's right to remuneration requires, *inter alia*, a binding sales agreement (23 § EAA). In the marketing phase, the estate agent typically advertises the property on both the agency's website and on an independent online marketplace. In Sweden, there is one dominant online marketplace. This online marketplace was previously owned by some of the larger real estate agency firms and the real estate agency organization. However, a couple years ago, it was sold to an international investor. To have one dominating online marketplace has both pros and cons. One pro is that everyone refers to it as 'The marketplace for homes', and if someone wants to know the supply, they know where to go. The cost of advertising on this marketplace is diversified in regard to asking price and on what additional service than just being put on the market is desired. One such 'add on' is called the 'rocket', where, for a certain amount of time, someone can buy the top place of the webpage. The cost for this varies, but for the median Swedish home, the basic charge would be € 240, and it is paid by the seller.

In Sweden, negotiating the price of the property is mostly done by a bidding process administrated by the estate agent. Prospective buyers may start a bidding process, the forms of which are not regulated by statutory law, although the agent's discretion is limited by case law (Jingryd, 2021). However, given that no sale is binding until a written contract is signed by both buyer and seller, the bids in the bidding process are not binding (4:1 LC, and 6:4 Tenant-Ownership Home Act, THA), whether made verbally or in writing. Similarly, a declaration from the seller that a bid is accepted is not binding. The seller is also under no obligation to sell to the highest bidder. However, the estate agent is required to inform the seller of all bids, even if they have reason to believe that a given bid will not be accepted.

In the process of marketing the home, EU legislation has also been implemented, in that all homes must have an energy declaration. This is something that the seller is obliged to do before the contract is finalized. If not, the buyer has the right to contract a charted inspector to perform the inspection and do the declaration on behalf of the seller.

Contract phase

Once the bidding is closed by the seller, a contract signing meeting is scheduled. A binding sale requires a sale contract that fulfils the requirements in 4:1 LC or 6:4 THA. The contract specifies all the terms for the sale, including any contingency clauses or liability waivers, and a down payment of at least 10% is paid no later than a few days. However, the contract often includes contingency clauses such as inspection clauses, which give the buyer the right to inspect the property after the contract signing but prior to possession. It is not unusual that the agreed sales price is somewhat renegotiated after the inspection, but only marginally.

If the transaction concerns a tenant-ownership home, then the buyer needs to apply for membership with the tenant-ownership association. This is usually done with the help of the estate agent at the same time as the contract is signed. The membership application can be denied only if the buyer has financial problems.

The possession date, when the rest of the contract sum is paid and the sales deed signed, is typically around three months later. The buyer then applies for title registration at the Land Registry, without which no binding mortgage can be accomplished. The estate agent or the bank usually assists the buyer with the application. Depending on the property's existing mortgage sum, the Land Registry may need to issue a new mortgage deed. This comes at a cost of 2% of the mortgage amount and an administration fee of € 40.

Registration phase

The Swedish Land Registry records the transfer of ownership and issues a title deed. The title deed is declaratory only, as the conveyance is already binding between the parties as soon as the contract is signed and any contingencies are met. However, as mentioned, it can be decisive in the case of a title dispute.

A stamp duty of 1.5% of the purchase sum as well as an administration fee of € 90 is tied to the title registration. These fees are paid to the Land Registry by the buyer. The estate agent usually assists the buyer with the registration of the title, in which case the buyer pays the sums to the agent on the date of possession. However, this assistance is included in the estate agent's commission, and the buyer does not pay any remuneration to the estate agent. For tenant-ownership homes, there is no land registration and thereby no stamp duty or administration fee for either the buyer or seller. Except for the stamp duty and energy declaration, no legal fees are tied to the transaction process in Sweden. The legal advice pertaining to the sale and the drawing up of the necessary deeds are all included in the estate agent's duties, and thereby also included in the commission to be paid by the seller.

Post-transaction phase

As mentioned, there are no fees for selling a home in Sweden. However, if the seller makes a profit, a 22% capital gains tax on the profit must be paid. If the sold property was the seller's place of residence, they may be eligible for a referral of the capital gains tax (§47 Income Tax Act). If such a referral is granted, the seller must pay an annual fee of 0.5% of the taxable profit.

4. Summarizing the conveyance process and its transaction costs

To summarize, the Swedish conveyance process for both real estate and tenant-ownership is arguably simple. The ownership is conveyed with a sales contract, or, where it is agreed that a deed of sale shall be signed at the day of possession, upon the signing of the deed. Title registration, while mandatory under §20 LC only for real estate, is merely declaratory. Notably, the Swedish transaction process has no pre-contract phase, reducing both time and risk for the parties because they can be certain from the signing of the contract that, barring any contingency clauses, the sale

Table 20.1 Fees and taxes during the conveyance process in Sweden

Phase	*Fees and taxes*	*Real estate*		*Tenant-ownership*	
		Seller	*Buyer*	*Seller*	*Buyer*
Marketing	Real estate agent's fees	2%	-	2%	-
	Marketing fees	€ 240	-	€ 240	-
	Energy performance certificate	€ 350	-	-	-
Contract	Legal fees	-	-	-	-
	Tenant-Association fee	-	-	-	€ 114
	Building inspection	-	€ 900	-	€ 300
Registration	Stamp tax	-	1.5%	-	-
	Land registration fee	-	€ 90	-	-
	Mortgages fees	-	2% (of new mortgages) + € 40 adm.	-	€ 47
Post-registration	Capital gains tax	22% (of profit)	-	22% (of profit)	
	Legal fees	-	-	-	-
	Sales tax	-	-	-	-
Total (EUR for the model case price)		**€ 2,590**	**€ 2,530**	**€ 2,240**	**€ 461**
Total (% share to the model case price)		**2.6%**	**2.5%**	**2.2%**	**0.5%**
Excluded costs from the model case		22% CGT	2% of new mortgages	22% CGT	-

Note. Model case price: € 100,000

Source: Author

is binding. Furthermore, the complexity of the conveyance is lessened and the costs are lower, as, in addition to the marketing of the property, the impartial estate agent fulfils duties involving both legal advice pertaining to the sale and the drawing up of the necessary deeds.

Table 20.1 summarizes the transaction costs in Sweden both for real estate and tenant-owner homes.

Note

1 1 Swedish Krona = 0.1 Euro (as of 1 October 2021).

References

Boverket. (2019, December). *Boverkets indikator no. 2*. Retrieved from https://www.boverket.se/globalassets/publikationer/dokument/2019/boverkets-indikatorer-december-2019.pdf

Jingryd, O. (2021). *Fastighetsförmedling*. Lund: Studentlitteratur.

Lantmäteriet. (2019). *Fastighetsfakta*. Retrieved November 20, 2021, from www.lm.se/fastighetsfakta/
Mäklarstatistik. (2020). *Bostadsmarknaden*. Retrieved November 20, 2021, from www.maklarstatistik.se
Statistiska centralbyrån (SCB). (2019a). *Bostadsbeståndet, updated last 190425*. Retrieved November 20, 2021, from www.scb.se/hitta-statistik/statistik-efter-amne/boende-byggande-och-bebyggelse/bostadsbyggande-och-ombyggnad/bostadsbestand/pong/statistiknyhet/bostadsbestandet-2018-12-31
Statistiska centralbyrån (SCB). (2019b). *Antal färdigställda lägenheter i flerbostadshus resp. Småhus, updated last 190508*. Retrieved November 20, 2021, from www.scb.se/hitta-statistik/statistik-efter-amne/boende-byggande-och-bebyggelse/bostadsbyggande-och-ombyggnad/nybyggnad-av-bostader/pong/tabell-och-diagram/antal-fardigstallda-lagenheter-i-flerbostadshus-resp.-smahus/Statistiska centralbyrån (SCB). (2020). *Bostadsbeståndet, updated last 190425*. Retrieved Fabruary 7, 2022, from https://www.scb.se/hitta-statistik/statistik-efter-amne/boende-byggande-och-bebyggelse/fastighetspriser-och-lagfarter/fastighetspriser-och-lagfarter/
Svensk författnings samling (SFS). (1970). Land Code, Chapter 4.
Svensk författnings samling (SFS). (2011). Estate Agents Act.

Common law countries

Canada
United States of America
England & Wales
Ghana
Nigeria
Kenya
South Africa
Thailand
Malaysia
Australia
Common law
Common law only
Mixed system using common law

21 AUSTRALIA

Clive Warren

1. The Australian housing market

Residential property in Australia[1] is held in two principal forms: freehold (ownership) and leasehold (tenant rental). In addition, a property may be part of a strata title or community title scheme, in which the freehold land is subdivided into strata. This is the case for most apartment blocks where owners hold the freehold in their apartment but are subject to a community title scheme which manages the common areas, lifts, corridors, car parks, and so forth and levies owners to cover maintenance and insurance costs. Leasehold property provides exclusive possession of the property subject to the terms of the lease. Legislation provides rights to the lessee (tenant) to enjoy possession and the lessor (owner/landlord) the right to receive rent. In the case of residential property the rights of tenants are further protected through legislation to limit the rights of landlords to evict tenants without reasonable grounds or notice period.

The law in Australia relating to real property is derived from English law and closely resembles that which pertains in the United Kingdom. Australia, however, is a federated country which comprises six states and two territories with each state governing subject to the British Crown and, as such, each state has its own statutes governing the transfer of rights in real property. While the statutes governing property are similar, there are subtle differences between the states and territories – some of which is discussed later in relation to conveyancing of property title.

The real estate market in Australia has shown considerable growth over recent years. The largest rate of growth over the past decade has occurred in the major capital cities; Sydney has gained 66%, Melbourne 43%, and Brisbane 19.6% (ABS, 2020) (see Figure 21.1). The current price increases are fuelled by record low interest rates with some predictions of up to a 20% increase over the next two years.

When analyzing the Australian real estate market over time some interesting patterns can be observed. Figure 21.2 illustrates a temporal perspective using a time period from 2003 to 2020 divided into major capitals. Prices diverge after the recovery from the 2007–2008 financial crises with Sydney, Hobart, and Melbourne experiencing very large increases in prices and on

DOI: 10.1201/9781003130383-25

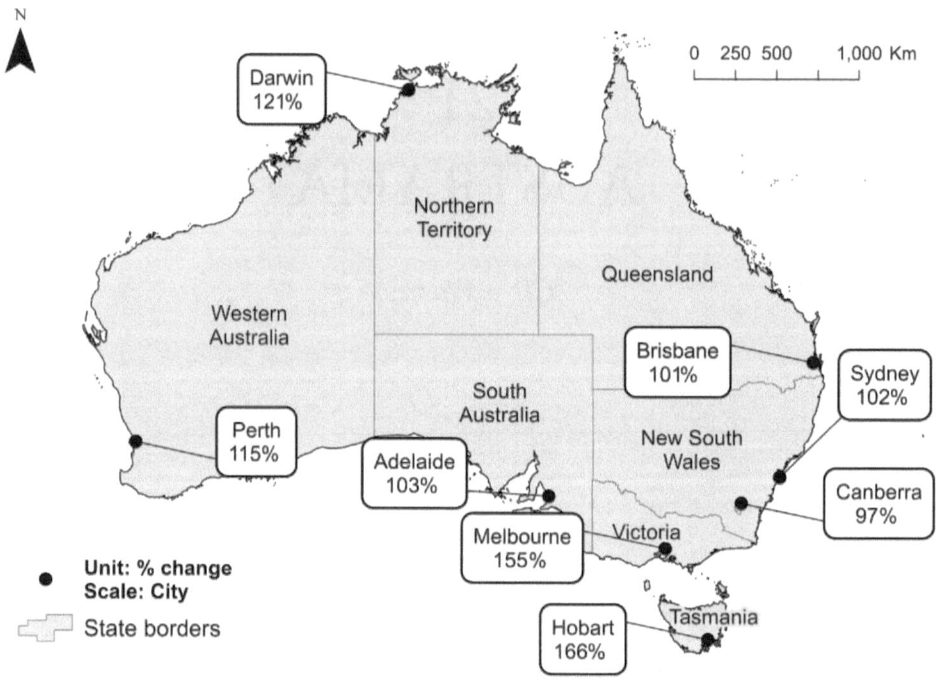

Figure 21.1 House price index change (in percent), 2003–2020

Note. 2011 December = 100

Source: ABS (2021) Australian Bureau of Statistics, Online Stat Database

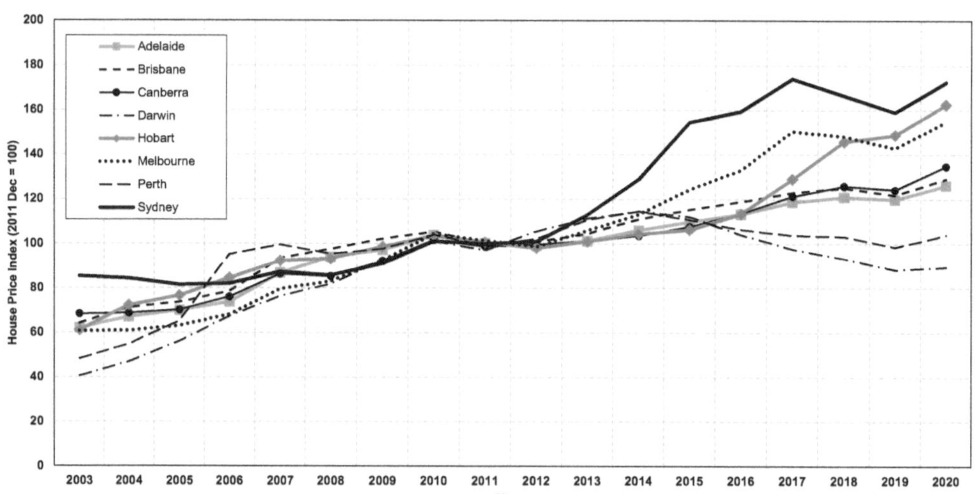

Figure 21.2 Change in house prices in Australia, 2003–2020

Source: ABS, Australian Bureau of Statistics, Online Stat Database

the other side Darwin and Perth with price levels today equal to or lower than those during the financial crisis.

The estimated value of Australian residential housing stock is € 4,661 billion, with a mean price of € 441,280 (ABS, 2020). Approximately 66% of Australians own their home with or without a mortgage, and 32% rent their home. One in five (20%) own one or more investment properties with 5% owning four or more properties (ABS, 2019). Detached housing comprises 75.5% of the stock with 9.9% being semi-detached or town houses, while 13.6% are flats or apartments (ABS, 2021). The average size of detached houses is declining and currently stands at 228.8 m^2, with the average apartment being at 128.8 m^2.

2. Regulatory framework

The Australian real property regulatory system was originally derived from English Law. Each state and territory have its own legislation covering the tenure and transfer of property. As in the United Kingdom, all property is held on a grant of title from the Crown and states administer all Crown land (Land Act, 1994, s14). While derived from a common legislative origin, there are subtle differences in each jurisdiction. In describing the title process this chapter focuses on the state of Queensland but highlights differences in other states where appropriate.

In respect of residential property the state grants a freehold title such that the owner of which can either sell the freehold title or grant a lease over the whole, or any part, of the title. The exception to this is the Australian Capital Territory where only 99-year leases are granted.

The process of transferring the title in a property is governed by the Land Title Act, 1994 (QLD), which prescribes the process for registering titles. All that is required for a legally binding contract for a sale to exist is a written note signed by the parties identifying the property and an agreed price (Property Law Act, 1974, s59). The Property Law Act, 1974 governs the process for the conveyance of title from the vendor to the purchaser. While there is no legal requirement for the parties to use a solicitor or registered conveyancer, in practice it is very unusual for each party not to have their own independent legal advisor. There is only a minimum requirement that the sale agreement be made in writing, but it is customary for a detailed sales contract to be entered into by the parties. There are standard form contracts produced by the Real Estate Institute in each jurisdiction which are generally used by real estate agents when negotiating a sale.

Australia was one of the first countries in the world to adopt the Torrens system of title registration. The system was first introduced in South Australia in 1858 and replaced the need for title deeds to be held by a vendor and produced as evidence of title when selling property. Torrens title system comprises a government-held register of every separate piece of land with a volume and folio number against which the owners' details are registered, together with any easements, rights, mortgages, or leases. The registered owner cannot have their title challenged or overturned as it is considered to be an 'indefeasibility title'.

The process of conveying a title from one party to another is governed by the Property Law Act, 1974 and by the Land Title Act, 1994, which require the title transfer to be recorded on the land registry. This process is completed by the legal representatives of both parties attending the office of the Land Title Registry to complete the transfer of title. This process is commonly referred to as the Date of Completion, when all rights in the property are transferred to the new owner and any new mortgagees rights are entered against the title.

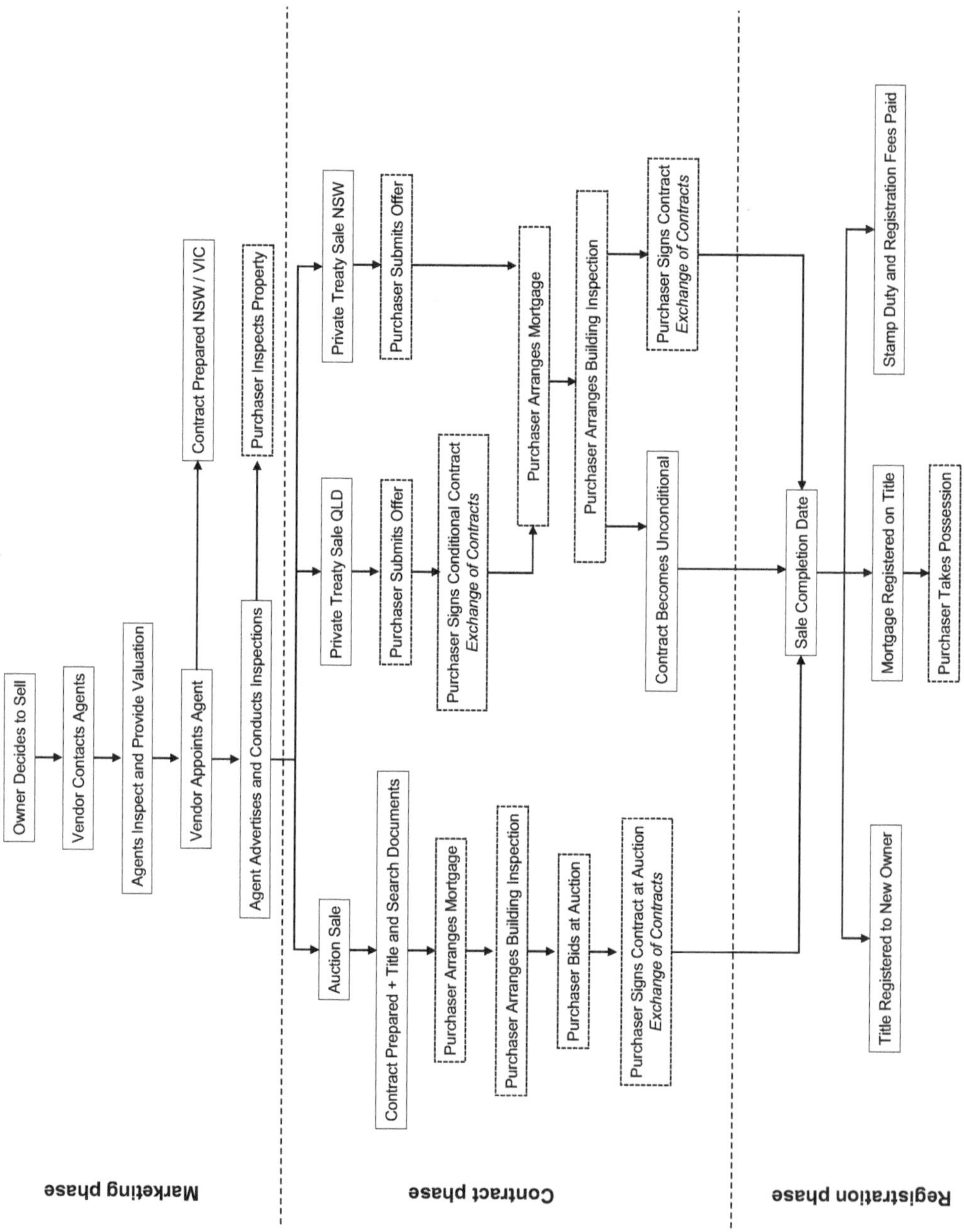

Figure 21.3 The conveyance process in Australia

Source: Author

3. The conveyance process

The process of selling a freehold property in Australia is highly regulated with slightly different practices followed in each state and territory. There is, however, a marked difference in the sales process between states and territories as a result of the differing legislation. Figure 21.3 shows the freehold sales process.

Each jurisdiction in Australia regulates the real estate agency industry requiring that each estate agent's office principal and each estate agent within that office is registered (Property Agents and Motor Dealers Act, 2000, QLD). The objective of this regulation of the profession is to ensure that consumers are protected from undesirable and unethical practices. Registration requires agents to complete education and experience requirements. The education of agents is primarily administered by the Real Estate Institute in each state.

Marketing phase

While there is no obligation on a vendor to engage an agent to sell their property, the majority of residential property transactions involve a vendors agent. Increasingly purchasers are also engaging a buyer's agent to search out potential properties on the market and to negotiate a price or to bid at an auction.

The majority of agency practices belong to a national organization with most offices being owned and operated as a franchise from the parent organization. This franchise arrangement gives the agency a much larger level of recognition by both vendors and purchasers and a degree of certainty in the level of service offered. Vendors select an agent based on their perceived efficiency on achieving a sale, together with the level of fee charged. Agent fees range from 1% to 5% of the sale price depending on the location and type of property. In addition, most agents charge for media advertising and for placing the property on an online portal. Many agents maintain their own online property list, but the online market is dominated by two major players that have developed out of the newspaper industry: domain.com.au and realestate.com.au. It would be unusual for any property for sale not to be listed on one, or both, of these portals.

When an estate agent is instructed to act for a vendor they are required to act in the best interest of the vendor and do not owe any duty to a potential purchaser. In reality, however, the agent attempts to steer a path between the parties to achieve an amicable outcome while obtaining the best price for the vendor. The agent's commission on sale of the property is paid by the vendor, but in the cases where a buyer's agent has been instructed by a purchaser then each party would pay their own agent's fees.

Agents usually seek an exclusive sales period, typically of three months, though some vendors opt for a multiple agency agreement. The advantage of a sole agency period is that the agent is more highly motivated to achieve a sale within that period and is, therefore, guaranteed a fee for the work carried out. Prior to engaging an agent vendors normally seek advice from several agents on the likely sales price for the property and the likely time that the property will remain on the market. Agents are required to act ethically and not mislead vendors by over stating their valuation in order to achieve an agency instruction.

The actual sales process differs between jurisdictions. In New South Wales (NSW) and Victoria there is a much higher reliance on sales by auction than in other states, where sale by private treaty is the norm.

Sale by auction requires the agent to undertake an advertising campaign and conduct viewings of the property. Potential purchasers must be prepared to bid at the auction, so they must have made all necessary enquiries regarding the property, including any building surveys, title searches, and obtaining any mortgage finance approvals. The agent will engage a licensed auctioneer to conduct the sale. The auctioneer may charge a fee of around € 640 or this may be paid from the agent's commission on the sale. At the auction the vendor can set a reserve price prior to the auction. The auctioneer is prohibited by law from taking fictitious bids in order to enliven the action process. If the property fails to reach the reserve price it will be 'passed in' and the auctioneer or agent may continue to negotiate with interested parties to reach the reserve price. If the property sells at auction the winning bid is binding on the purchaser and they will be required to sign the sales contract immediately following the fall of the hammer. The sales contract will set out the deposit required to be paid at this time, typically 10% of the purchase price, and the date for completion of the sale, which is usually 28 days from the date of contract.

Sale by auction is increasing in popularity as it is seen as an efficient and open sales process. In a tight market a good auctioneer can often achieve a higher sale price as a result of competing parties bidding against each other. The rate of auction success, as measured by the number of sales achieved under the hammer, is often used as an indicator of the activity in the market and potential for rising prices. In early 2021 the property market has seen some significant price increases with auction clearance rates reaching 85% in Sydney.

A sale by private treaty is a less certain process from a vendor's perspective but also has less risk from a purchaser's perspective as it does not require significant expenditure on building inspections prior to agreeing a purchase. The agent will undertake marketing of the property and arrange regular 'open houses' at which the agent will show potential buyers through the property as well as arranging private viewings. Most potential purchasers who require finance to purchase a property will seek pre-approval from their lender before looking at properties. This process ensures the purchaser is able to borrow sufficient funds based on their income level.

Once a potential purchaser is identified and is willing to make an offer, the agent is required by law to pass all offers onto the vendor for consideration. If the offer is at, or close to, the asking price, agents will often ask purchasers to sign a purchase contract which is presented to the vendor. If the vendor agrees to the offer, then they sign the contract of sale which then becomes binding on both parties. If the vendor makes a counter-offer, by altering the signed contract, this is then presented back to the purchaser to agree the change in order for the contract to become binding.

Contract phase

Private treaty contracts often take the form of a *pro forma* contract prepared by the Real Estate Institute. The contract contains typical terms and conditions which include a period within which the purchaser may obtain a satisfactory building condition inspection, a pest inspection, and finance. In respect of these clauses the purchaser must meet the time constraints imposed, typically 14 days. Reasonable steps to gain any finance have to be made and these clauses cannot be used to rescind the contract without reasonable grounds. Once these conditions are met, or the period outlined under the contract expires, then the contract is said to be 'unconditional' and will proceed to completion on the date specified in the contract. The previous contract process differs in NSW, in that, before marketing a property the vendor is required to prepare a sales contract including title documents, a drainage diagram, and current planning zoning. Contracts are not exchanged until the purchaser has arranged finance and carried out any building inspections. On exchange the contract is binding on both parties, subject to a five-day cooling off period for the purchaser (Conveyancing Act, 1919).

Registration phase

The registration of title in Australia is undertaken by the Land Registry in each state and territory and the title is held against a registered plan of the land. Transfer of title occurs at the Registry when representatives of both parties attend to conduct the completion of the sale when the purchase price is transferred to the vendor and the purchaser registered as the new owner. Stamp duty and other registration fees must be paid prior to registration. The estate agent and conveyancer deduct their fees prior to making the final remittance to the vendor.

4. Summarizing the conveyance process and its transaction costs

The process to market and transfer of residential property in Australia is relatively simple. While each state and territory have its own legislation and tax regimes, they are all similar in the processes involved. In all cases a written contract is required and the title is transferred via a government-held land registry. Table 21.1 outlines the fees and charges applicable to the sale of residential property in each of the larger states.

The rate of stamp duty on the sale of land varies between states and territories; some also offer discounted stamp duty to first-time home buyers within certain thresholds. The rate of stamp duty is also different for owner occupiers to that for investors. A significant additional stamp duty is also charged to foreign investors who are not permanent residents of Australia. In addition, residential property that is not owner occupied can be subject to an annual land tax, the rate of which varies from state to state. In Queensland the tax applies to the value of land only and not the buildings on it. For an individual investor the rate varies from 1% to 2.25% based on the value of the land over € 384,000.

5. Residential rental property

In Australia rental occupancy represents about 32% of homes. Most rental property freeholds are held by individual investors with no large corporate investors active in the market. The letting process is regulated with legislation providing considerable protection to tenants from unlawful eviction and actions by the landlord. In Queensland the Residential Tenancies and Rooming Accommodation Act, 2008 provides these protections and establishes a disputes resolution process and a rental bond regime to hold any security deposits.

Residential lettings are usually arranged on behalf of landlords by managing agents. The agreement must contain the required statutory clauses or in the case of NSW a statutory form of agreement is required (Residential Tenancy Act, 2019). The tenancy agreement is typically for a period of one year but may be for any period, although a lease over 3 years is required to be registered with the Land Registry. At the end of the term the tenant may hold over from month to month or seek to enter a new tenancy agreement. The statutes require a landlord to give between 30 and 90 days' notice to terminate, to a tenant who is holding over, depending on the reasons for requiring possession. Notice of 30 days is required at the end of a tenancy agreement if the tenant is not holding over. Landlords require a court order to regain possession if a tenant does not leave the property after the notice period.

Table 21.1 Fees and taxes during the conveyance process in Australia

Phase	*Fees and taxes*	*Seller*	*Queensland*	*NSW*	*Victoria*
			Buyer	*Buyer*	*Buyer*
Marketing	Real estate agent's fees	1.5–5%	-	-	-
	Marketing fees	€ 320–1,280	-	-	-
	Auction fees	€ 640	-	-	-
Contract	Legal fees	€ 960–3200	-	-	-
	Bank mortgage application	€ 0–384	-	-	-
Registration	Title registration	-	€ 125 + € 24 / € 6,400 above € 115,200 purchase price	€ 93	€ 62 + € 1.5 for every whole € 640 of purchase price to a maximum of € 2,310
	Stamp duty (owner occupier)	-	< € 3,200 = 0 € 3,200–48,000 = 1.5% + € 48,000–345,600 = 3.5% + € 345,600–640,000 = 4.5% + > € 640,000 = 5.75%	€ 0–8,960 1.25% + € 8,960–19,840 = 1.5% + € 19,840–53,120 = 1.75% + € 53,120–198,400 = 3.5% + € 198,400–661,120 = 4.5% + > € 661,120 = 5.5%	€ 0–16,000 = 1.4% + € 16,000–83,200 = 2.4% + € 83,200–281,600 = 5% + € 281,600–352,000 = 6% + > € 352,000 = 6%
	Stamp duty (non-owner occupier)	-	-	-	€ 0–16,000 = 1.4% + € 16,000–83,200 = 2.4% + € 83,200–614,400 = 6%% + > € 614,400 = 5.5%
	Stamp duty (foreign investor)	-	Additional 7%	Additional 8%	Additional 8%
Post-registration	Sales tax	-	-	-	-
Total (EUR for the model case price)		**€ 3,420–10,504**	**€ 2,665**	**€ 2,592**	**€ 2,743**
Total (% share to the model case price)		**3.4–10.5%**	**2.7%**	**2.6%**	**2.7%**
Excluded costs from the model case		-	Stamp duty is considered as owner occupier Not considering stamp duty 7% or 8% for foreign investor		

Note. Model case price: € 100,000

Source: Author

Property managers will typically charge a letting fee equivalent to one week's or two weeks' rent plus any advertising costs. Tenants are required to lodge a security deposit of one month's rental which is held by the government Rental Bond Authority. Ongoing management fees in Australia range from 7% to 10% of the rent collected depending on the size and location of the property.

Note

1 1 Australian Dollar = 0.64 Euro (as of 1 October 2021).

References

Australian Bureau of Statistics (ABS). (2019). *Housing occupancy and costs*. Canberra: Author.

Australian Bureau of Statistics (ABS). (2020). *Residential property price indexes: Eight capital cities*. Canberra: Author.

Australian Bureau of Statistics (ABS). (2021). *Online stat database*. Retrieved March 6, 2021, from http://stat.data.abs.gov.au/Index.aspx?DataSetCode=ABS_CENSUS2011_B32

Conveyancing Act 1919 No. 6 NSW.

Land Act (1994) QLD.

Land Title Act 1994 (QLD).

Property Agents and Motor Dealers Act 2000 QLD.

Property Law Act 1974 QLD.

Residential Tenancies and Rooming Accommodation Act 2008 QLD.

Residential Tenancy Act 2019 NSW.

22
CANADA

Yaëll Emerich

1. The Canadian and Quebec housing markets

Understanding Canada's[1] housing market requires study of its three levels of government – federal, provincial, and municipal – and its two legal traditions of common law and civil law, as they all individually impact its functioning. By virtue of these political and legal interactions, it is necessary to consider not only the current policies, laws, and regulations that shape Canada's housing market but also the historical, political, social, and legal forces that have influenced its development (Hulchanski, 2006).

The two main forms of tenure in Canada are homeownership and rental housing, which are both regulated by separate legal regimes. Home ownership is more prevalent than rental housing. Two-thirds of all households own their home, while about one-third of all renters will eventually purchase a house. Against a backdrop of low non-market housing supply, only 5% of Canada's homeowners live in non-market social housing (public housing, non-profit housing such as cooperatives). The remaining 95% of homeowners purchase their housing on the private market (Balakrishnan, 2021).

Nonetheless, globally, Canada fits the liberal-welfare regime type with low social spending, which means that housing is provided almost exclusively through the market and social housing is only residual (Chisholm & Hulchanski, 2019). It has been described as an 'intermediate between those of Northwestern Europe and the more residual US model' (Suttor, 2016, p. 4).

The rate of homeownership, which currently sits at 68% nationwide, is relatively stable in Canada. Direct and indirect government subsidies ensure that homeownership rates have never fallen below 60%. Due to these high rates of ownership, rental housing is often seen as a residual part of Canada's housing system. However, Canada's main urban settings tell a different story: homeownership levels in 2016 dropped to 58.7% in Toronto, 55.8% in Vancouver, and 50.5% in Montreal (The spatial division in Vancouver and Toronto is presented in Figure 22.1). Low-income earners, singles, and minorities are overrepresented in the rental market.

For over a decade, rising Canadian real estate prices have elicited serious public concern. For example, in 2020, a total of 551,392 houses were sold in Canada, which marks an increase of 12.6% over 2019 sales as illustrated in Figure 22.2.

DOI: 10.1201/9781003130383-26

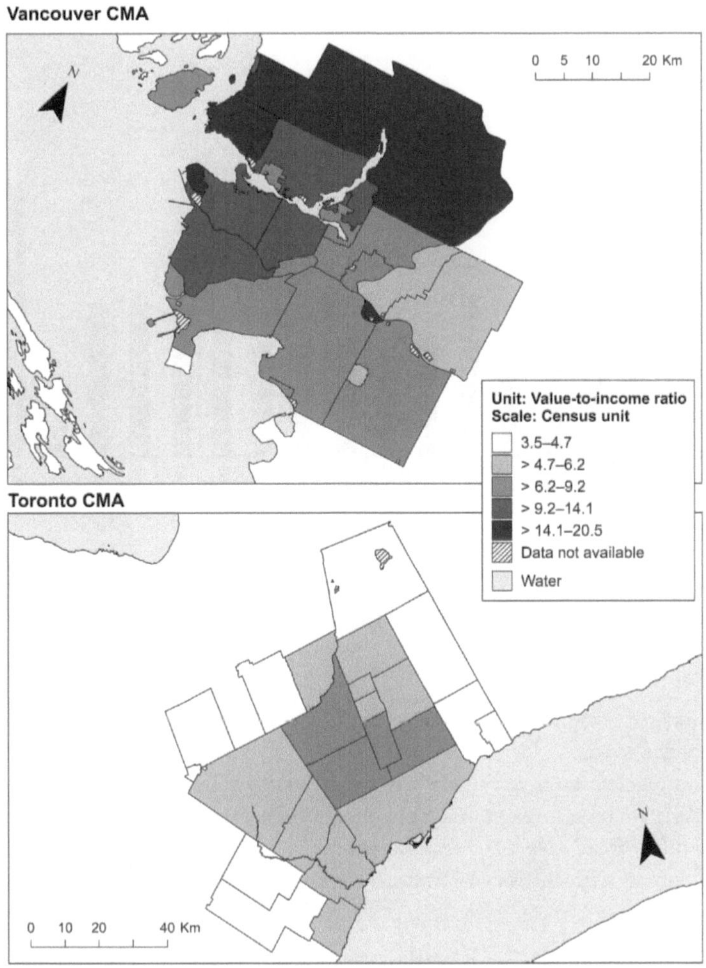

Figure 22.1 Value-to-income ratio ranges in the Vancouver and Toronto census metropolitan areas by census subdivision, 2018

Note. The Metro Vancouver A census subdivision (CSD) includes several unincorporated areas within the Vancouver census metropolitan area (CMA), including the University Endowment Lands and the University of British Columbia, the area north of West Vancouver and North Vancouver, and several islands.

Source: Statistics Canada, Canadian Housing Statics Program (CHSP) 2018, Table 46-10-0049-01

The average selling price was € 419,023 – a 17.1% jump from the previous year. However, excluding Canada's priciest markets from this analysis, as in, Toronto and Vancouver, lowers the median selling price by € 89,700.

The Canada Mortgage and Housing Corporation (CMHC) – whose mandate includes the promotion of housing affordability and choice and the protection of the availability of adequate funding for housing at low cost – was created in 1946.

In an effort to slow down the market, the federal government has introduced some constraints on mortgage loans. For example, there is a minimum 20% down payment to buy. However, buyers have the option of making a minimum down payment of 5%, which is then

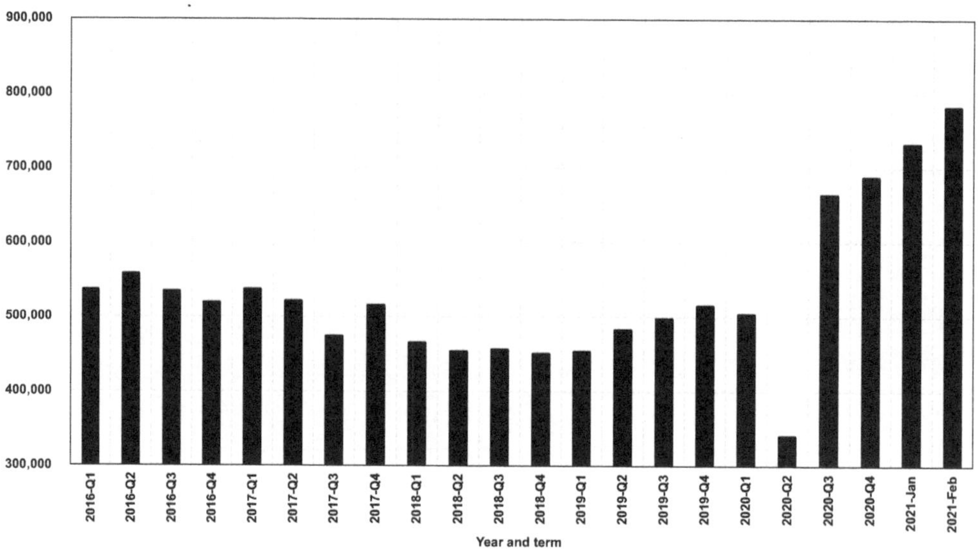

Figure 22.2 Residential sales activity – seasonally adjusted data at annualized rates

Source: The Canadian Real Estate Association, National Statistics (2021 March)

assorted by mandatory insurance with the CMHC, used to protect banks against the risk of non-payment of the loan.

The provinces regulate their rental markets using different approaches. In Quebec, the *Tribunal administrative du logement* regulates rent in the province. When a tenant refuses a rent increase, the landlord can turn to the *Tribunal administrative du logement*, who will decide on the proposed rent increase. This approach contrasts with British Columbia, where maximum annual rent increases are legislated.

2. Regulatory framework

The sale and development of real estate is essentially a matter of provincial jurisdiction. However, the conveyancing process is different in Quebec (which is a civilian jurisdiction) and in the rest of Canada (common law jurisdictions).

Canada's conveyancing process requires the intervention of a professional (a lawyer or a notary in Quebec) and is subject to an obligatory registration process both in Quebec and in the rest of Canada. The effect of the registration differs depending on the jurisdiction and can be either declaratory or constitutive of title.

As for renting, there are two types of rental agreements in Canada: written lease or verbal rental agreement. A written lease is always preferable, as it clearly specifies the agreement between a landlord and a tenant. Both written and oral rental agreements define the legal rights and responsibilities for landlord and tenant. The contract outlines the rent, the responsibilities of the parties, what is included or not, and the conditions for terminating the lease.

3. The conveyance process

The conveyance process in Quebec differs from the rest of Canada and is illustrated in Figure 22.3.

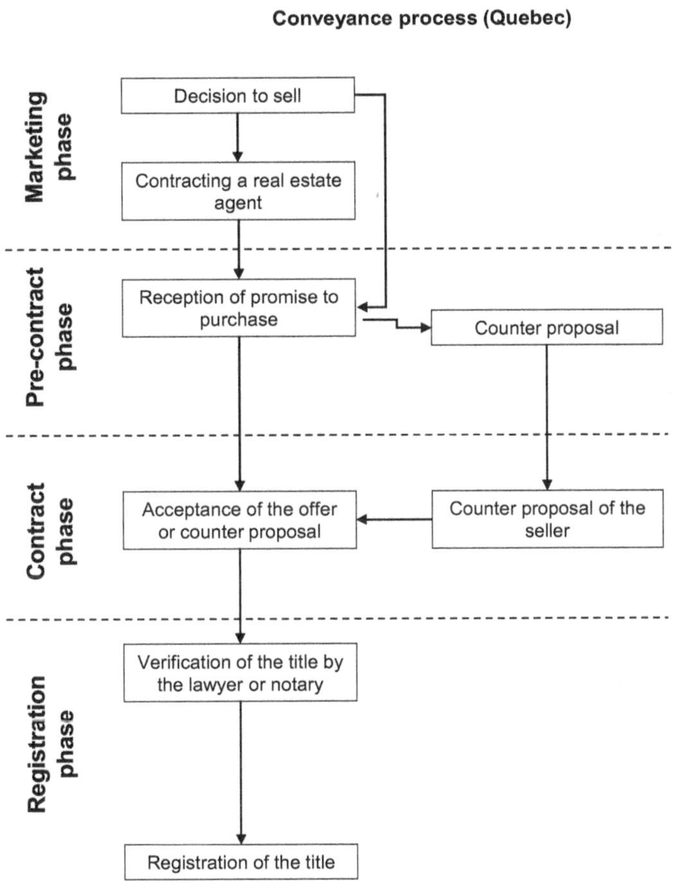

Figure 22.3 The conveyance process in Quebec

Source: Author

3.1 Quebec

Marketing phase

The first phase of the conveyance process often involves an estate agent. As an intermediary, the agent's fee is paid by the seller, usually by means of a percentage of the sale price. An increasing number of sales are transacted directly between sellers and buyers – for example, in Quebec, it is through a website called Proprio Direct, where vendors and buyers can find models of contracts and other relevant documentation.

Pre-contract phase

In Quebec, there is often a pre-contract phase where the buyer offers a price and proviso the offer to some conditions (usually the obtention of a mortgage from a bank institution and the inspection of the immovable).

The offer to buy must contain the following elements: the name of the buyer, the name of the vendor, the address of the property, the price of the sale, the amount of the account, any elements included in the sale, the date of possession (usually 30 to 60 days after the signature of the offer), the demand for a certificate of localization to date, the date of expiration of the offer, and any other conditions (e.g. upon a satisfactory inspection of the property or the approval of a loan by a financial institution).

The pre-contract phase is crucial, as it is binding. According to Art. 1396 of the Civil Code of Québec (C.c.Q.):

> An offer to contract made to a determinate person constitutes a promise to enter into the proposed contract from the moment that the offeree clearly indicates to the offeror that he intends to consider the offer and reply to it within a reasonable time or within the time stetted therein.
>
> *A mere promise is not equivalent to the proposed contract, however, where the beneficiary of the promise accepts the promise or takes up his option, both he and the promisor are bound to enter into the contract*, unless the beneficiary decides to enter into the contract immediately. (Italics added)

Also, in the case of a promise of sale, the law gives a real remedy called an 'action in execution of title' to the beneficiary: 'Failure by the promisor, whether seller or buyer, to execute title entitles the beneficiary of the promise to obtain a judgment in lieu thereof' (Art. 1712 C.c.Q.).

Contract phase

The buyer's agent will typically prepare the offer and submit it to the seller's agent, although some buyers will negotiate the offer directly with the seller. When acting for buyers, notaries or lawyers will examine the title and verify some issues (e.g. realty taxes or environmental matters) and prepare closing documents. According to the Civil Code of Québec, the vendor is bound to deliver the property and to warrant the ownership and quality of the title (Art. 1716 C.c.Q).

Registration phase

The intervention of a professional (lawyer or notary in Quebec) is necessary to enter into the transaction. The lawyer (or notary) verifies the title (in particular, that the property is free from any mortgage or charge) and proceeds to the registration of the title. While the rights are created by the contract, the registration gives effect to the contract vis-à-vis third parties.

3.2 Canada

Other provinces in Canada have enacted statutes that govern the acquisition, ownership, use, and development of real estate. In each province except Quebec, the real estate laws are based on common law and are illustrated in Figure 22.4. An estate refers to an interest in real property. It can be a freehold estate (e.g. an estate of indefinite duration) or a leasehold estate (e.g. an estate of definite duration). The fee is simple, as a freehold estate in these terms is equivalent to ownership.

Marketing phase

As with Quebec, real estate is often marketed through a real estate broker.

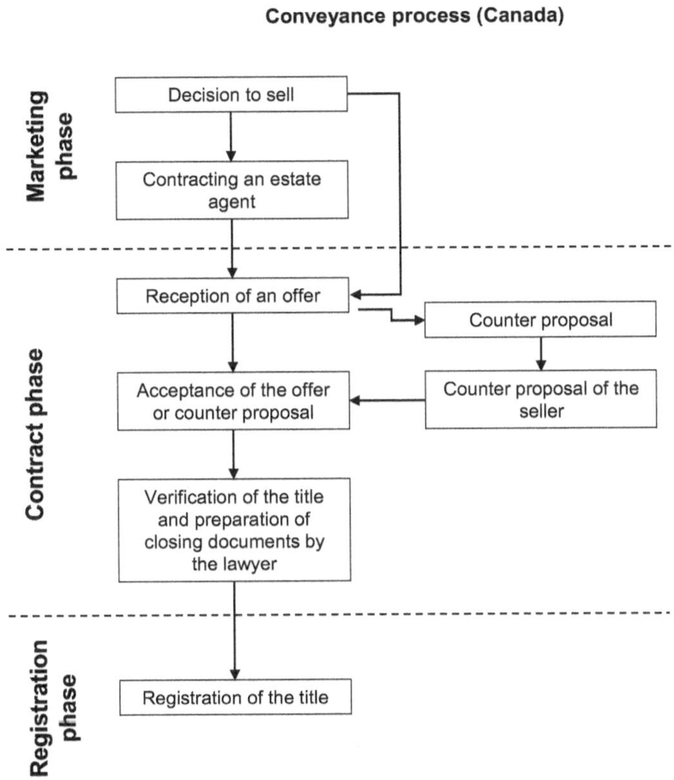

Figure 22.4 The conveyance process in Canada

Source: Author

Contract phase

The buyer's agent will typically prepare the offer and submit it to the seller's agent, although some buyers will negotiate the offer directly with the seller. When acting for buyers, lawyers will examine the title, verify some issues (e.g. realty taxes or environmental matters), and prepare closing documents.

Registration and transfer of title

All privately owned property in Canada is registered. Each province administers its system of registration of interests in land. Canada recognizes the registry system and the land titles (or Torrens) system. Each province uses either one or both systems. While the traditional registry system provides only for the public recording of title (similar to Quebec), in the land title system, the title is guaranteed by the government.

4. Summarizing the conveyance process and its transaction costs

To summarize, in the Canadian conveyance process for real estate as outlined in Table 22.1, ownership is conveyed with a contract of sale. The registration is mandatory and can be either

Table 22.1 Fees and taxes during the conveyance process in Canada

Phase	*Fees and taxes*	*Seller*	*Buyer*
Marketing	Real estate agent's fees	2–6% of purchase price	-
Contract	Inspection fees	-	€ 345
	Home appraisal cost	-	€ 242
	Mortgage insurance (down payment of less than 20% of property's asking price)	-	0.5–2.9% of mortgage loan
Registration	Notary fees	-	€ 828
Post-registration	Real property transfer tax	-	Higher amount between purchase price and municipal evaluation
	Federal taxes (newly constructed home)	-	5% of purchase price
	Provincial taxes (newly constructed home)		9.975% of purchase price
	Adjustment and designation of school and municipal taxes	-	Dependent on date of sale (days of the year remaining at the time of transaction)
	Sales tax	-	-
Total (EUR for the model case price)		**€ 2,000–6,000**	**€ 16,390**
Total (% share to the model case price)		**2–6%**	**16.4%**
Excluded costs from the model case		-	0.5–2.9% mortgage insurance fee. Real property transfer tax, school and municipal taxes.

Note. Model case price: € 100,000

Source: Author

declaratory (in Quebec and in the common law provinces that use the registry system) or constitutive of title (in common law provinces that use the land titles system). While a real estate broker is not mandatory in the marketing phase, the intervention of a professional (lawyer or notary) is mandatory.

Fees and charges apply when purchasing property. For example, if mortgage insurance is required because of down payment of less than 20% of the property's asking price, buyers can expect an additional cost between 0.5% and 2.90% of the total mortgage loan. In Quebec, a real property transfer tax also applies based on the higher amount between the price paid at purchase or the municipal evaluation. In the case of new homes, a federal tax (goods and services tax) and provincial tax (Quebec sales tax) must be paid, amounting to approximately 15% of the purchase price. If the property is inspected, an additional € 345 can be expected. Finally, the buyer will also pay notary fees of around € 828.

Note

1 1 Canadian Dollar = 0.69 Euro (as of 1 October 2021).

References

Balakrishnan, A. (2021, janvier 15). Les ventes d'habitation à un record en 2020. *La Presse Canadienne*.

Chisholm, S., & Hulchanski, D. (2019). Canada's housing story. In D. Maclennan et al. (Eds.), *Australia, Britain, Canada: Shaping futures changing the housing story final report* (p. 21). Glasgow: Policy Scotland.

Hulchanski, D. (2006). What factors shape Canadian housing policy? The intergovernmental role in Canada's housing system. In R. Young & C. Leuprecht (Eds.), *Canada: The state of the federation 2004 municipal-federal-provincial relations in Canada* (p. 221). Montreal: McGill-Queen's University Press.

Suttor, G. F. L. (2016). *Still renovating: A history of Canadian social housing policy*. Montreal: McGill-Queen's University Press.

23
ENGLAND AND WALES

Jill St George and Julian Sidoli

1. The English and Welsh housing market

Despite being two separate nations within the United Kingdom[1] the law and regulation of the housing market in England and Wales has historically been identical. Even now divergences are minor and none effect the actual conveyancing process itself. The residential housing market in England and Wales consists of home ownership, privately rented and publicly or socially rented property (see Table 23.1). Home ownership can be subdivided into those who own the property outright and those who are buying the property with the help of a mortgage. The market itself is buoyant with house prices rising in all regions. There remains a shortage of supply of property of all types, and this continues to push both house price inflation and increases in rent in the private rented sector. When it comes to the spatial distribution a clear north-south division can be observed. Figure 23.1 indicate higher sales price of houses in the south.

This chapter aims to provide an overview of how property transactions are undertaken in England and Wales, what issues may arise, and what remedies are available. It may come as a surprise that prior to the formality of exchange of contract there is no remedy for a buyer or seller changing their mind or for the seller to subsequently accept a higher offer from a third party, known colloquially (albeit now sanctioned by august dictionaries) as 'gazumping'.

A brief historical overview

Notionally, all land is owned by the Crown although this does not mean even in theory that Queen Elizabeth II owns it. It is an odd remnant of feudalism harking back to the Norman conquest of 1066 where William I distributed estates in land to his nobles who in turn taxed and administered the land and the people thereon. The system in England and Wales, in stark contrast to many civilian jurisdictions, has no real use of the notary. Indeed, whilst 'notary public' exists it is something of an esoteric category rarely touched upon by the average purchaser. Civilian property lawyers have sometimes seen – fairly or not – a randomness and lack of certainty with regards to the registration of ownership and title. The traditional approach was based on showing 'good title' – in effect a bundle of documents showing that the current owner had bought it from the previous owner who had in turn bought it from the previous owner and so

DOI: 10.1201/9781003130383-27

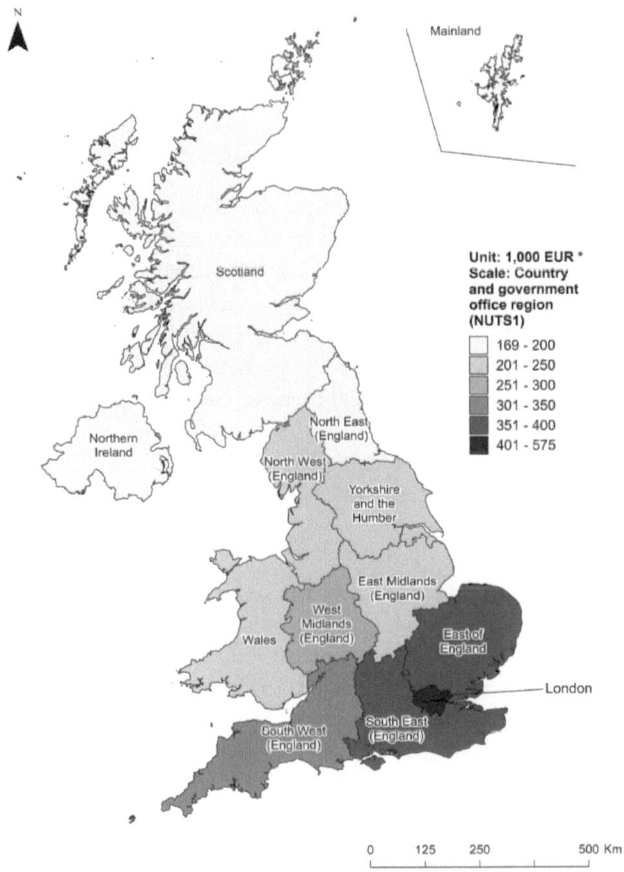

Figure 23.1 Average house price by country and government office region, 2021 April

Note. *1 Euro = 0.8475 Great Britain Pound (as of 2021 September)

Source: HM Land Registry, National Statistics UK House Price Index summary: April 2021 Published 16 June 2021

Table 23.1 Households by tenure FYE 2007–FYE 2017

FYE	*Owned outright*	*Buying with a mortgage*	*Private rented*	*Social rented*
2007	31	37	13	18
2008	32	35	14	18
2009	32	35	15	18
2010	32	34	16	18
2011	33	32	17	18
2012	32	32	17	18
2013	33	31	18	18
2014	33	31	19	18
2015	33	30	19	18
2016	34	29	20	18
2017	34	28	20	17

Source: Family Resources Survey 07–17

on. 'Good title' meant that there were no gaps in the chain so that no one could claim 'better title'. This position was amended by the Land Registration Act 2002 which established that all property when sold had to be centrally registered.

2. The legal framework

The legal transfer of home ownership from buyer to seller in England and Wales is facilitated through private contract. This is supplemented by several standard industry protocols, many of which are recommended but not mandated by law. Where the law does intervene, it is often only with a broad framework, leaving the parties free to decide how to conduct their own transactions, albeit with a standard form contract (*Standard Conditions of Sale*, 5th edition) mandated for an effective disposition. Common parties to the transactions (although only the buyer and seller are mandated by law) include estate agents, surveyors, mortgage lenders, and solicitors or licensed conveyancers.

Summarized, the process of conveyancing in England and Wales can be broken down into three stages. It is only at the third stage that the contract becomes binding and formal legal redress becomes available.

1 Properties are marketed for sellers by estate agents and the seller accepts an offer (subject to contract).
2 Once the offer is accepted, both parties instruct either a solicitor or a licensed conveyancer to conduct the conveyance (Conveyancing Association). The buyer is responsible for gathering information on the property and initiating enquires
3 Contracts are signed and exchanged, at which point the sale becomes binding on both parties. A completion date is entered into the contract; and change of ownership is registered with the Land Registry and monies due paid. In reality, exchange of contracts and completion usually happen at the same time.

3. The conveyance process

Marketing phase

In England and Wales, although not mandatory, most properties are marketed for sale by estate agents either online or offline. The property and accompanying price are advertised by the estate agent as an invitation to tender (a formal invitation to make an offer to buy) as opposed to a fixed price for sale (*Pratt Contractors Ltd v Transit New Zealand* [2005] 2 NZLR 433). Estate agents are not required by law to be licensed or qualified, and they are primarily regulated by four pieces of legislation – the Estate Agents Act, 1979, Consumer Protection from Unfair Trading Regulations, 2008, the Consumers, Estate Agents and Redress Act, 2007, and the Money Laundering Regulations, 2007. Since 2014, the National Trading Standards Estate Agency Team has been responsible for regulating estate agents. Estate agents must also describe the property fairly (Property Misdescriptions Act), but most estate agents will include a disclaimer on their documentation which places the responsibility back onto the buyer.

A buyer will contract with an estate agent for the sale of their property, with contracts typically taking one of two forms. The first is to have a sole agency agreement, wherein a single estate agent is appointed to market a property for sale. Ordinarily with such an agreement, unless expressly included to the contrary, if the sellers find a buyer themselves, they are not liable for the estate agent fees. The second is a sole selling rights agency, wherein a single estate agent is appointed to market a property for sale and the estate agent is entitled to a fee regardless of who finds a buyer. Brazg (2021a) notes:

- The current average high street estate agency fee is 1.18% (+ Value Added Tax – currently 20%)
- High street estate agents remain the most popular form of estate agents, with only 3–4% of all house sellers currently using online estate agents

The online sector holds only 5% of the market (Brazg, 2021b). Property portals are very popular, specifically Rightmove and Zoopla, however, these are not estate agents, and instead act as an advertising aggregator, listing properties from a variety of estate agents. Rightmove reports receiving 127.5 million visits per month, whilst Zoopla reports receiving 50 million (Home Owners Alliance, 2020).

The role of estate agents is however more limited in England and Wales than in many jurisdictions. Their role is essentially to present to buyers available property and negotiate the terms of the sale. A memorandum of sale is produced which has no legal effect but from that point it is the conveyancer who is responsible for the contract, enquiries, transfer, and registration.

Pre-contract phase

Much of the current structure governing conveyancing in England and Wales is attributable to two statutes – the Law of Property Act, 1925 and the Land Registration Act, 1925. The Law of Property Act, 1925 remains in force today, while the Land Registration Act, 1925 has been repealed and replaced with the Land Registration Act, 2002. The preliminary stages to the conveyancing process are not binding; parties are bound and subject to penalties only on withdrawal post the exchange of contracts.

Both solicitors and licensed conveyancers are qualified and regulated to act for either a buyer or seller in a property transaction. Under industry regulation, a licensed conveyancer can act for both parties, whereas a solicitor can act only for one side. The term 'conveyancer' is used colloquially for both. Once an offer to buy has been accepted by the seller, both parties appoint a conveyancer to act on their behalf during the conveyancing process. The law does not mandate the use a legal professional for a conveyance, however most mortgage lenders do.

The volume of work required of a conveyancer when acting for the seller is significantly less than when acting for the buyer owing to the application of the maxim *caveat emptor* (discussed in further detail ahead). In brief, a seller's conveyancer is responsible for confirming the identity of the buyer, verifying that they have the funds for purchase (from information provided by the buyers conveyancer), obtaining copies of the title deeds from the Land Registry if applicable, responding to any questions about the property from the buyer's conveyancer, preparing the contract for sale, exchanging contracts with the buyers conveyancer, transferring funds (most commonly to the lender and/or estate agent) and preparing and sending the transfer deeds to the buyers conveyancer to update the Land Registry. Sellers are required to complete a Fixtures and Fitting Form (TA10) which details any fixtures, fittings, and contents included with the sale, and a Property Information Form (TA6). The TA6 requests the seller provide information on, amongst others, boundaries, details of any neighbour disputes, any building works and associated approvals, planning applications affecting the property, any electrical or gas work which has been carried out, and details on those occupying the property and the date upon which they will vacate.

The buyer's conveyancer is responsible for verifying the identity of the buyer and ensuring funds are in place, most commonly a mortgage offer and checking the lenders associated terms. The buyer's conveyancer must be registered on an approved panel provided by the lender. The title deeds are checked when received from the seller's conveyancer to ensure that they own the property in question and have the right to sell it, as is a draft copy of the contract. Following receipt of the draft contract, the buyer's conveyancer will raise any queries which they may have with the seller's conveyancer.

The common law maxim of *caveat emptor* applies in property transactions – 'let the buyer beware' – which places burden on the buyer to investigate and ask, as opposed to on the seller to produce and disclose, anything over and above that included in the Property Information Form (TA6). It is therefore prudent, although not mandatory, for the buyer to conduct as much due diligence as possible on the prospective property. In addition to the enquiries discussed earlier, this traditionally includes surveys (valuation and building – often requested by lenders) and searches. Searches are enquiries made on the buyer's behalf by their conveyancer to various registries who hold information about the property, the surrounding land/area, and any other factors which may directly or indirectly affect it, in particular for a lender, its marketability. Searches are not generally mandated by law (see New Build requirements of SuDS and LLC1 as an exception) but are required by most lenders where a property is being mortgaged. The most conducted searches are as follows:

- Official or Regulated Local Authority (with focus on planning, building control, highways, and pollution issues)
- Environmental (with focus on flooding, landslides, subsidence)
- Water and Drainage (with focus on ownership and maintenance of sewers and drains, how water supply is provided, etc.)

Depending on the property and its location, searches can include Chanel Repair Liability (whether the property is liable for repairs of the parish church), Canal and River (in particular where a property backs on to a river or stream), and Coal Mining and Brine (where a property is located in an area where historic mining took place which may affect the land which the property sits on) amongst others. Once all enquires have been submitted and responded to and searches investigated, the buyer's conveyancer provides a single, comprehensive report on all matters concerning the conveyance. This is known as the RoT, or Report on Title.

Contract phase

Contracts for the sale of land are private contracts governed by the Law of Property (Miscellaneous Provisions) Act 1989. Under the Act, contracts for the purchase of property must be made in writing (S2 (1)) and must be signed on (or on behalf of) each party to the contract (S2 (3)). It is common, although not mandated by law, for the buyer to pay a 10–20% deposit on exchange of contracts, providing the buyer with an equitable interest in the property but not a legal interest, which arises on completion and registration of title with the Land Registry. To secure this equitable interest, a notice of exchanged contracts can be provided to the Land Registry (*Williams v Broadoak Private Finance Ltd* [2018] EWHC 1107 (Ch)). Included within the exchanged contracts will be a date upon which completion takes place. On completion date, the purchase price will be transferred to the seller. Under S4 (1) Land Registration Act, 2002, the transfer of a freehold estate or leasehold in excess of seven years must be registered with the Land Registry. Under S6 (4), this registration of title must occur within two months of the transfer. Where more than one person buys a property, they will be required to specify their equitable ownership – either as joint tenants or tenants in common. The preferred option will be ascertained by the conveyancer prior to exchange and will be included on the Land Registry. Under S36 Law of Property Act, 1925, joint tenants both hold the equity in its entirety, whereas tenants in common own specified equitable shares.

If the buyer is satisfied with the contents of the final report, including responses to all enquiries by the seller and the results of any searches carried out, both the buyer and seller are required to sign a contract which is held by their respective conveyancer. The seller's conveyancer will also hold a signed transfer deed for the Land Registry (TR1 form) which will be provided to the buyer's conveyancer upon exchange. As noted earlier, a completion date will be included in the signed contract, and the buyer's conveyancer would be in receipt of the required funds. Historically, the 'exchange' of contracts would occur in person; however, it is now primarily conducted over the phone, wherein the conveyancer from both sides confirm that they have all necessary to complete the transaction and they 'exchange' contracts. Once contracts have been exchanged, the purchase is legally binding. Where there exists a chain of purchases occurring contemporaneously, the conveyancer at the start of the chain will contact the next conveyancer up to confirm receipt of all necessary and a time will be established for all in the chain to confirm receipt. Confirmation must travel up the chain, and then back down before the time allocated to allow for exchange to take place at the time specified. Once exchange has taken place, the buyer will receive a 'completion statement' from their conveyancer which details which monies have been paid, those due to be paid, and that necessary to complete on the date stated in the contract.

On completion day, the purchase funds are transferred from the buyer's solicitor to the seller's solicitor, at which point the purchase is complete. As, upon exchange on contract the sale is binding, failure to complete on the date agreed is a breach of contract with penalties attached. Under S49 (1) Law of Property Act, 1925, the court has jurisdiction to make an order on claim for compensation which they deem 'just' and both parties are able to rely on remedies under the law of contract, which include rescinding the contract, deposit return/forfeit (depending on which party failed to complete), and possible damages and/or costs.

As noted earlier, freehold properties and leaseholds with a lease in excess of seven years must be registered with HM Land Registry. The buyer's conveyancer completes their corresponding sections of the TR1 form, which transfers the registered title from the seller to the buyer and submits the signed transfer request to the Land Registry. Additional documents to be sent to Land Registry include the following:

- The witnessed transfers signed by the seller and the buyer
- A Revenue Certificate from HMRC
- Evidence of any charges to be registered against the property (commonly the interest of a mortgage lender)
- Evidence of any charges that have been removed

It is the buyer's responsibility to ensure that their ownership is registered and that the fee to do so is paid. Following this registration, a Title Information Document is produced by the

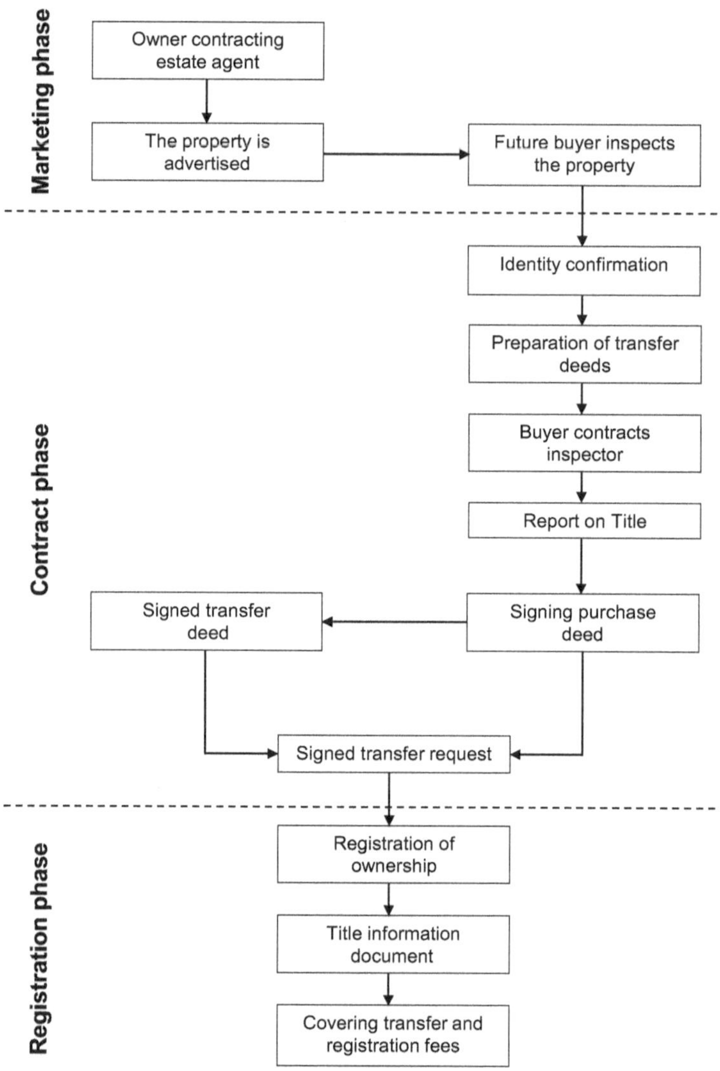

Figure 23.2 The conveyance process in England and Wales

Source: Department for Communities and Local Government

Land Registry which confirms the change to the register. Stamp Duty Land Tax (SDLT) is a tax payable to HM Government by the buyer of a property. Liability for SDLT is divided into bands, with the amount payable dependent on whether the home is a primary residence or second property and the value of the property. A Stamp Duty Land Tax form (SDLT1) must be completed when stamp duty is paid, and it is the responsibility of the buyer to ensure that this has taken place within 30 days of completion. Stamp duty must be paid in full and cannot be paid using a mortgage.

The conveyance process is illustrated in Figure 23.2, and the estimated fees and taxes during the conveyance process are collected in Table 23.2.

Table 23.2 Estimated fees and taxes during the conveyance process in England and Wales

Phase	*Fees and taxes*	*Freehold*		*Leasehold*	
		Seller	*Buyer*	*Seller*	*Buyer*
Marketing	Estate agent fees	0.5–3%	-	0.5–3%	-
	Value added tax	20%	-	20%	-
	Energy performance certificate	€ 41–177	-	€ 41–177	-
Contract	Conveyancing fees	€ 649–1180	€ 1,003–1,770	€ 649–1,180	€ 1,003–1,770
	Value added tax	20%	20%	20%	20%
Registration	Stamp duty land tax	-	Between 0% and 12% depending on the property value	-	Between 0% and 12% depending on the property value
	Registration of ownership	-	Between € 24 and € 1,074 depending on the value of the property, how registration is applied for, and whether it is the first registration	-	Between € 24 and € 1,074 depending on the value of the property, how registration is applied for, and whether it is the first registration
Post-registration	Sales tax	-	-	-	-
Total (EUR for the model case price)		**€ 1,420–5,193**	**€ 1,228–14,278**	**€ 1,420–5,193**	**€ 1,228–14,278**
Total (% share to the model case price)		**1.4–5.2%**	**1.2–14.3%**	**1.4–5.2%**	**1.2–14.3%**
Excluded costs from the model case		-	-	-	-

Note. Model case price: € 100,000

Source: Author

Note

1 1 Great Britain Pound = 1.18 Euro (as of 1 October 2021).

References

Bragz, G. (2021a). The definitive guide: Estate agent fees and commission. *The Advisory*. Retrieved May 4, 2021, from www.theadvisory.co.uk/estate-agents/fees-what-should-you-pay/

Bragz, G. (2021b). The truth about: Online estate agents, special report (part one). *The Advisory*. Retrieved May 4, 2021, from www.theadvisory.co.uk/estate-agents/online-estate-agents/

Consumer Protection from Unfair Trading Regulations. (2008). Retrieved from https://www.legislation.gov.uk/uksi/2008/1277/contents/made

Consumers, Estate Agents and Redress Act 2007.

Estate Agents Act 1979.

Home Owners Alliance. (2020). Rightmove, Zoopla and the rest: Which is best? *Homes Owners Alliance*. Retrieved May 4, 2021, from https://hoa.org.uk/advice/guides-for-homeowners/i-am-selling/rightmove-zoopla-which-is-best/

Land Registration Act 1925.

Land Registration Act 2002.

Law of Property Act 1925.

Law of Property (Miscellaneous Provisions) Act 1989.

Money Laundering Regulations 2007.

Pratt Contractors Ltd v Transit New Zealand [2005] 2 NZLR 433.

Williams v Broadoak Private Finance Ltd [2018] EWHC 1107 (Ch).

24
GHANA

Emmanuel Kofi Gavu, Wilfred K. Anim-Odame and Omokolade Akinsomi

1. The Ghanaian housing market

The housing sector in Ghana[1] is segmented into formal and informal. The formal sector consists of public and private developers that focus mainly on building houses for the middle- and high-income households; while informal categorization consists of self-builders and small-scale builders that focus mainly on low-income households (Arku, 2009). Ghana's housing sector has evolved since the government's implementation of the Structural Adjustment programmes of the 1980s. This has seen a shift from government being directly involved in housing provision to more private sector involvement with government providing an 'enabling environment' as a facilitator. Housing in the private sector is provided by individuals and corporate entities as owner-occupiers or landlords, who provide detached and semi-detached houses, apartments, and 'compound' houses. Compound houses refer to single-storey houses with two or more rooms that serve a household with a courtyard where multiple households share common utilities, for example, bathroom and toilet facilities (Gavu and Owusu-Ansah, 2019). Tenants typically pay between six and 36 months of rent in advance to landlords in lieu of occupation.

Legislation on ownership, rental, and other forms of tenure

The 1992 Constitution of the Republic of Ghana is the primary legislation on ownership of landed assets. Article 18 of the Constitution provides that every person has the right to own property either alone or in association with others. Under Article 18(2), no person shall be subjected to interference with the privacy of his home, property, correspondence, or communication except in accordance with law and as may be necessary in a free and democratic society for public safety or the economic well-being of the country for the protection of health or morals, for the prevention of disorder or crime, or for the protection of the rights and freedoms of others.

The formal laws governing land administration and development include the Land Act, 2020 (Act 1036), Real Estate Agency Act, 2020 (Act 1047), Lands Commission Act, 2008 (Act 767), Stamp Duty Act, 2005 (Act 689), Land Use and Spatial Planning Authority Act, 2016 (Act 925).

DOI: 10.1201/9781003130383-28

House prices and rentals

The range of prices and rentals in the high-end market in the capital, Accra, is shown in Table 24.1.

Yield averages between 9% and 10%, and gross floor area ranges from 51 m^2 (studios) to 900 m^2 (townhouses) in urban centres.

Supply of housing is mainly from private investors (formal and informal). The formal sector is well-organized and targets high-end market. Efforts of the informal sector are mostly unco-ordinated and generally targeted at the lower end of the market.

Mortgage market and rates

The mortgage debt to GDP ratio is 1.62 in Ghana; very small compared with averages within the sub-region. Mortgage is available to a limited segment of the population – mainly high-income earners in urban Ghana as well as those in the diaspora. However, the market size has since 2015 increased by 19%. Only a small number of households (0.8%) finance their homes with a mortgage (Ghana Statistical Service, 2014; Gyamfi-Yeboah, 2020).Ghana's total housing finance was about 10% of GDP in 2018 (Centre for Affordable Housing Finance in Africa, 2020). The First National Bank (formerly Ghana Home Loans) is the principal specialized mortgage provider in the country. Table 24.2 details the mortgage characteristics in Ghana.

Transaction volumes

The property market is dominated by residential followed by commercial properties. The residential market reveals three housing categories: luxury (high-end), mid-end, and low-end.

Table 24.1 The range of prices and rentals in the high-end market in the capital, Accra

Description	*Prices (€)*	*Rentals (€/month)*
Studio	73,100–86,000	1,118–1,462
1 bedroom	129,000–180,600	1,290–1,892
2 bedroom	223,600–301,000	2,150–3,010
3 bedroom	361,200–498,800	3,010–3,870
Penthouse	516,000–602,000	3,870–4,730

Source: Author

Table 24.2 Mortgage market characteristics

Typical mortgage rate | Term | Deposit	23% | 15 years | 20%
Ratio of mortgages to GDP	1.62%
Number of mortgage providers	8
Number of microfinance providers	180
Annual housing growth rate	4.40%

Source: Centre for Affordable Housing Finance in Africa (2020)

The residential market, which is the most vibrant, registers about 85,000 transactions per year for both sales and rentals (Centre for Affordable Housing Finance in Africa, 2020).

Price variations within different parts of the market

Submarkets exist within the housing sector with differentiated prices and rentals (see Gavu & Owusu-Ansah, 2019). The submarkets constitute a very important element in the household portfolio. These submarket classifications have implications for housing affordability, economic development, and individual wealth management. Figure 24.1 provides house price movements over the period. A cursory inspection of the information provided in Figure 24.1 also shows the volatile nature of prices within the housing market in Ghana.

Characteristics of the housing stock

Due to excess demand over supply, the housing stock is characterized by shortages and crowding mostly within the informal market. The aggregate housing stock comprises detached, semi-detached, self-contained houses, gated communities, and compound houses. Three main types of accommodation are available in compound houses: *Single room*, *hall-and-chamber*, and *hall-and-chamber self-contain* units. *Single room* units consist of one room that serves as both living room and bedroom. The *hall-and-chamber* unit has two rooms: one used as living room (which can also be converted to a bedroom) and the other as a bedroom. In both cases, all other facilities are shared by households on the same compound. With the *hall-and-chamber self-contain* units, the tenant has a private space for kitchen, toilet, and bath. A detached house is a free-standing residential building (sometimes referred to as a single-family home). A semi-detached refers to a house that is connected to another property on at least one side. A self-contained house is the local term for a studio apartment in Ghana; spaces such as kitchen and bathroom are not

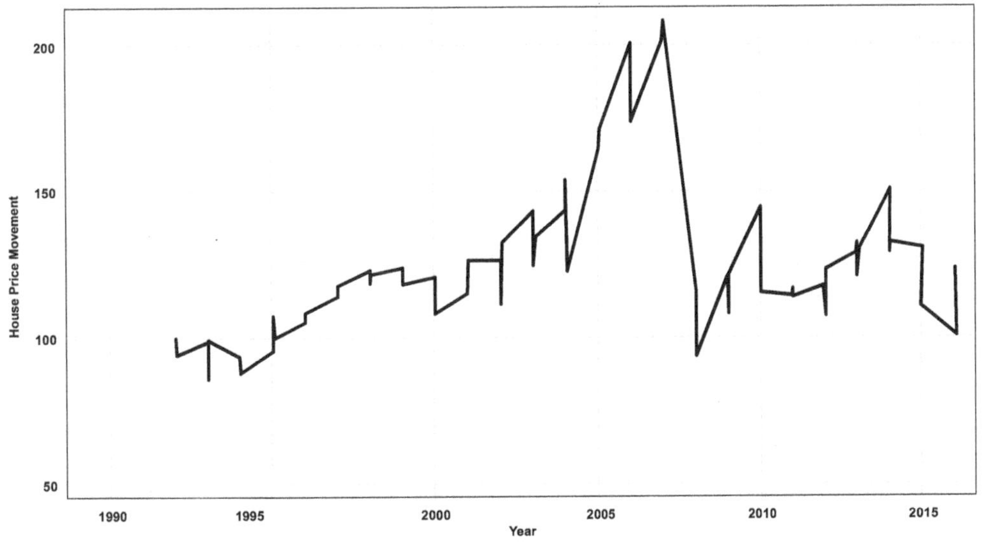

Figure 24.1 House price movements in Ghana, 1990–2015

Source: World Bank

shared with others on the compound. Gated communities refer to housing estates with strict controlled access for residents and non-residents and often are characterized by closed perimeter walls and fences.

Who is constructing new stock? A brief overview of the construction industry and its performance

The majority of new housing stock is being supplied by the private sector where individuals or companies employ equity or remittances (informal sector) and equity or debt (formal sector) to deliver housing units for the market (both formal and informal). Notable formal supply in the market is provided by property developers including Devtraco and Devtraco Plus, Taysec, Clifton Homes, Goldkey Properties, Trasacco, and Manet, among others. Investor opportunities exist in the low-end market because of the large housing deficit of about 2 million units across the country (Ministry of Works and Housing, 2019; Salifu Osumanu, Aigbavboa, & Thwala, 2018).

2. Regulatory framework

Specific regulatory framework/processes that guide the market

There are several legal and regulatory frameworks which guide the Ghanaian housing market. Landownership forms a fundamental factor to a flourishing housing market. The Constitution of the Republic of Ghana specifies several clauses that guide the ownership of Land. Chapter 21 of the Constitution discusses land matters. Clause 257 on the other hand specifies that all public lands are vested in the president on behalf of and in trust for the Ghanaian people. The Constitution defines public land as any land vested in the Government of Ghana for the public service of Ghana and land acquired in the public's interest. All lands in the Northern, Upper East, and Western Regions of Ghana before the enactment of the Constitution which were vested in the Government of Ghana are not deemed to be public lands.

Chapter 21, Clause 266 of the Constitution discusses the ownership of land by non-citizens. A person who is not a citizen of Ghana is not eligible to hold freehold interest in land. Any non-citizen of Ghana who held freehold interest on land in Ghana before the 22 August 1969 would have that right converted to a leasehold of 50 years which would then revert to the state after the expiration of the lease.

Chapter 20, Clause 267 is specific on land and property pertaining to the stool and skin. The clause specifies that all stool lands in Ghana are vested in the relevant stool which is held in trust on behalf of the subjects of the stool in accordance with customary law. For the disposition and development of any stool land to take place, this must be certified by the regional lands commission of the region and must be consisted with the development plan authorized by the planning authority concerned. The collection of rents, dues, royalties, revenues, and any payments is to be paid in the stool account and disbursed as follows: 10% to the Office of the Administrator of Stool Lands (OASL), 25% to the stool, 20% to the traditional authority, and 55% to district assembly.

The Land Act, 2020 (Act 1036) is intended to harmonize and consolidate laws on land to ensure sustainable land administration, and effective and efficient land tenure, among others. A number of customary land secretariats have been established and are in operation across the country to provide facilities for traditional landowners to manage their lands effectively.

Dispute resolution is a factor to a successful and thriving housing market. Land disputes may occur due to multiple sales of the same piece of land to more than one prospective purchaser. This practice may be attributed to several factors including boundary disputes, joint inheritance, multiple allocations of land by chiefs, and complexity in title transfer (UN Habitat, 2019). Redress is through the courts; however, the customary land secretariats also engage in settling disputes through alternative dispute resolution mechanisms.

3. The conveyance process

When a valid contract for the sale or lease of landed property is finalized, the vendor (property owner) becomes a trustee for the vendee (purchaser) to whom the beneficial ownership passes in equity, regardless of his legal title. This makes the vendor responsible for the due care over the property as a trustee must over properties held in trust. The vendor becomes liable to pay compensation to the vendee if the subject property is injured or damaged.

Prior to the completion of the sale, the death of the vendor will not affect the enforcement of the contract. The deceased personal representatives will step into his shoes and complete the transaction. Similarly, on the death of the purchaser before completion, his personal representatives can sue and be sued on the contract of sale.

The vendor is obliged on a sale of land to prove his title. Indeed, the burden is on him to satisfy the vendee that he has title to the land in question. It is absolutely essential for the vendor to produce and deliver an abstract of title at his own expense to the vendee. The abstract of title could be an official search from the Lands Commission (both Public and Vested Lands Management Division, and Land Registration Division) that summarizes the chronological history on the land, which is sometimes referred to as the root of title. The purchaser should inspect the originals of all documents and/or court judgement(s) in respect of the land or property; these must all be in favour of the vendor. Nonetheless, the purchaser may wish to conduct an independent search at the Lands Commission for further assurance.

A land economist, valuer, or lawyer could act on behalf of either party to conduct the search and advise his client appropriately. These professionals will serve a critical role in interpreting the search report, which sometimes is not very straightforward to comprehend. The Ghana Institution of Surveyors and Ghana Bar Association regulate the conduct of their members and also prescribe fees chargeable for such services.

Conveyancing styles may vary but the Conveyancing Act, 1973 (NRCD 175) provides a general framework that contains four principal ingredients: party, property, price, and signature (PPPS). The first 'P' refers to the name of the parties. Names of the vendor and vendee under the conveyance shall be written in full without abbreviation and together with their respective address. In addition, each party must state the capacity in which he is executing the conveyance. Both parties must be of legal age; in Ghana the law pegs it at 18 years or above. The *nemo dat* rule applies strictly. This is the Latin phrase *Nemo dat quod non habet* which means 'no one can give what they do not have'. The second 'P' refers to the property's description. This is usually provided as a schedule that describes the dimensions of the polygon (shape of the land) in feet or metric in relation to the adjoining land or property. A cadastral or parcel plan is provided in addition to depict the shape of the polygon and also provide other details including the acreage of the land. The description should as far as possible be consistent with that in the previous deeds or documents in order to avoid any dispute.

The price of the land is the third 'P'. This amount is also termed as the consideration of the transaction. It has to be stated in words and figures, as a lump sum and/or period payment. The date and frequency of the periodic payment must explicitly be stated in the conveyance to avoid any doubt. Ground rents are typical periodic payments, usually on an annual basis. For leases, the commencement date and expiry date are very essential and must always be stated. The term certain should also be prominently stated.

The 'S' represents the attestation where the parties execute the deed in the presence of at least one witness for each party. In the case of stool, skin, and family lands, it is recommended that the processes of the transaction involve at least two principal elders, who should sign as witnesses for the grantor (chief or family head).

The Real Estate Agency Act, 2020 (Act 1047) seeks to regulate real estate agency practice in Ghana. It is also intended to regulate the conduct of practitioners and commercial real estate transactions including sale, purchase, letting, and other real estate transactions. In order to practice as a real estate agent in the country, a person must have a license issued by the proposed Real Estate Agency Council. This will go a long way to regulate and promote the real estate market in the country.

3.1 Marketing phase

Agency/professional fees – price ranges

There are agency fees charged to either prospective tenants, landlords, buyer, or vendor when professionals such as lawyers, valuers, or 'agents' are engaged in the transaction. Their fees range between 5% and 10% of the transaction cost for sales or rentals, but in most cases are in line with fees and charges of the relevant professional bodies: Ghana Institution of Surveyors (for valuers and estate managers) or Ghana Bar Association (for lawyers). It is common to find non-professional 'agents' charging a fee for conducting a prospective client on an inspection of a property. The fee chargeable is often subject to negotiation.

Advertisements

Properties available for sales or letting are usually advertised through professionals in the built environment, agents, dedicated websites, or sometimes through personal contacts, or print media.

Negotiation process and who controls the market

The market is pro-landlord. It favours landlords because demand generally exceeds supply and also because of the rising population and high urbanization pressures. Nonetheless, the pro-landlord rental market in the country explains the fundamentals because there are no effective rent controls (Anim-Odame, 2016); parties in all transactions freely agree on the rent or price payable.

3.2 Contract phase

Signing contract and payment terms

As in all property markets the cardinal principle of *caveat emptor* applies at all times. Arm-length negotiations are common, and parties have the right to dictate terms and conditions. Payment terms are usually subject to mutual agreement and for rent, payments are typically made in advance. It is strongly advised to engage professionals – valuer or lawyer – in all land-related and property-specific transactions.

Registrable instruments

Registrable instruments under land title registration in the cities of Accra, Tema, and Kumasi are provided under section 19 of the Land Registration Act, 1986 (PNDCL 152) to include allodial title, common law freehold, customary law freehold, usufructuary interest, leasehold interest, and customary tenancy. Instruments covering land in all other regions are registered in line with the provisions of the Deeds Registry Act, 1962 (Act 122).

Rent advance payments and pricing in Ghana

In Ghana, landlords usually demand lump sum payments of between six months and five years rent in advance from prospective tenants; this is known as rent advance payment (Ehwi, Asante, & Gavu, 2021; Gavu, 2020). This payment precludes the obligation to make periodic payments until the lease expires and another agreement is renegotiated.

Residential leases, in general, put the responsibility for routine maintenance and all utilities costs on tenants. This leaves landlords' outgoings limited to management costs estimated at 7% of annual rental income, property rates which vary with location at up to 1.4% of annual rent, and ground rents at up to 0.3% of annual rent. Total deductions from gross rents therefore are estimated at 8.7% (Anim-Odame, 2012).

3.3 Registration phase

There are two systems of land registration in Ghana: land title registration and deeds registration. The Lands Commission records the transfer of ownership and instruments that affect land and also issues a land title certificate (see Figure 24.2). The land title registration applies to transactions in Accra, Tema, and part of Kumasi Metropolitan; all other regions are under the deeds registration system. The land title registration process is briefly described as follows: An applicant picks up appropriate registration forms from the Client Service Access Unit (CSAU) and then fills out and submits the forms together with all relevant land documents and an approved barcoded parcel or cadastral plan. The submitted document is subjected to a vetting process against an appropriate checklist. When the contents of a document are found to have complied with the checklist, a bill is generated for fee payment at the on-site banking at the CSAU, with an indication of a possible date of collection of the completed registered document. The applicant pays the required administrative fees and is given a receipt of acknowledgement as well as a job number for the application. The documents are forwarded from the CSAU to the respective back offices of the divisions of the Lands Commission for the other registration processes. The applicant's documents are batched to, first, the Survey and Mapping Division (SMD) and the Public and Vested Lands Management Division (PVLMD) for records verification (searches), whilst the main application is batched and forwarded to the Land Registration Division (LRD), the division responsible for the issuance of land title certification. Second, when search results are received from SMD and PVLMD, the application is forwarded to the technical unit for verification of records. Third, if there are no adverse findings from the search results, a request is sent to SMD for the preparation of a cadastral or parcel plan depending on the subject neighbourhood. However, if there are queries, objections, or conflicting interests/recorded transactions, then an inspection of the subject property may be required to resolve same. Fourth, when the parcel or cadastral plan is received, LRD proceeds to publish the application in the daily newspapers to give notice to the general public for 14 days. Fifth, if no objections are received within the 14 days, the application is forwarded to approval officers for vetting and approval. Sixth, if application is recommended for approval, the land title certificate is printed out, vetted, and forwarded to the land registrars for final vetting and signature. Details on the Land Register comprises folio, volume, certificate number,

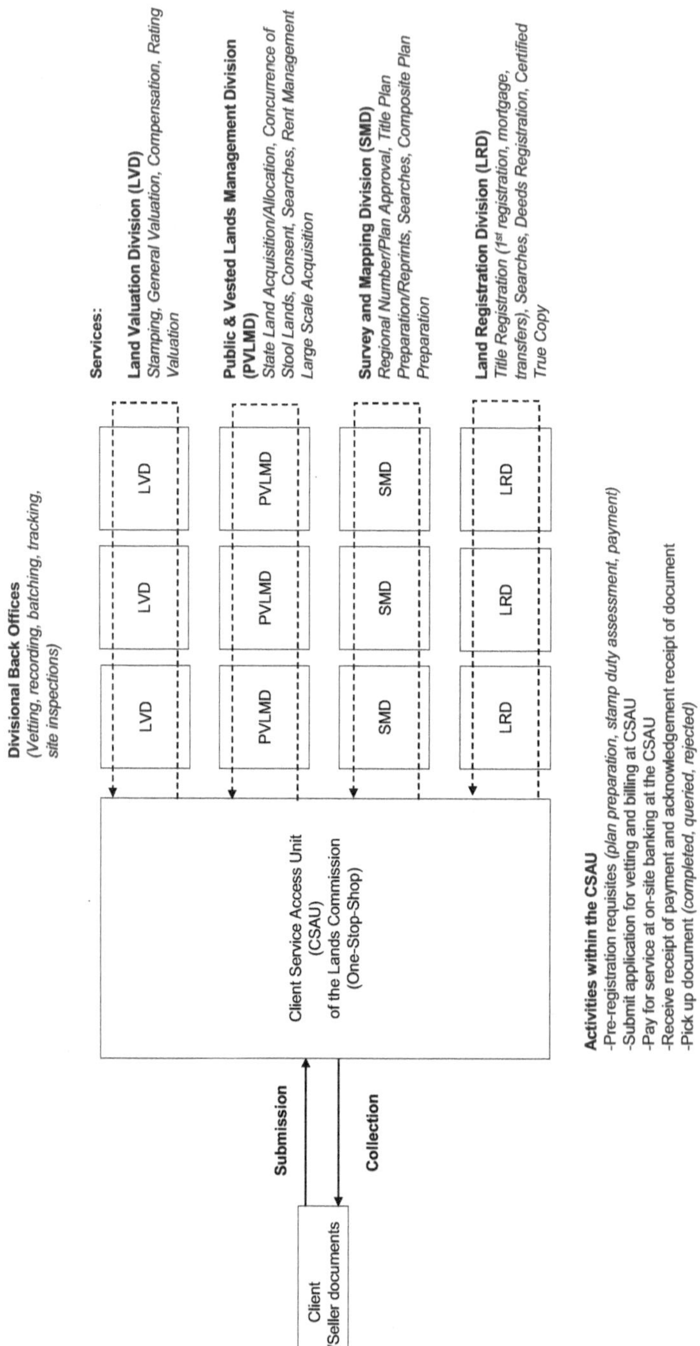

Figure 24.2 The conveyance process and interface with the Lands Commission

Source: Author

description of parcel, parties, date of issuance, and date and nature of instrument, among others. Seventh, the signed land certificate is sealed and forwarded to the technical unit for plotting and recording of the particulars of the application in the database. Finally, the land certificate is bound, embossed, and forwarded to CSAU for collection by the applicant. In all these processes, the only interface an applicant has with the Lands Commission is CSAU. On completion, the applicant is notified by a text message to visit the CSAU and sign and collect the registered document. However, if there are queries, objections, conflicting interests, further particulars needed, or the need for an inspection of the subject property, the client through a text message may be invited by the officer-in-charge.

Deeds registration provides for the registration of instruments affecting land (e.g. leases, assignments, mortgages). The registration is based only on the deed and not the land itself. This registration does not confer title on any land. As discussed earlier, the deeds registration applies to most parts of the country. The registration procedure is similar to the land title registration process discussed earlier; however, in the end the applicant receives the registered lease document (registered deed).

Figure 24.2 provides clients' interface with the Lands Commission (and the types of services that can be accessed). Currently, the commission has introduced an online platform to receive various applications and requests from clients.

3.4 Post-transaction phase

Capital gains tax is assessed by the Ghana Revenue Authority and payable by the vendor to the Authority. Ground rents are paid by tenant in accordance with the provisions in the lease directly to the landlord (for family or individual lands or properties) or to the Office of the Administrator of Stool Lands (for stool and skin lands or properties). Property rates are forms of property taxes paid in Ghana on annual basis to the Metropolitan, Municipal, or District Assembly where the property is located. The rateable value of the property is assessed by the Land Valuation Division of the Lands Commission as the basis for the Assembly to apply an appropriate rate impost to arrive at the property rate payable by the owner. Landlords are expected to file their returns annually to the Domestic Tax Division of the Ghana Revenue Authority and disclose rent charged as the basis for the determination of rent income tax payable.

4. Summarizing the conveyance process and its transaction costs

The conveyancing process for land tenure (ownership and transfer) in Ghana is seemingly complicated due to the duality of land owning and land administration systems. The enactment of the Land Act, 2020 (Act 1036), which has consolidated and brought together under one umbrella several laws dealing with land, seeks to address a number of the challenges. More so, the one-stop-shop provided by the Lands Commission and the ongoing digitization and automation of land services delivery will by and large ensure efficient and effective land services delivery. Would-be purchasers or tenants are cautioned to conduct due diligence at all times so that they do not breach the legal principle of *caveat emptor*. It is highly recommended that a professional such as a valuer or a lawyer is engaged to achieve value for money. Appropriate strategic interventions are expected to be implemented by government and key stakeholders to entirely formalize the real estate market in Ghana to boost investors' confidence. Table 24.3 provides a summary of transaction costs during the conveyancing process in Ghana.

Table 24.3 Fees and taxes during the conveyance process in Ghana

		Real estate	
Phase	*Fees and taxes*	*Seller*	*Buyer*
Marketing	Real estate agent's fees	5–10% of total sale or rent amount	5–10% of total sale or rent amount
	Marketing fees	varies	varies
Contract	Legal fees	-	3–10%
	Bank mortgages fees	-	2.5% of loan amount (minimum € 70)
Registration	Stamp tax	-	0.25–1%
	Land registration fee	-	€ 98
Post-registration	Capital gains tax	15% (of profit)	-
	Legal fees	-	3–10%
	Sales tax	-	-
Total (EUR for the model case price)		**€ 5,000–10,000**	**€ 11,348–31,098**
Total (% share to the model case price)		**5–10%**	**11.3–31.0%**
Excluded costs from the model case		15% CGT	2.5% mortgage fee

Note: Model case price: € 100,000

Source: Author

Note

1 1 Ghanaian Cedi = 0.14 Euro (as of 1 October 2021).

References

Anim-Odame, W. K. (2012). The nascent residential investment market in Ghana. *Journal of Real Estate Literature, 20*(2), 289–303.

Anim-Odame, W. K. (2016). Developing real estate markets in Sub-Sahara Africa: The fundamentals. *Real Estate Finance, 33*(1), 26–32.

Arku, G. (2009). Housing policy changes in Ghana in the 1990s. *Housing Studies, 24*(2), 261–272. doi:10.1080/02673030902719763

Centre for Affordable Housing Finance in Africa. (2020). *2020 yearbook: Housing finance in Africa*. Retrieved from http://housingfinanceafrica.org/

Ehwi, R. J., Asante, L. A., & Gavu, E. K. (2021). Towards a well-informed rental housing policy in Ghana: Differentiating between critics and non-critics of the rent advance system. *International Journal of Housing Markets and Analysis*. doi:10.1108/IJHMA-12-2020-0146

Gavu, E. K. (2020). Conceptualizing rental housing market structure in Ghana. *Housing Policy Debate*. doi:10.1080/10511482.2020.1832131

Gavu, E. K., & Owusu-Ansah, A. (2019). Empirical analysis of residential submarket conceptualisation in Ghana. *International Journal of Housing Markets and Analysis, 12*(4), 763–787. doi:10.1108/IJHMA-10-2018-0080

Ghana Statistical Service. (2014). *Ghana living standards survey round 6 (GLSS 6)*. Retrieved from www.statsghana.gov.gh/gssmain/fileUpload/Living conditions/GLSS6_MainReport.pdf

Gyamfi-Yeboah, F. (2020). Dr. Frank Gyamfi-Yeboah: The mortgage industry landscape in Ghana. *Myjoyonline*. Retrieved from www.myjoyonline.com/opinion/dr-frank-gyamfi-yeboah-the-mortgage-industry-landscape-in-ghana/

Ministry of Works and Housing. (2019). *2018 statistical report*. Retrieved from https://mohua.gov.in/index.php

Salifu Osumanu, I., Aigbavboa, C. O., & Thwala, D. W. (2018, July 21–25). Examining the relationship between lean adopotion and housing finance in Ghana. In S. Fukuda (Ed.), *Proceedings of the AHFE 2018 international conference on affective and pleasurable design* (pp. 301–311). Cham: Springer. doi:10.1007/978-3-319-94944-4_33

UN Habitat (2019). Ghana Housing Profile. Retrieved from https://unhabitat.org/sites/default/files/documents/2019-07/ghana_housing_profile.pdf

25

UNITED STATES

Rebecca Lewis and Robert Parker

1. The U.S. housing market

The United States (U.S.)[1] housing market for homes consists of two main forms: investment and ownership. In this chapter, we focus on the ownership process. Several forms of ownership exist: investment vs. resident and cash vs. mortgage transactions. Investment properties are purchased by individuals and corporations to lease to tenants. In the United States, it is important to distinguish cash vs. mortgage transactions because the conveyance process and transaction costs differ.

The real estate market and housing market has a variety of financial regulations that govern lending eligibility at a federal level, while procedural aspects and transaction costs vary across states and are governed by a suite of federal, state, and local policies.

The vast majority of housing construction occurs by private industry. Government investment supports 10% of units in various forms, but the federal government no longer builds and maintains public housing. Since the Great Recession in 2008, rental and sales prices have increased steadily while housing production has decreased below historic levels. Reforms were instituted following the 2008 recession which led to a drop in sales prices, but construction also slowed during this period as shown in Figure 25.1. The Great Recession revealed challenges in the lending market associated with loan qualification and mortgage terms; this led to reforms in mortgage and disclosure processes under the Truth in Lending Act (12 CFR part 1026).

The yearly sales transaction volume averages 4% of the total units (5.34 million in 2019) (National Association of Realtors, 2021). Figure 25.2 shows the yearly change in consumer prices and house prices over a period of 30 years. In 2020, the average price of homes was approximately € 106 per square foot but varied tremendously across the country, with average prices ranging from € 21 per square foot in Detroit to € 697 per square foot in San Francisco as illustrated in Figure 25.3 (National Association of Realtors, 2021).

The distribution of homes is as follows:

- 69% detached or attached homes (82% owner occupied, 18% rental)
- 26% multi-units (two units + in structure) (13% owner occupied, 87% rental)
  - Within rentals (two+ units) (ownership of properties is 24% private individual, 60% non-individual, 4% non-profit)
    - 10% of rentals are subsidized by federal programs
- 6% other types of housing

DOI: 10.1201/9781003130383-29

Figure 25.1 Housing starts vs. household formation, 1990–2020

Source: Federal Reserve Economic Data, 2021

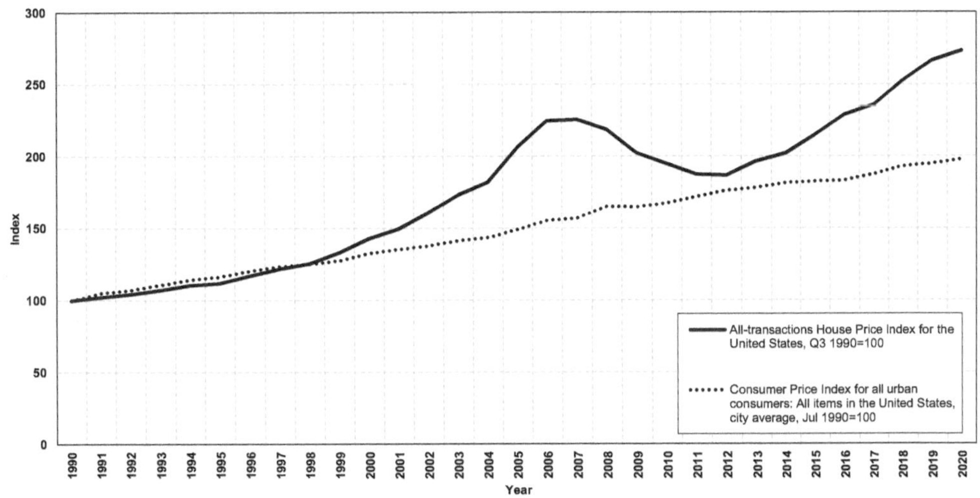

Figure 25.2 Yearly change in consumer price and house prices in the United States (in percent), 1990–2020

Source: Federal Reserve Economic Data, 2021

(American Community Survey, 2015–2019; United States Census Bureau, n.d.c)

The average size of an owner-occupied housing unit is 1,900 ft^2, while the average rental unit is 980 ft^2 (American Housing Survey, 2019). The size of owner-occupied units has grown steadily from 1,500 ft^2 in the 1950s to 2,300 ft^2 in the latest American Housing Survey in 2016–2017. Rental units have stayed consistent in size at approximately 950 ft^2. The majority of multi-units were built between 1960 and 1979 (28%) or 1980 and 1999 (29%), while most detached or attached houses were also built between 1960 and 1979 (29%) or between 1980 and 1999 (28%) (American Community Survey, 2015–2019). In the United States, housing

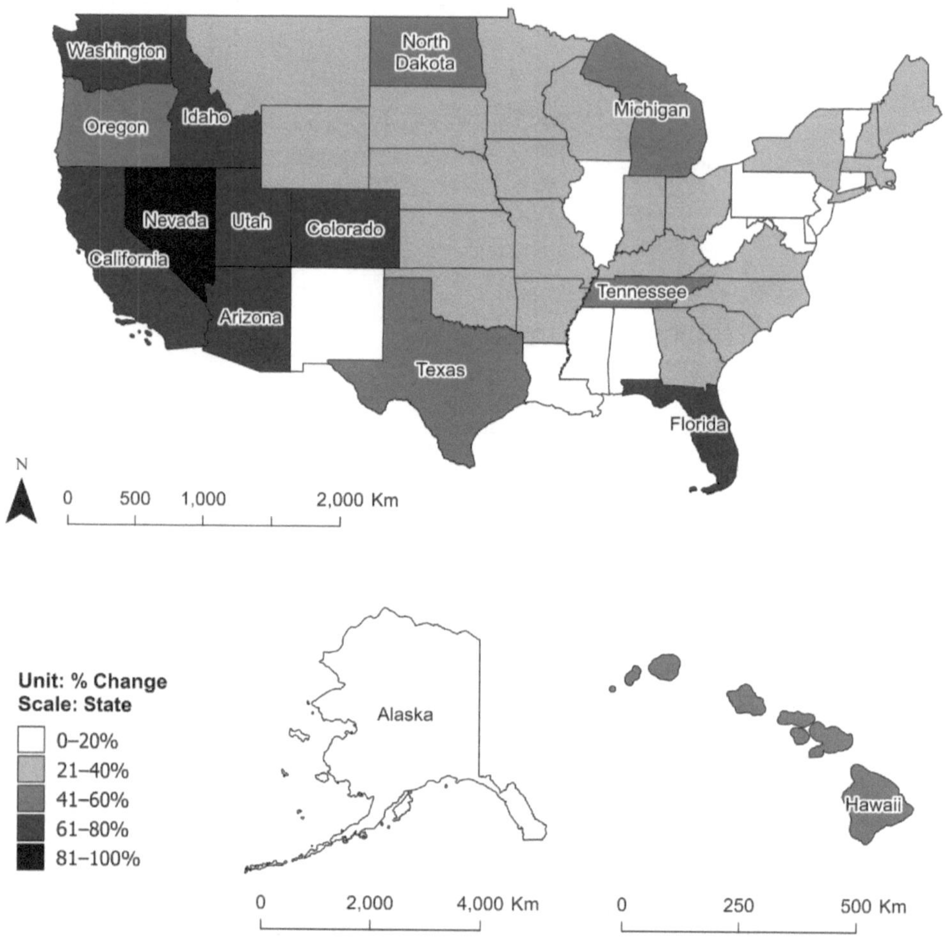

Figure 25.3 Change in house prices in the United States, 2010–2020

Source: Federal Housing Finance Agency Price Index

construction fell during the Great Recession and has not yet recovered as the number of new housing starts was 0.72 units per new household from 2010 to 2016 – far lower than the 1.1 units per household recommended in a healthy market (Up for Growth, 2018).

2. Regulatory framework

Federal, state, and local regulations

Real estate transactions are subject to federal, state, and local laws. Most states follow a combination of statutory and common law. The implication of this approach is that adjudicated disputes may impact how future transactions are structured or the content of legal documents that accompany the transaction. International laws do not govern real estate assets in the United States, and few restrictions exist on ownership of real property, although foreign owners must comply with certain reporting requirements.

Real estate is thought of as a bundle of rights – a set of legal privileges conveyed to the owner. These include right of possession, control, exclusion, enjoyment, and disposition. In general, homeowners receive the entire bundle of rights upon purchase. Moreover, these rights do not supersede the boundaries of other federal, state, and local laws. Many homes are subject to rules of homeowner associations (HOAs) that limit right of control. HOAs may impose restrictions on everything from pets to paint colors to where vehicles may be parked.

System of conveyance

Property rights are recorded through deeds, a form of legal document used to transfer property ownerships. Recording protects property rights and establishes priority in most jurisdictions. Mortgages become effective only by recording and must be recorded within ten days of execution. Property recording is a county (or parish, as in Louisiana) function in the United States. Requirements for recording are established by each state and vary by state. Counties may also have specific requirements.

The federal government has no requirements that land titles be registered. Several states initially used the Torrens system, and it is voluntary in the few jurisdictions that still use the system. Over time a unique land recording system evolved in the United States based on property deeds. Six types of deeds exist in the United States, but only two are commonly used for residential transactions: warranty deeds and mortgage deeds. Warranty deeds guarantee that the property is free of encumbrances (i.e. debts, liens) and that the seller has the legal right to sell the property. Most residential sales have warranty deeds. Buyers who take out loans for a property will also sign a mortgage deed giving the lending institution the right to put a lien on the property if the loan is not repaid or to foreclose on the property if payment stops.

Title to a house passes upon delivery of the deed. The deed is then recorded by the county in which the property is located. Many jurisdictions or lenders require title insurance to protect lenders and homeowners from claims regarding legal ownership of the home. This protects the owner from issues related to the title from the seller or previous owner(s). It is customary to conduct a search to identify any issues with the title. Title searches typically include researching local real estate records, property surveys, and affidavits from the seller.

Ownership is not complete until transferred by a conveyance or sales deed. A sales agreement is a promise to transfer a property in the future but is not complete without execution of a conveyance deed. A sales deed is also a conveyance deed and is used for cash purchases. Conveyance deeds can also include mortgage, lease, gift, or exchange deeds. The conveyance deed typically includes a property survey, a list of rights included with the property, the full chain of titles up to the current seller, the delivery method to the buyer, a statement of consideration (e.g. price), additional terms and conditions, a clause of property ownership, and the signatures of both parties. Most residential sales use a fee simple absolute form of conveyance that states the holder has present and future interest in the property and the ability to sell part or all of the property at any time.

3. The conveyance process

In this section we outline how the conveyance process works for home sales, focusing on owner occupation rather than investment properties.

Marketing phase

According to the National Association of Realtors, 88% of home sales in 2020 were aided by a real estate agent and/or broker. Real estate agents help people sell, purchase, or rent property and the United States has about 2 million licensed agents. The primary role is to facilitate transactions between buyers and sellers. They carry offers and counter-offers between the parties to inform clients of the many requirements of a real estate transaction. A listing agent represents the seller and helps establish listing prices as well as making suggestions about how to best market the property.

Real estate agents and brokers are paid by commission. Commissions are negotiable and typically are between 5% and 6% of the sales price. The commission is split up among agents and brokers with each of the parties (listing agent, buyer's agent, listing broker, and buyer's broker) receiving 25% of the total commission (or 1.5% to each party for a 6% commission).

Licensing is governed at the state level, and all 50 states have training requirements and 40 states require licenses. Many sellers find agents through referrals from family or friends or based on reputation. The seller typically signs a listing agreement with the selected agent that provides the agent rights to market the property. Many listing agreements are exclusive where the agent is the sole representative of the buyer.

The next step is to establish a sales price. This typically involves the listing agent and broker conducting a market analysis and review of comparable properties and recent sales. Once the sales price is established, the property is typically registered in the local or regional Multiple Listing Service (MLS). The purpose of an MLS is to publish a 'unilateral offer of compensation' by listing brokers to all other brokers in the MLS. The database includes a broad array of information about each property. The key benefit is that subscribers have access to the most comprehensive list of available properties. In 2020, the United States had 580 MLS listings. It is worth noting that data providers like Zillow and Redfin have become significant players in marketing real estate and allow buyers access to data on properties for sale in nearly every market.

MLS listings include a listed price from which the 'unilateral offer of compensation' is derived (based on the assumed commission). An offer is considered a contractional obligation but can be renegotiated (just as purchase price can be negotiated) at any time between the seller's and buyer's broker.

Once the terms are negotiated with the listing agent, the property is put on market. Marketing typically includes placing a for sale sign in the yard and conducting 'open houses' where anyone is invited to tour the property. Marketing increasingly includes virtual marketing – high-resolution photos and often 3-D virtual tours posted online. This allows buyers much more flexibility and the opportunity to screen properties in the comfort of their home and without the assistance of their agent. Buyer's agents work closely with buyers to understand what type of property and amenities they are looking and will provide information on properties they think meet the buyer's criteria.

When a buyer finds a property they like, they submit an offer to the listing agent through the buyer's agent. All subsequent negotiations occur through the buyer's and seller's agents. The offer depends on a broad range of factors – macroeconomic conditions, interest rates, local market conditions, neighborhood amenities and demographics, schools, parks, size of the home and lot, and amenities are some of the factors and may be more or less than the listing price. The offer may include conditions such as the property appraising for the offer price or passing inspection. Typically the offer is countered by the seller until both parties get to agreement on the price and terms. These are called 'arm's length' sales where both parties agree to terms that meet their interests. Such sales are considered market rate sales.

Most buyers will take out a mortgage; in the past decade the percentage of cash sales has varied between 20% and 50%. Cash buyers skip the complicated and lengthy process of applying for a mortgage and can sometimes get a reduced price. Purchasers who take out mortgages have a broad array of lenders to choose from and will typically pre-apply to be pre-screened for credit worthiness and maximum loan amount. Qualifications vary by lender but typically include factors such as credit scores (the VA requires applications have a FICO score of at least 580), debt-to-income ratios (which can range between 28% and 44%), credit history, down payment, and collateral. The Fair Housing Act (24 CFR Part 100) makes it unlawful to discriminate based on race, color, national origin, religion, sex, familial status, or handicap (disability).

Contract phase

After a seller and buyer agree to an offer price, the property goes into escrow. At this point, the buyer puts a small amount into an escrow account as an earnest payment (€ 2,580) to show that they are serious buyers. This small escrow payment goes towards the down payment when the sale closes; if a buyer walks away, the seller keeps this sum of money. Following reforms after the Great Recession, the practices around down payments have shifted. Buyers must offer 20% down or pay for mortgage insurance on top of principal and interest payments.

During the 'pending' phase between offer and closing, several additional steps typically occur. The mortgage lender works with the buyer to prepare a uniform residential loan application with the buyer's assistance. The buyer arranges a home inspection with a private home inspector. Based on the results of a home inspection, the buyer may ask for additional terms to the sales offer conditions or price. If the buyer is financing a portion of the house with a mortgage, the bank typically requires an appraisal. If the appraisal is lower than the than agreed upon offer price, the mortgage provider may require additional negotiation of the sales price or the buyer may choose to pay more than the appraised value if the down payment is large enough. Once an offer and all conditions are accepted by a seller, the process moves towards a closing date. A few days before closing date, the buyer and realtor have an opportunity to 'walk through' the property to ensure that repairs and renovations have been completed as agreed upon.

The closing date and possession date is typically four to six weeks after offer. At this date, the buyer and seller sign all legal paperwork with a title company. The title company handles the legal paperwork for both the buyer and seller while filing necessary forms to transfer the deed and record the sale. The down payment is due to the seller at the time of signing (usually a cashier's check) and the mortgage lender pays the remaining sum to the seller. The title company files the paperwork to register the change in property ownership with the county assessor's office. When the property is deemed 'recorded', the buyer has ownership and can begin occupying the property.

Each step in the contract phase requires payment to various private and public actors in the process. Table 25.1 summarizes some of these costs.

Registration phase

The county assessor's office records the transfer of ownership and issues a title deed. If the transaction is financed through a mortgage, the lender holds the deed until the loan is paid in full. In the case of a cash transaction, the buyer obtains the deed.

Some counties require real estate transaction taxes or stamp taxes in addition to flat-rate administrative fees for title registration. These fees are paid by the mortgage lender on behalf the buyer during the closing process. The mortgage lender and the title company assist the buyer with the process. The legal fees are paid through the mortgage process; all fees are aggregated and included in the total mortgage amount on top of the sales price.

Post-transaction phase

After the transaction, the seller may be required to pay a capital gains tax on the sale of their home to the state or federal government. Sellers are exempt from the tax for € 215,000 of profit (single person) or € 430,000 of profit (married persons) if the sellers owned the home for more than two years. Sellers are completely exempt from the capital gains tax if they apply the full value of the home sale into another home within two years. Federal capital gains taxes on home sales range from 0% to 20%, depending on income level.

Table 25.1 Fees and taxes during the conveyance process in the United States

Phase	*Fees and taxes*	*Real estate (mortgage based)*		*Real estate (cash transactions)*	
		Seller	*Buyer*	*Seller*	*Buyer*
Marketing	Real estate agent's fees (split among buyer/seller agent and broker)	3–6%	0–3%	3–6%	0–3%
	Marketing fees (MLS)	-	-	-	-
Contract	Legal fees	-	-	-	-
	Mortgage lender fees	-	€ 1,290 (varies)	-	
	Appraisal	-	€ 516 (varies)	-	Waived
	Inspection	-	€ 430 (varies)	-	Waived
	Title company fees	-	€ 430–1,290 (varies)	-	€ 430–1,290
Registration	Transfer taxes	-	1.7–2.0% of price (varies by state)	-	-
	Recording fee	-	Varies by locality (€ 198 in Oregon; € 97 in Florida)	-	-
Post-registration	Capital gains tax	0–20% of profit (if over € 215,000 or € 430,000)	-	0–20% of profit (if over € 215,000 or € 430,000)	-
	Legal fees	-	-	-	-
	Sales tax	-	-	-	-
Total (EUR for the model case price)		**€ 3,000–6,000**	**€ 3,045–11,213**	**€ 3,000–6,000**	**€ 430–4,290**
Total (% share to the model case price)		**3–6%**	**3.0–11.2%**	**3–6%**	**0.4–4.3%**
Excluded costs from the model case		CGT 0–20% of profit (if over € 215,000 or € 430,000)	-	CGT 0–20% of profit (if over € 215,000 or € 430,000)	-

Note. Model case price: € 100,000

Source: Author

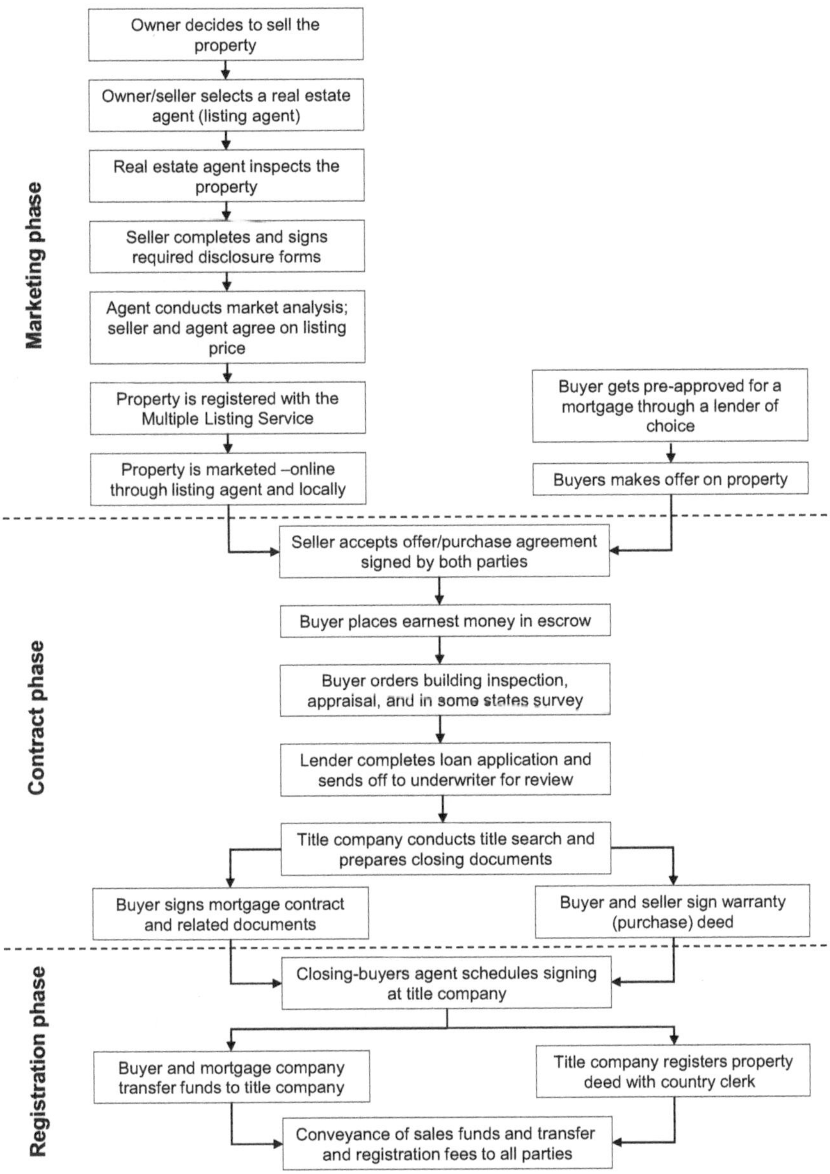

Figure 25.4 The conveyance process in the United States

Source: Author

Many states (37 currently) impose real estate transfer taxes on the transfer of the title of real property (National Association of Realtors, 2005). The taxes and fees include a transfer fee and a percentage rate charged on the value of the home. Local governments sometimes charge their own transfer fees as well. The taxes vary from a flat rate of € 1.7 in Arizona to a percentage of the sales price (up to 2%).

Homeowners are responsible for paying property taxes on an annual basis. Some homes are required to pay HOA fees monthly. Property taxes are paid to the local government and the rules vary by state and locality in terms of homestead exemptions, property tax relief, and how taxable value is determined. The effective real estate tax rate varies from 0.28% to 2.49% of the home value. Real estate taxes on a median home range from € 505 in Alabama to € 7,191 in New Jersey.

4. Summarizing the conveyance process and its transaction costs

To summarize, the conveyance process involves several private and public sector actors and taxes and fees that vary by state and locality as outlined in Figure 25.4. Ownership is conveyed when closing paperwork is signed and the title has been recorded by the county assessor's office. The pre-contract phase can be lengthy and complex. Buyers are served by real estate agents and mortgage lenders who help navigate the process. The inclusion of several private sector actors, including the real estate agents, home inspectors, appraisers, mortgage lenders, and a title registration company add to the complexity and expense in the process. Upon sale, favourable federal laws make it easy to bypass capital gains taxes on sales. Each property is responsible for paying annual property taxes to the local government. Transfer taxes vary by state and locality as well.

Note

1 1 US Dollar = 0.86 Euro (as of 1 October 2021).

References

National Association of Realtors. (2005). *Real estate transfer taxes*. Retrieved November 20, 2021, from www.nar.realtor/smart_growth.nsf/docfiles/transfertaxrates(8-05).pdf/$file/transfertaxrates(8-05).pdf

National Association of Realtors. (2021). *Real estate statistics*. Retrieved November 20, 2021, from www.nar.realtor/research-and-statistics/quick-real-estate-statistics

United States Census Bureau. (n.d.a). *American community survey (2015–2019)*. Retrieved November 20, 2021, from www.census.gov/programs-surveys/acs

United States Census Bureau. (n.d.b). *American housing survey (2019)*. Retrieved November 20, 2021, from www.census.gov/programs-surveys/ahs.html

United States Census Bureau. (n.d.c). *Rental housing finance survey (2019)*. Retrieved November 20, 2021, from www.census.gov/programs-surveys/rhfs.html

Up for Growth. (2018). *Housing underproduction in the United States*. Retrieved November 20, 2021, from www.upforgrowth.org/sites/default/files/2018-09/housing_underproduction.pdf

26
KENYA

Raphael Kieti, Peter Mburu and Omokolade Akinsomi

1. The Kenyan housing market

The Kenya[1] housing market contains a blend of formal and informal markets. The informal market consists of slums and squatter settlements (shanties) and is characterized by houses of poor structural quality, overcrowding, tenure insecurity, and poor sanitation. These areas are home to approximately 61% of Kenya's urban population (Cytonn Research, 2018). The informal market is unregulated, and transactions take place outside formal institutions and frameworks. The formal housing market, on the other hand, is vibrant and consists of three main forms: rental, ownership (cash purchase and mortgage), and tenant-purchase schemes. Transaction practices and procedures in the formal market are fairly standardized, take place in formal institutions, and are governed by laws, policies, and administrative guidelines issued by the Government of Kenya (GOK) and the Law Society of Kenya (LSK). The procedures and processes, together with the governing legislation, differ between rental, ownership, and tenant-purchase schemes, with implications on both the conveyance process and transaction costs, as we show throughout the chapter.

According to the Centre for Affordable Housing Finance in Africa (CAHF, 2019), 64% of Kenyan households own their accommodation and 36% rent. Of the 32% of households that live in urban areas, 70% rent and 30% own, whereas in rural areas, 88% own and 12% rent. The rate of ownership in the capital city, Nairobi, is quite low, with over 90% of households living in rented accommodation. Due to rapid urbanization in Kenya, the demand for housing far outstrips the supply, not only in major towns and cities but also in secondary towns and rural areas. The predominant forms of housing in urban areas are detached units (41%), medium- and high-rise flats (21%), swahilis (21%), and shanties (13%) (CAHF, 2019). With an estimated housing deficit of 2 million units in Kenya (World Bank, 2017), many lower income urban households live in single-room accommodation and shanties (informal housing). The effective housing demand is constrained by many factors, among them low household income and high house prices. According to the Kenya National Bureau of Statistics (KNBS), 74.5% of formal sector workers earn less than Kshs. 50,000 (€ 390) per month. A total of 53% of urban renters pay less than Kshs. 2,000 per month (€ 16) in rent, 26% between Kshs. 2,000 and Kshs. 4,000 (€ 16 to € 31), 16% between Kshs. 4,000 and Kshs. 10,000 (€ 31 to € 78), and only 5.5% pay above Kshs. 10,000 (€ 78).

DOI: 10.1201/9781003130383-30

House prices have increased rapidly over the last decade. In the late 2000s to 2016, the real estate sector in Kenya enjoyed a boom, with the sector's contribution to the gross domestic product (GDP) growing from 4.8% in 2010 to 8.4% in 2016. This was driven by the entry of private and institutional developers, characterized by modern high-end house designs and an increase in gated communities (Cytonn Research, 2019). The value of properties rose rapidly, with rental rates increasing by 200% between 2007 and 2017, according to the Hass Consult rental index (2013). According to the Hass Consult price index, the average value of a detached unit (bungalow, townhouse, or maisonette) in Nairobi rose by 330% from Kshs. 7,100,000 (€ 55,380) in 2000 to Kshs. 30,100,000 (€ 234,780) in 2017, with the average price for 1-, 2- and 3-bedroom detached units currently at Kshs. 13,700,000 (€ 106,860), and the average price for 4-, 5- and 6-bedroom detached units at Kshs. 40,100,000 (€ 312,780). Overall, house prices in metropolitan Nairobi (Nairobi and satellite towns) average Kshs. 140,000 (€ 1,092)/m^2 for detached units and Kshs. 99,000 (€ 772)/m^2 for apartments/flats. Rents average Kshs 577 (€ 5)/m^2 per month for detached units and Kshs. 462.6 (€ 4)/m^2 per month for apartments in the Nairobi Metropolitan area (Cytonn research, 2019). In other towns in Kenya, house prices are lower compared with Nairobi, averaging Kshs. 6,100,000 (€ 47,580) in 2016, according to the Kenya National Bureau of Statistics. Figure 26.1 shows the yearly changes in consumer price and house prices in Kenya.

The high house prices in Kenya and especially the capital city, Nairobi, are due to high land cost, infrastructure costs, high construction costs (materials, design fees, and labour), and the high cost of financing. Due to the scarcity of serviced land in good locations, land contributes approximately 25% to 40% of house development costs in urban areas, which consequently impacts on end-user prices. In Nairobi, for example, land prices have been growing at a six-year compound annual growth rate (CAGR) of 17.4%. Mid-level construction costs in Kenya range from Kshs. 44,000 (€ 343) to Kshs. 64,000 (€ 499)/m^2, depending on the level of finishes, height, and other related factors, and account for 50% to 70% of house development costs. Several parts of Kenya lack the requisite infrastructure for development, such as proper access roads

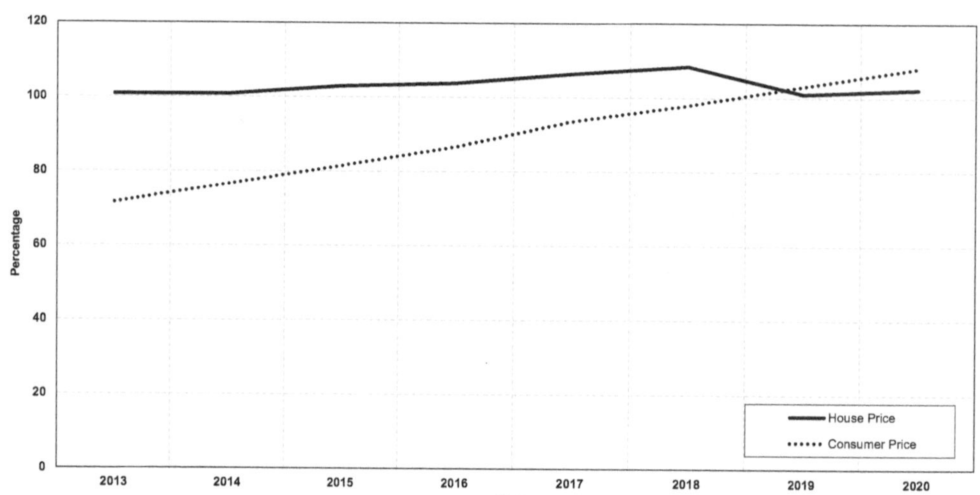

Figure 26.1 Yearly change in consumer price and house prices in Kenya (in percent)

Source: Kenya National Bureau of Statistics

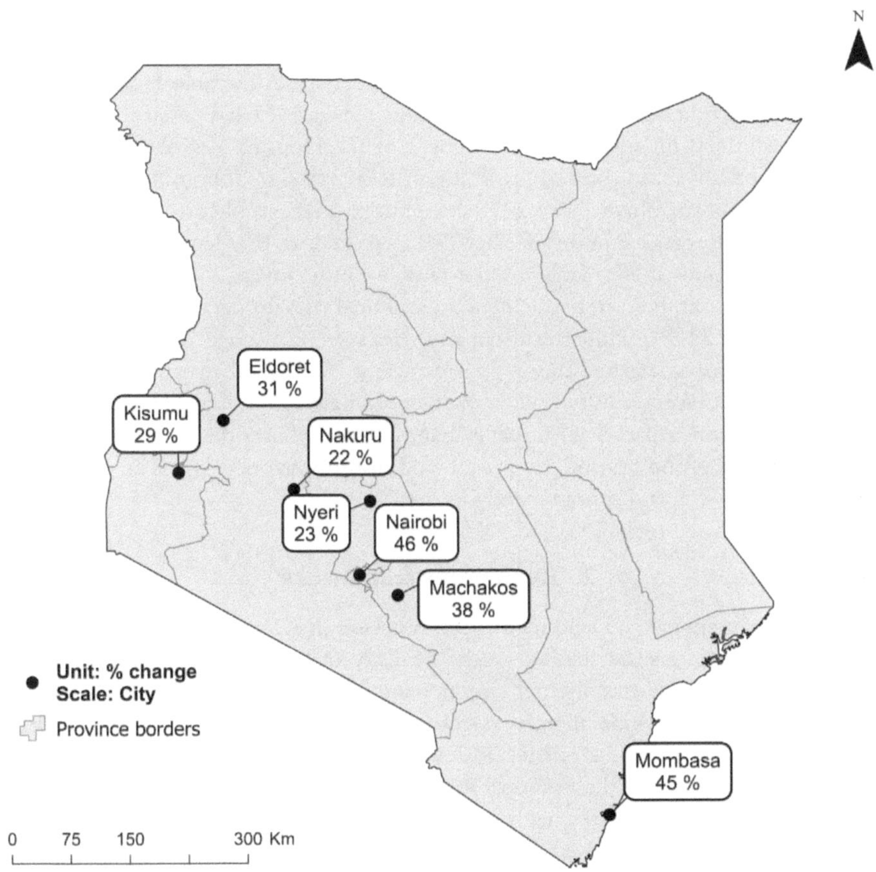

Figure 26.2 House price index change in Kenya (in percent), 2009–2020

Source: Author

and power, water, and sewerage services. Developers are forced to incur these costs, which are then passed on to the end buyer. According to research conducted by the Centre for Affordable Housing Finance in Africa in 2015 and 2016, infrastructure contributes to approximately 15% of total house development costs in Kenya. The house price index change in Kenya is provided in Figure 26.2.

House development is highly capital intensive, and developers have to look for alternative funding sources, which, in Kenya, is mainly from banks and financial institutions. Unfortunately, the cost of borrowed funds from banks has been high and averaged 18% in 2012, ranging from 11% to 25%, and in 2013, the average interest rate was 16.89%, ranging between 15.5% and 19% (Central Bank of Kenya Annual Reports, 2012, 2013). As a result of the high interest rates, the GOK in 2016 enacted the Banking (Amendment) Act 2016 that capped interest rates at a maximum 4% above the Central Bank of Kenya (CBK) base rate with a view to stabilize interest rates and house prices. Consequently, interest rates stabilized at 13.5%, but the interest rate capping regime had negative effects on the housing market and the economy, leading to reduced credit advancement to the private sector due to tightened underwriting standards by the banks. The Act was subsequently scrapped in 2019, heralding a return to a free-market interest rate regime in Kenya. The current interest rates in Kenya in a free-market environment averaged 12% in 2020, which is still high and unaffordable, and raises the cost of developments and impacts on house prices. Recently, the GOK under the Affordable Housing Programme (AHP) in the government's Big Four Agenda launched in 2017 has taken measures aimed at reducing house prices, including the amendment of the Housing Act to create the National Housing Development Fund (NHDF) to address the supply side of housing. Another measure was the enactment of the Finance Act 2019 that created the Kenya Mortgage Refinance Company (KMRC), a mortgage liquidity facility (MLF) to address the demand side of housing by availing long-term funds at affordable interest rates to house buyers.

According to Cytonn Research (2019), average annual sales for detached units in the Nairobi Metropolitan area is 22.5%, while the apartment sales volume average 23.9% per annum.

The average size of a detached unit is approximately 167m^2, and apartments average 130m^2. Tenant purchase units average 60m^2, while the size of single rooms with shared common facilities is on average 18m^2. Apart from municipal and council houses that were built in the 1950s and 1960s, the bulk of the housing by private and institutional developers in Kenya was built between 1973 and 1983, and more recently in the 2000s.

2. Regulatory framework

The property conveyancing and land registration processes in Kenya are functions of the government through land registries established in the Ministry of Lands and Physical Planning in the capital city, Nairobi, and in several other counties and sub-counties across the country. The primary statute governing conveyancing process is the Land Registration Act (LRA), which, among other things, establishes the land registries (S. 7; LRA). Other legislation include the Registration of Documents Act (RDA) and the Sectional Properties Act (SPA). It is important to mention that the RDA deals with not only agreements relating to the sale of land but also any other item, for instance, the sale of a car (Mburu, 2017). The SPA enacted in 2020 provides a means for dealing with sectional properties, such as apartments (flats), offices, and other units within a building.

The formal requirements to accomplish a legally binding conveyance are laid down in parts III and IV of the Land Registration Act (LRA) No. 3 of 2012. The binding conveyance requires the execution of an instrument of transfer in the prescribed form, usually a transfer of land/lease/charge form, duly filled with the details of the property (land reference number, size, location, etc.), names of the transacting parties and their conveyancing advocates, including their addresses and the transaction price. The execution of the instrument of transfer consists of appending the signatures of the parties (buyer and seller) on it or affixing their thumbprints or any other mark as evidence of personal acceptance of the transaction and this is done in the presence of conveyancing advocates (S. 44; LRA). It is this instrument of transfer which is presented to the land registry for registration and issuance of title in favour of the buyer.

Under S. 44(5) (LRA), the transacting parties, in addition to executing the instrument, shall attach a copy of identity card or passport, copy of personal identification number (PIN) certificate issued by the Kenya Revenue Authority, passport size photographs, and where applicable, a marriage certificate. Where the transfer relates to a lease, the consent of the lessor has to be obtained and a rent clearance certificate issued certifying that no rent is owed to the National or County government in respect to the land. Further, if the property is situated within the area of a rating authority, a rates clearance certificate by the respective rating authority is required, certifying that all outstanding rates, penalties, and other charges payable to the authority in respect of the land, including rates and other charges up to the date of request for transfer, have been paid in full. These clearances and consents (collectively the completion documents) are however not required by the land registrar. They are a part of the due diligence on the part of the buyer or their advocates to ensure that she/he does not buy a property that has unpaid liabilities and utility bills.

Sometimes there is the requirement of spousal consent to put into effect the instrument of transfer found at S. 12 of the Matrimonial Properties Act, although the LRA does not require this consent, most advocates and banks still do require spousal consent to transact, especially when dealing with registration of charges. Section 46 of the LRA requires the instrument of transfer to be stamped in accordance with the Stamp Duty Act, as evidence that all government stamp duties in respect to the transaction have been paid. Section 40 of the LRA provides that transfers should take effect immediately but does not provide timelimes for the conveyancing process.

3. The conveyance process

The registration of property transactions in Kenya take place at the land registries spread across the 47 counties in Kenya. It is important to note that not all lands in Kenya are registered and titled. Only about 40% of lands in Kenya are registered with the majority, and the sparsely populated lands remain unregistered, as they fall under the category of community lands and public lands. There are also lands that fall under the category of informal or squatter settlements, which are mostly untitled, and ownership is through informal documents such as share certificates issued by land-buying companies or property developers, allotment letters issued by local authorities, sale agreements witnessed by local administrative leaders, and so forth. In some informal settlements, especially in the slums, people have no recognizable legal rights to the land they occupy; they simply moved in, built houses, and stayed. Most of the land that people occupy in this manner is government land and land reserved for public purposes which had remained idle and unoccupied for a long time. However, today there are concerted efforts to bring all lands in Kenya under the purview of registration.

Private lands in most urban areas on which there is the development of houses are however registered, and it is safe to assume that the subject of this chapter's discussions is registered properties that are titled. Figure 26.3 shows the conveyancing and property registration processes.

The conveyancing process and the registration of outright transfers for houses is almost the same as that of the registration of mortgages. There are slight differences, which are highlighted in the steps shown in Figure 26.3.

Marketing phase

At this stage, it is important to have a *willing buyer* and a *willing seller* and to bring them to a meeting. To get a good house, or any property for that matter, it is important to take time to get the best in terms of budget, location and size, as well as considering other attendant costs and attributes in terms of security, rents, rates, availability of water, roads, power, sewer services and service charges, if any.

Estate agents come in handy to help the parties go through this step of sourcing for the house/property. The estate agents in Kenya are registered with the Estate Agents Registration Board (EARB) under the Estate Agents Registration Act Cap 533 of the laws of Kenya. While not mandatory, registered estate agents in Kenya are members of the Institution of Surveyors of Kenya (ISK), and some of them are practicing property valuers registered under the Valuers Act Cap 532 of the laws of Kenya. However, most estate agents are neither registered valuers nor members of ISK, and in this case, they are bound by the sub-contracts that they enter into with their principals. In fact, according to a Ministry of Lands report, while there are over 80,000 estate agents practicing in Kenya, only about 380 estate agents are registered by the EARB and

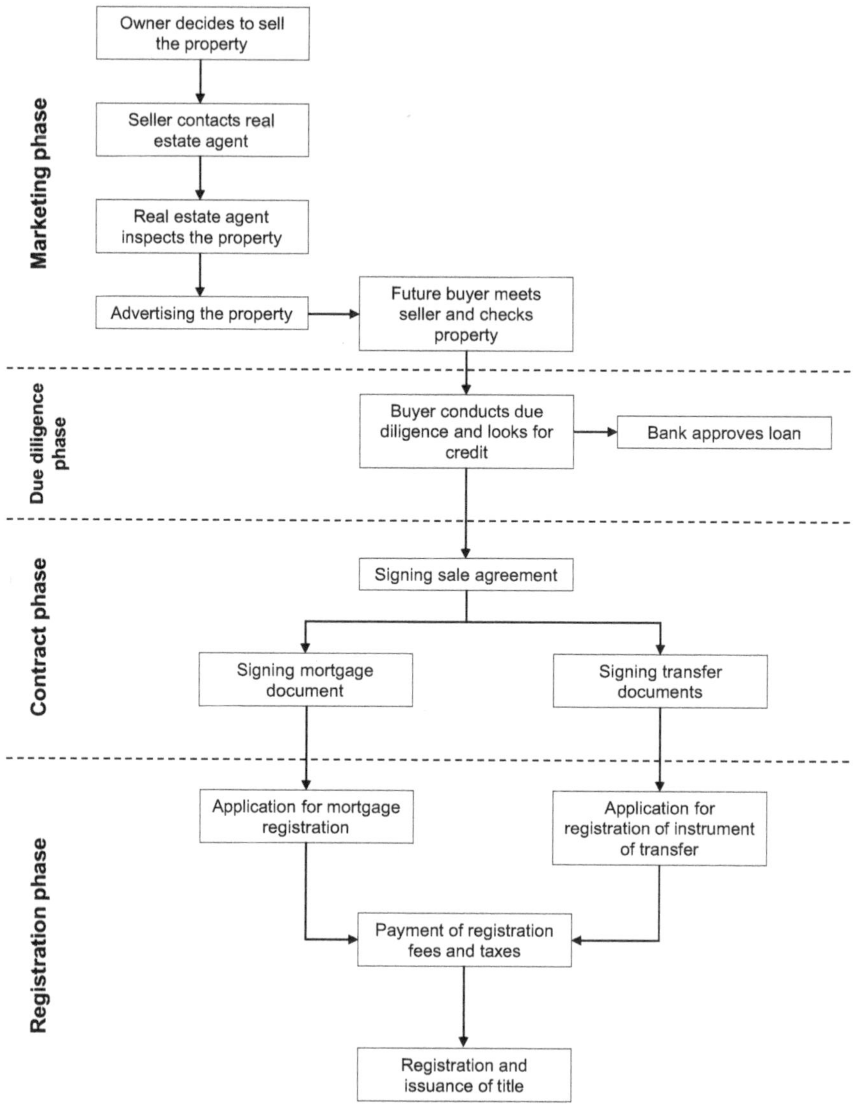

Figure 26.3 The conveyancing and property registration process for titled real estate in Kenya

Source: Author

only a handful are members of ISK. To secure their interest, it is advised that transacting parties (seller/buyer/mortgagor/mortgagee) deal only with registered estate agents who are members of the ISK and are registered and practicing property valuers. Estate agents are paid a commission as stipulated in the Estate Agents (Remuneration)(Amendment) Rules of 2002. Most agents prefer to have a written agreement deposited with a lawyer so as to secure their pay. Most agents' role ends at the point of the meeting of the buyer and seller; however, in other instances, they are also helpful in other steps of the sale, but most of these are best handled by advocates (lawyers).

Due diligence assures both the buyer and the seller of complete authenticity of the transaction they are getting into. The buyer is advised as much as practically possible to personally handle, or at least oversee, the carrying out of all necessary steps so as to ascertain the correctness of the title by themselves with the help of their agent, lawyer, surveyor, or valuer. This is because sometimes there can be serious defects in the title, in which case, the buyer needs the services of truthful advocates or agents. There are many instances of fraudulent titles finding their way into the property market in Kenya, for instance, in *Martha Chelal v Elijah Kipkemoi*. In this case, the property of the estate of Kipyegomen Chelal (deceased) had fraudulently been registered in the name of a third party who later sold it.

The court held that Section 26 of the Land Registration Act provides for indefeasibility of title and protection of title holders, but it also provides for impeachment of titles that have been procured fraudulently, by mistake or by misrepresentation. The court found and issued a declaratory order that the registration of *LR NO. MOIBEN/MOIBEN BLOCK 4(TUGEN) 115* in the name of the first defendant was fraudulently obtained and directed the land registrar to have it cancelled. An apparently innocent third-party purchaser in this case was also not spared, and they were also found not to have a valid claim on the title to the land.

Due diligence involves several factors, such as viewing the land, the title, and the identity of the owner; meeting the owner; carrying out official title searches from the Lands Office, companies registry (where applicable), registrar of persons, and from the survey and physical planning office; checking survey maps and development plans; and carefully examining the grounds, with a keen ear on the neighbours. In the case of *National Land Commission v Afrison Export Import Limited*, the Environment and Land Court (ELC) observed that an official search is not a conclusive act of due diligence. The court puts into consideration competing interests in the title, among other things, that a buyer needs to consider.

In high-risk areas, the buyer could also demand that the vendor swears that the land is not subject to any legal holdups in court or elsewhere, that there is no dispute over ownership notwithstanding positive search results, and that the land is not part of a public utility. As to whether the land/property was initially a public utility, the report of illegal allocations of public land commonly known as the 'Ndungu Report' is a useful reference document. Further, a careful examination of the spatial land use plans or development plans for the area could tell whether the land was initially set aside for public purpose. The search should also reveal any outstanding loans, liens, caveats, matrimonial claims, and annual rent payable.

Contract phase

Only after ascertaining that the title is clean and in good standing should the parties proceed to the next stage – that of concrete negotiations and the drawing up of the sale agreement. The agreement is drawn by the vendor's lawyer and fees should be borne by the vendor. When all the parties and their lawyers make the relevant amendments and make a final agreement, they then sign it and it is attested to by their respective lawyers. It is also common to have one lawyer acting for both parties; however, some people think that this may give rise to conflict of interests.

Upon signing the agreement, a commitment fee is payable according to the stipulations of the agreement; this is usually 10%, although some buyers take a big risk by paying a huge percentage of the purchase price, even up to 100%. The agreement basically lays down the rules of carrying out the transfer from start to finish. It describes the parties, the parcel subject of the agreement, the purchase price, and the mode of payment, among other things.

It is important for either party to demand that anything they feel is pertinent to the transfer be incorporated in the sale agreement in order to avoid the emergence of conflicts.

No assumptions should be made. Each party then keeps a copy of the sale agreement and makes steps or acts as per the same. The agreement also stipulates the completion period and which fees are payable when and by which party. Either party may opt to register the agreement under the Registration of Documents Act (RDA), although there is no strict requirement for that. It is important to note that the agreement is not registrable against the title. The seller then moves in quickly to secure the completion documents which include the consent to transfer (or to charge), a rates clearance certificate, and rents clearance certificates (if applicable). Utility bills also have to be cleared, including water and electricity consumption.

The buyer's lawyer, in the meantime, drafts the instrument of transfer and sends it over to the vendor's lawyer for comments and approval. The transfer, upon approval by both parties and the legal counsel, is signed and the respective lawyers attest to their client's signature, marks, or thumb imprints.

The vendor's lawyer then forwards all completion documents plus the instrument of transfer and the original title to the purchaser's lawyer, usually upon an undertaking or as per the agreement. When all the documents including the signed transfer document are in the safe custody of the purchaser and their advocates, it is likely that the vendor feels vulnerable. This is usually remedied by agreement provisions that an undertaking to pay must be issued by the buyer's lawyer to the vendor's lawyer. Further provisions could be that the balance of the purchase price be deposited with the vendor's lawyer upon their undertaking not to release a penny until a successful registration of the transfer.

Registration phase

After having cushioned both parties at this critical stage, the buyer – personally or through their lawyer or estate agent – then moves to the Ministry of Lands office to get the property valued by a government valuer for transfer, assessment, and payment of stamp duty. More often than not, the government valuers must visit the unit of land or property in question in order to ascertain the market value so as to come up with the right amount of tax payable as stamp duty and capital gains tax. The stamp duty assessment is carried out by the collector of stamp duties, usually a government land registrar or registrar of titles, and is currently assessed at 4% (of the value of the property) for properties within municipalities and cities and 2% for properties within rural areas. The document (original transfer) is then assessed, as mentioned earlier, and the amount is thereafter payable to the Tax Authority (Kenya Revenue Authority) bank account.

The original transfer and the counterparts (usually two) are stamped with the full amount, while counterparts are franked at Kshs. 20 (€ 0.16). The application for registration follows with the forms available at the land registries. These forms are filled in quadruplicate. The application includes the original title, transfer in triplicate duly signed and franked, stamp duty receipt and bank slip, valuation form, and relevant completion documents. The applicant is then issued with one of the booking forms bearing the A-book number, basically the application number. The applicant should always produce this number at the registry as the tracking number so as to see the progress of the application.

The registration process takes ten working days, according to the Ministry of Lands Service Charter, and thereafter, the name of the purchaser is entered in the land register and a title is issued accordingly. When there is a mortgage, it is entered in the register immediately after the entry of the purchaser as the new owner.

Post-registration phase

The registered title documents are then handed over to the purchaser's lawyer, and in the case that there is a mortgage, the documents are handed over or collected from the registry by the lender's lawyers. Within the stipulated timelines as per the agreement, the final payment is released to the seller while the purchaser is handed the title in their name. In the case of a mortgage, the title documents as well as the mortgage papers are kept in the safe custody of the bank while the bank releases the final payment to the seller.

4. Summarizing the conveyance process and its transaction costs

To summarize the conveyancing and property registration process for titled real estate in Kenya, the process is fairly simple and straightforward. The ownership is conveyed through an instrument of transfer duly signed by the transacting parties and witnessed by an advocate. Careful examination of title documents and due diligence is necessary to be certain of the authenticity of the transaction. The conveyancing and registration processes should be carried out with the input of professionals such as advocates and estate agents. Due to the involvement of the professionals, the process, although simple, can also be expensive, especially given that their fees are based on the market value of the property or the transaction price. The Government of Kenya, through the Ministry of Lands and Physical Planning, recently launched a Land Information Management System (LIMS) which will automate all land transactions, and this is expected to enhance transparency in the conveyancing processes, expedite the process, and lower transaction costs. Table 26.1 summarizes the transaction costs for titled real estate in Kenya.[2, 3, 4, 5]

Table 26.1 Fees and taxes during the conveyance process in Kenya

Phase	*Fees and taxes*	*Real estate*		*Tenant-purchase*	
		Seller	*Buyer*	*Seller*	*Buyer*
Marketing	Estate agent's fees	Graduated scale but say 3%	Graduated scale but say 1.5%	-	-
	Marketing fees	€ 31		-	-
	Advocates fees	-	Graduated scale but say 2%	-	-
Contract	Advocates fees	Graduated scale but say 2%	Part of advocates fee paid above under due diligence	Graduated scale but say 2%	Graduated scale but say 2%
	Bank mortgages fees	-	Advocates fee 2% of mortgages € 4.3 Registration fee	-	-
Registration	Stamp duty tax	-	4%	-	4%
	Registration fee	-	€ 4.3	-	€ 4.3
Post-registration	Capital gains tax	5% of profit	-	-	-
	Sales tax	-	-	-	-
Total (EUR for the model case price)		**€ 5,031**	**€ 7,508**	**€ 2,000**	**€ 6,004**
Total (% share to the model case price)		**5.0%**	**7.5%**	**2.0%**	**6.0%**
Excluded costs from the model case		5% CGT	2% mortgage fee	-	-

Note. Model case price: € 100,000

Source: Author

Notes

1 1 Kenyan Shilling = 0.0078 Euro (as of 1 October 2021).
2 Advocates fees are usually on a graduated scale – see schedule 1 in the Advocates(Remuneration) (Amendment) Order, 2014, available online.
3 Advocates fees basically cover drafting of documents and advice. In case due diligence will require other professionals, say land surveyors, these are charged separately.
4 Estate agents fee are usually on a graduated scale – see the Estate Agents(Remuneration) Rules, 2002, available online.
5 Tenant-purchase schemes are usually government housing projects.

References

CAHF. (2019). *Housing the economy: Assessing Kenya's affordable housing market.* Midrand: Fin Mark Trust.

Central bank of Kenya (CBK). (2012). *Bank supervision annual report 2012.* Retrieved from https://www.centralbank.go.ke/images/docs/Bank%20Supervision%20Reports/Annual%20Reports/bsd2012-r.pdf

Central bank of Kenya (CBK). (2013). *Bank supervision annual report 2013.* Retrieved from https://www.centralbank.go.ke/images/docs/Bank%20Supervision%20Reports/Annual%20Reports/BSD2013AnnualReport.pdf

Cytonn Research. (2018). *Affordable housing in Kenya.* Cytonn Investment Ltd. Retrieved from https://cytonnreport.com/research/affordable-housing-in-kenya-cytonn-weekly-162018

Cytonn Research. (2019). *Nairobi metropolitan area property report.* Cytonn Investment Ltd. Retrieved from https://cytonn.com/uploads/downloads/nma-residential-report-2019.pdf

The Estate Agents Act, cap 533, Kenya.

HassConsult Ltd. (2013). *Second quarter property index report.* Kenya. Retrieved from https://hassconsult.co.ke/images/Q22013/TheHassPropertyIndexQ2_2013.pdf

Mburu, P. N. (2017). *Strategies to modernize the land registration system in Kenya.* Groningen: University of Groningen.

National Land Commission v Afrison Export Import Limited eKLR 2020.

Republic of Kenya Estate Agents Registration Act Cap 533 of the laws of Kenya, The Government Printer Nairobi.

Republic of Kenya Registration of Documents Act 1902. The Government Printer Nairobi.

Republic of Kenya Land Registration Act, No 3 of 2012. The Government Printer Nairobi.

Republic of Kenya Sectional Properties Land Act, No 21 of 2020. The Government Printer Nairobi.

The Valuers Act cap 532, kenya.

World Bank. (2017). *World bank annual report 2017.* Retrieved November 2, 2021, from documents.worldbank.org

27
MALAYSIA

Muhammad Najib Razali, Wong Woei Chyuan and Omokolade Akinsomi

1. The Malaysian housing market

The housing market in Malaysia[1] has been performing well in terms of pricing and transaction volume since independence in 1957. The Malaysian housing sector constitutes more than 50% of total property transactions. Since 2001 the contribution of the housing sector to the property industry has been firmly between 62% and 73% (NAPIC, 2020). This indicates the importance of this sector to Malaysian economic growth. The boom in the residential sector through the increase in demand has made it valuable for property investors and developers alike, as it provides an opportunity for them to gain more profits. Malaysia witnessed a property boom during the early 1990s which prompted property players to participate in this sector. Nevertheless, the Asian financial crisis (AFC) of 1997 resulted in contraction growth in the economy which led to the housing sector sharply declining. However, since the recovery from the AFC, the housing sector in Malaysia has been progressing actively, which can be seen from the upward trend of property transactions since 2000. The records show that since the year 2001, the total transactions of the residential sector display a gradually increased trend. However, it is evident that from 2005 the total number of transactions has decreased. This is because during this period, the housing sector in Malaysia experienced an over-supply coupled with a significant period of high price, which resulted in unaffordability in the housing market. The house price index change in Malaysia is seen in Figure 27.1.

According to the Malaysian House Price Index Report 2020 by the National Property Information Centre, Malaysia (NAPIC), there has been a substantial increase in house prices since 2009 (NAPIC, 2020). This is due to the housing market becoming an investment asset for investors. As an investible asset, this leads to the property market being one of the main contributors to the economic growth of Malaysia. The property development sector has become a major investment vehicle as the demand for residential property is always consistent. Since residential property is the strong backbone in the Malaysia property market, the changes in house prices will impact the whole property market and economy. In Malaysia, house pricing is determined by a free market with limited government control. The government plays a role in establishing the policy to ensure the residential property sector functions according to rules and regulations. In recent years, Malaysia has

DOI: 10.1201/9781003130383-31

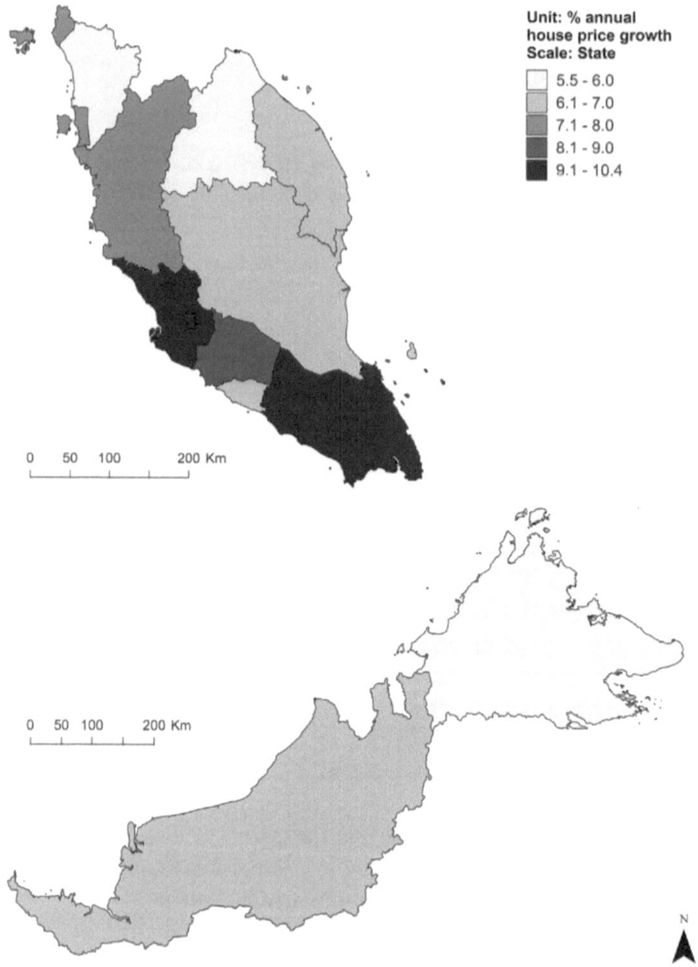

Figure 27.1 House price index change in Malaysia (in percent), 2009–2020

Source: Author

seen significant increases in house prices. House prices represent a significant portion of household wealth.

House price trend

The real estate market in Malaysia has been recovering from the global financial crisis (GFC) as an outcome of the government's economic policies. Consequently, the good economic environment has accelerated the growth of the housing sector, which triggered a significant increase of house prices across all states in Malaysia. The positive growth shown by the housing price index is beneficial for Malaysia as the overall real estate sector is one of the largest industries that drives economic growth (Jarad, Yusof, & Shafiei, 2010; Nuruddin, Bakar, & Jaafar, 2015). The contribution of the housing segment can be seen by the average total of contribution towards the overall transactions of property in Malaysia, which is approximately 65% since 2000. This

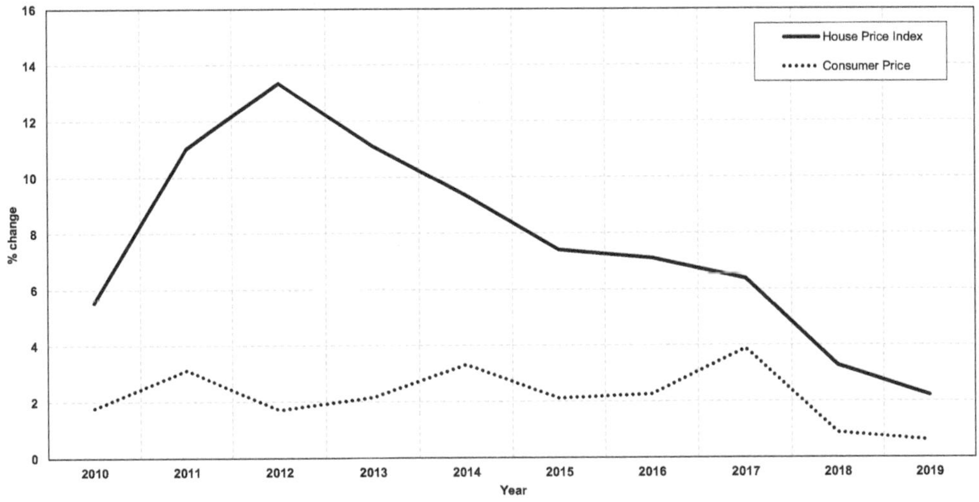

Figure 27.2 Yearly change in consumer price and house prices in Malaysia (in percent), 2010–2019

Source: Author

further indicates the significance of this sector towards the growth of the economy. However, in recent years, there has been a decreasing trend in the value of the property sector, especially after it reached its peak in 2014. In 2019, the total value of residential property stood at € 15,205 million which is slightly increased from the previous year. The trend in transactions and value of the residential sector seems to show some steady performance. However, in terms of house prices, the record shows the trend is moving upwards, where the house price average is currently at RM 430,786 (€ 90,465) compared with just RM 200,000 (€ 42,000) in 2009. According to a report by Khazanah Research Institute (KRI) (2015), most of the states in Malaysia are facing conditions of 'severely unaffordable' and 'moderately unaffordable' in terms of house prices. The statement has been strengthened by the NAPIC (2020) report, which mentioned that the Malaysia House Price Index (MHPI) has experienced continuous growth, although at the moderate level in recent years. Based on the year-on-year changes of Malaysian house price over the period 2010 to 2019 (see Figure 27.2), the significant increased changes in patterns were seen from 2010 to 2012. Nevertheless, from 2012 the trend has shown some decline in changes which indicate the house prices in Malaysia over this period have not seen much movement. During this period the average house price was between € 56,910 and € 89,670, which is considered high. The consumer price index (CPI) in Malaysia has also seen changes between 0.6% and 3.87%, which is considered stable. This indicates differences in the changes between the price of durable goods and housing. The Government of Malaysia has taken several measures to reduce the impact of the high increase in house price. Among the initiatives that have been introduced are PR1MA (1 Malaysia People's Housing Project), which is established under PR1MA Act 2012. This initiative is aimed at delivering more affordable homes. Other measures include the abolishment of the developer interest bearing scheme (DIBS) in which the purpose is to avoid property bubbles caused by high increases in house price. The involvement of institutional investors also plays a major role to overcome the house price issue in Malaysia. Among others is sustainable lending policy, which is based on the prudent and responsible lending that enables borrowers to pay and promote sufficient credit risk.

In term of housing stocks in Malaysia, there is a total of 5,775,095 units of housing scattered over all the states in Malaysia as of 2019 (NAPIC, 2020). Selangor and Johor were the leading states with the highest number of residential units. These states are also amongst the developed states in Malaysia with a high population. Less populated states such as Perlis, Kelantan, and Terengganu showed less residential stocks over the similar period. According to the NAPIC (2020) report, the majority of the exiting supply was composed of landed property which constitute more than half of the total stocks. This is followed by stratified properties which have seen increasing trend in recent years.

House price changes

When observing the monetary perspective, the volume of residential property transaction does not reflect heavily on the value of the property itself. For instance, the volume of property transacted was the highest in 2012; however, the transaction value for this year is lower than 2017. This suggests that the high-end segment made a significant contribution to the value of property transacted in 2017. Price change activities were actively occurring during the period of 2009 to 2012. Nevertheless, the downward trend has been seen post the 2012 period, which also saw house prices increase over the period. The descending trend coupled with an increase in house price patterns signal the unaffordable housing issues in Malaysia. As the change became lower in recent years with prices significantly increasing over the last decade, this signifies that the unaffordable housing issues remain unresolved. Furthermore, there is still concern on the imbalance of the market that deters the value of transacted property. Based on a recent study by Ling, Almeida, Shukri, and Sze (2017), the number of unsold residential properties is at its highest due to the mismatch between affordability and the price of newly launched properties. It also revealed that most of the unsold properties were priced at € 52,500 and above. This has created a problem in catering to the demand for property below € 52,500.

2. Regulatory framework

Malaysia has a clear regulatory and legal framework on the following components of real estate regulations in JLL Global Real Estate Transparency Index: real estate tax, land use planning control, eminent domain, real estate debt regulation, and land and property title registration. This is evidenced by Malaysia's 'Transparent' ranking status in the index in 2020. This ranking places Malaysia in the same tier as Singapore and Hong Kong. This chapter focuses on the property title component of real estate regulations.

Like many countries around the world, land matters are under the purview of states or local authorities in Malaysia according to land tenure laws, dictated by the National Land Code (NLC), 1965. Malaysia adopted Australia's Torrens system where landownership is transferred using the title and registration system. Land titles need to be registered with the land office (Land Registry) in the state or district where the subject property is located to complete the transfer of property rights from the seller to the buyer. Land titles in Malaysia can be divided into freehold and leasehold, based on the duration of ownership entitlement. A freehold title gives its owners the perpetual rights over a property. A leasehold property on the other hand comes with a maximum leasehold period of 99 years. Landowners of leasehold land can however apply for an extension of the lease period from the state government for a fee that is calculated based on the extension period, land size, and the market value of the land.

Ownership restrictions exist on properties that are classified as *Bumiputera Lot* and *Malay Reserved*. These properties can be sold or leased only to Bumiputera buyers who are inclusive of Malays and Orang Asli or indigenous peoples of Malaysia. *Bumiputera Lot* arises from a development quota system that mandates certain percentages of new units to be allocated as *Bumiputera Lots* (e.g. 30% in Kuala Lumpur). Developers need to apply for the release of these Bumiputera quota units to non-Bumiputera buyers. Malay reserved land known as *Tanah Rizab Melayu* are reserved only for Malay Muslims. Properties built on Malay Reserved Land strictly cannot be transferred or leased to non-Malay buyers. Ownership restrictions also exist in low- and medium-cost housing units or housing schemes sponsored by the government in the primary market where these houses are available for sale only to low- to-medium-income householders. Owners of these subsidized units are subjected to a typical 5–10 years moratorium holding period. Foreigners are not allowed to buy properties with restrictions as previously stated. Moreover, they are not allowed to buy properties valued less than € 0.21 million in most of the states in Malaysia.

3. The conveyance process

This section of the chapter describes the conveyance of properties in Malaysia. The conveyance process in Malaysia can be divided into the following phases: marketing phase, pre-title registration phase, title registration phase, and post-title transaction phase.

Marketing phase

Most of the real estate transactions are carried out with the help of real estate agents or negotiators in Malaysia, with 'real estate agents' referring to fully accredited (academic qualification and experiences) agents who registered under the Board of Valuers, Appraisers, Estate Agents and Property Managers (BOVAEP). They are allowed to set up their own real estate agency. Real estate negotiators are agents who are registered under a real estate agency recognized by BOVAEA. A person can become a real estate negotiator by attending a full two-day course related to real estate agency and laws. Note that the terms 'real estate agents' and 'real estate negotiators' are used interchangeably in this chapter.

In Malaysia, real estate agents are usually hired by property sellers instead of buyers. Agency contracts can be exclusive or non-exclusive depending on the negotiation between the property agents and the sellers. Sellers who are desperate to dispose their properties and those who expect aggressive marketing activities from the property agents may be willing to offer exclusive selling rights to the agents. This exclusivity period can go up to a period of three months.

Real estate agents are remunerated based on a percentage of property selling price. The maximum fee is 3% as dictated by BOVAEP. This fee is usually paid by the sellers. In terms of marketing strategies, online property portals such as *PropertyGuru.com.my* and *iProperty.com.my* are common advertising tools used by property agents in Malaysia. ProperyGuru.com claims to have access to 12 million online property seekers every month and is offering advertising packages to property agents. The subscription price ranges from € 140 to a maximum of € 817 depending on the number of concurrent listings made available to the agents and other services provided by the portal. Real estate agents also advertise through their company website, social media, and on banners and sign boards at the property site.

Pre-title registration phase

This phase begins with the signing of an offer to purchase form between the buyer and the seller at the agent's office. This form specifies all the sales terms such as offer price, encumbrances, vacant possession, and the date to sign the sales and purchase agreement. This contract is binding where typically a 1% of the purchase price is paid by the buyer to seller as earnest deposit. The remaining 9% is to be paid upon signing of the Sale and Purchase (S&P) Agreement and the balance 90% is paid within three months from the S&P Agreement. For property purchase transactions that are financed by bank loans, buyers will sign into the S&P Agreement only after successfully obtaining a property loan from the banks. The 1% earnest deposit is refundable if buyers fail to obtain a property loan.

In Malaysia, there exists two types of property transactions. The first type of property sale involves the purchase from a property developer – in this instance, this involves property under construction, and this is governed by the Housing Development (Control and Licensing Act, 1966). Alternatively, property sale and purchase are between the buyer and seller (owner), and this is governed by a private contract which includes terms and conditions in the form of an S&P Agreement as agreed by the buyer and the existing owner. Buyers and sellers are advised to secure the services of their respective lawyers during the transaction process. Once the S&P Agreement is agreed between the buyer and owner, a number of documents are to be signed including the S&P Agreement, Memorandum of Transfer Form (Form 14A) under the National Land Code (NLC) 1965, (this is for individuals with a property title), or a Deed of Assignment (this is for properties without an individual title).[2]

Title registration phase

In Malaysia, once the seller and owner are in agreement and have signed the S&P Agreement, the property is registered under two phases. The first phase is with the stamp office, when the adjudication Form 14A/Deed of Assignment will be completed and submitted to the stamp office. This can be done manually or online.[3] Once this has been accepted by the stamp office, the lawyer will be provided with an adjudication number. The next phase is that the property will be valued by the Valuation and Property Services Department, also known as Jabatan Penilaian dan Perkhidmatan Harta (JPPH), and a report will be prepared and sent to the stamp office for stamp duty evaluation. This process usually takes one to eight working days depending on the complexity of the subject property. The stamp duty is between 1% and 3% of the value of the property and must be paid within 30 days from the notice of assessment date issued by the stamp office unit.

For property without an individual title, the transfer of property rights from seller to buyer is considered done once the Deed of Assignment is duly executed and stamped. The Deed of Assignment is a legal instrument that conveys the transferring of property rights from the seller (assignor) to the buyer (assignee). For property with an individual title, the next phase is to formally register the new ownership on the property at the state- or district-level land office. Documents needed at this phase include the original land title document, certified copies of the buyer's and seller's national registration identity card/passport (company Memorandum and Articles of Association for corporate buyers/sellers), stamped Form 14A, and a copy of current quit rent and assessment receipt. The land office will endorse the name of the buyer on land title document. This land title document will be kept in the strong room at the land office. A new

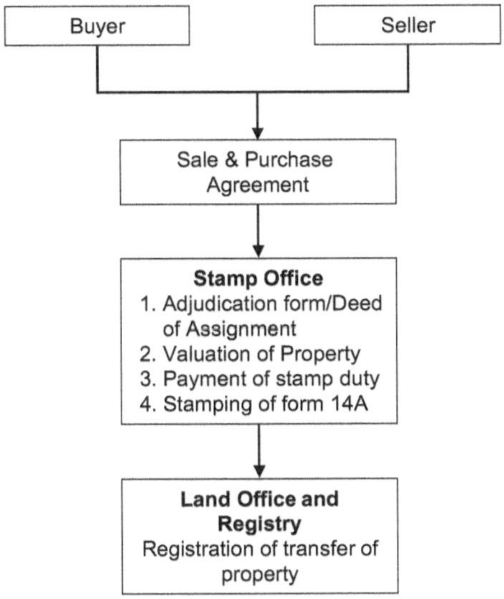

Figure 27.3 Registering of residential properties in Malaysia

Source: Author

land title document will be issued to the buyer. This registration process is completed within one working day from the date of complete document submission.

Post-title transaction phase

Sellers are liable to capital gains tax; this differs from citizens, permanent residences, and non-citizens. For citizens and permanent residences, the real property gains tax rate of 5% is chargeable if the property asset is disposed after the fifth year after acquisition. Thereafter if the property is disposed in the fifth year a 15% property tax rate applies, if sold in the fourth year a 20% tax rate will apply, and within three years a 30% tax rate applies. For individual non-citizens the tax rate differs from that of Malaysian citizens. The real property gains tax rate of 10% is chargeable if the property asset is disposed after the fifth year after acquisition. Thereafter, if the property is disposed in the fifth year or below, a 30% property tax rate applies. Figure 27.3 illustrates the registering of property with titles in Malaysia.

4. Summarizing the conveyance process and its transaction costs

In summary, the land title entitlements are clearly defined in Malaysia. The Malaysian government has striven to make the land registration process efficient using an online registration system at the submission phase. The duration of transfer generally takes two working days for less complicated properties with all the required documents ready. The transaction costs for a € 60,900 (median house price in 2019) property financed by bank loan are around 7.2%, which consists of 2.2% of stamp duty fees, 2% of legal fee, and 3% of property agent commission. Table 27.1 presents the fees and taxes during the conveyance process in Malaysia.

Table 27.1 Fees and taxes during the conveyance process in Malaysia

		Real estate	
Phase	*Fees and taxes*	*Seller*	*Buyer*
Marketing	Real estate agent's fees	3%	-
	Marketing fees	€ 63 for 90 days at iproperty.com.my portal	-
	Energy performance certificate	-	-
Contract	Legal fees	-	2%
	Bank mortgages fees	-	
Registration	Stamp tax	-	2.3%
	Land registration fee	-	Depends on states [1] Flat rate: € 21 (Kuala Lumpur) Variations: Up to € 420 (Johor)
Post-registration	Capital gains tax	5–30%	-
	Sales tax	-	-
Total (EUR for the model case price)		**€ 3,063**	**€ 4,321**
Total (% share to the model case price)		**3.1%**	**4.3%**
Excluded costs from the model case		5–30% CGT	-

Note. Model case price: € 100,000. The flat rate in Kuala Lumpur (€ 21) was used as the model case.

Source: Author

Notes

1 1 Malaysian Ringgit = 0.21 Euro (as of 1 October 2021).

2 Properties without individual titles include high-rise properties and residential units located in a gated community. These properties carry strata titles with shared common areas such as swimming pool, gym, elevators, and lounge.

3 Submission of the Form 14A/Deed of Assignment can be submitted online. Retrieved from http://pinharta.hasil. gov.my under the subpage http://stamps.hasil.gov.my under stamps.

References

Jarad, I. A., Yusof, N. A., & Shafiei, M. W. M. (2010). The organizational performance of housing developers in Peninsular Malaysia. *International Journal of Housing Markets and Analysis, 3* (2), 146–162.

Housing Development (Control and Licensing Act). (1996). Retrieved from http://www.commonlii.org/my/legis/consol_act/hdala1966416/

Khazanah Research Institute (KRI). (2015). *Making housing affordable, Perpustakaan Negara Malaysia, Kuala Lumpur.* Retrieved from https://www.scribd.com/document/467225474/FINAL-Full-Draft-KRI-Making-Housing-Affordable-with-hyperlink-220815-1

Ling, C. S., Almeida, S., Shukri, M., & Sze, L. L. (2017). Imbalances in the property market. *BNM Quarterly Bulletin, Quarter, 3*, 26–32.

National Property Information Centre. (2020). *Property market report 2009–2019.* Retrieved November 20, 2021, from www.napic.jpph.gov.my

Nuruddin, A. R., Bakar, S. P. S. A., & Jaafar, M. (2015). Unveiling the challenges faced by Malaysian housing developers through government policy changes. *Journal of Construction in Developing Countries, 20*(2), 37–52.

28
NIGERIA

Omokolade Akinsomi, Emeka Eleh and Olusegun Olanrele

1. The Nigerian housing market

The total number of formal residential dwellings in Nigeria[1] consists of 10.270 million (CAHF, 2020). Nigeria with a population of 202 million (World Bank, 2020) faces housing deficit problems. Recent estimates as at 2019 put the housing shortage in Nigeria at 22 million houses (Moore, 2019), which requires the construction of 700,000 units annually to reduce this housing shortage. The National Bureau of Statistics (NBS) reported an annual addition of 100,000 units to the housing stock (NBS, 2015), which is 14.3% of the estimated annual supply to offset the shortage.

As at 2019, the outstanding mortgage in Nigeria stood at 32,260 and the size of the mortgage market to GDP in Nigeria stands at 0.17% (CAHF, 2020). Access to finance is an obstacle in developing new houses and fast-tracking homeownership; the interest rates on mortgage is between 17% and 25%. However, the Central Bank of Nigeria (CBN) in conjunction with the Family Homes Fund Limited in September 2020 approved a sum of 200 billion Naira (€ 0.42 billion) to construct 300,000 homes targeted at low-income earners with interest rates at no more than 5% (Nairametrics, 2020). The 5% interest rate agreed was between the CBN and Family Homes Fund Limited, and it is expected that the low interest will impact future homeowners as well.

Moore (2019) highlights the issues affecting housing delivery in Nigeria as shortage in housing development funds, housing finance, urbanization, and the explosion of rural-urban migration. Ekpo (2019) attributes this housing shortage to insecurity in the land tenure system, high land cost, infrastructure, government bureaucracy, lack of housing information system/database, multiple taxation, high interest rates, high inflation, high costs in property development, corruption, and the continuous devaluation of the naira.

The Nigerian housing market is sub-divided into rental markets where rent is determined as well as the owner or asset market where capital values are formed. The last report by the National Bureau of Statistics in 2010 highlighted that the owner-occupied household percentage stood at 68% while renter status was estimated at 16.1%.

There is a lack of accepted house price index in Nigeria. For this analysis we employed rental data from Ubosi Eleh + Co. – a leading real estate consultancy in Nigeria.[2] As seen in Figure 28.1, this shows the rental percentage changes and inflation percentage changes

DOI: 10.1201/9781003130383-32

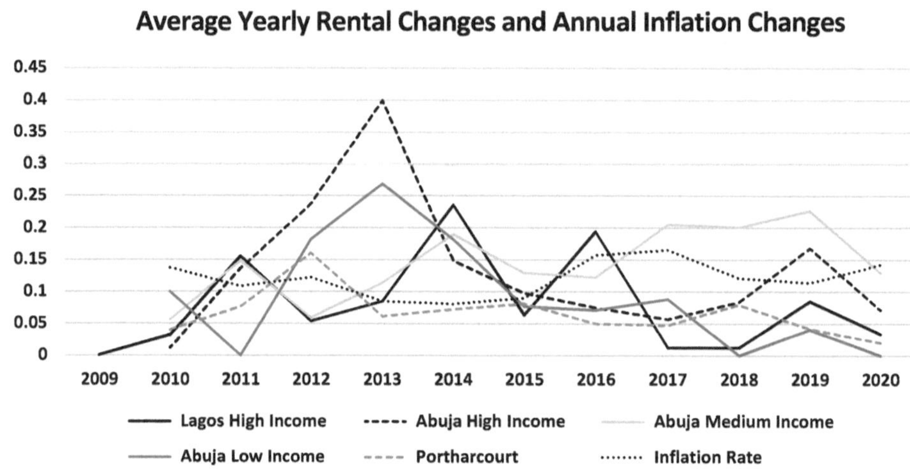

Figure 28.1 Rental value changes and inflation changes in Nigeria

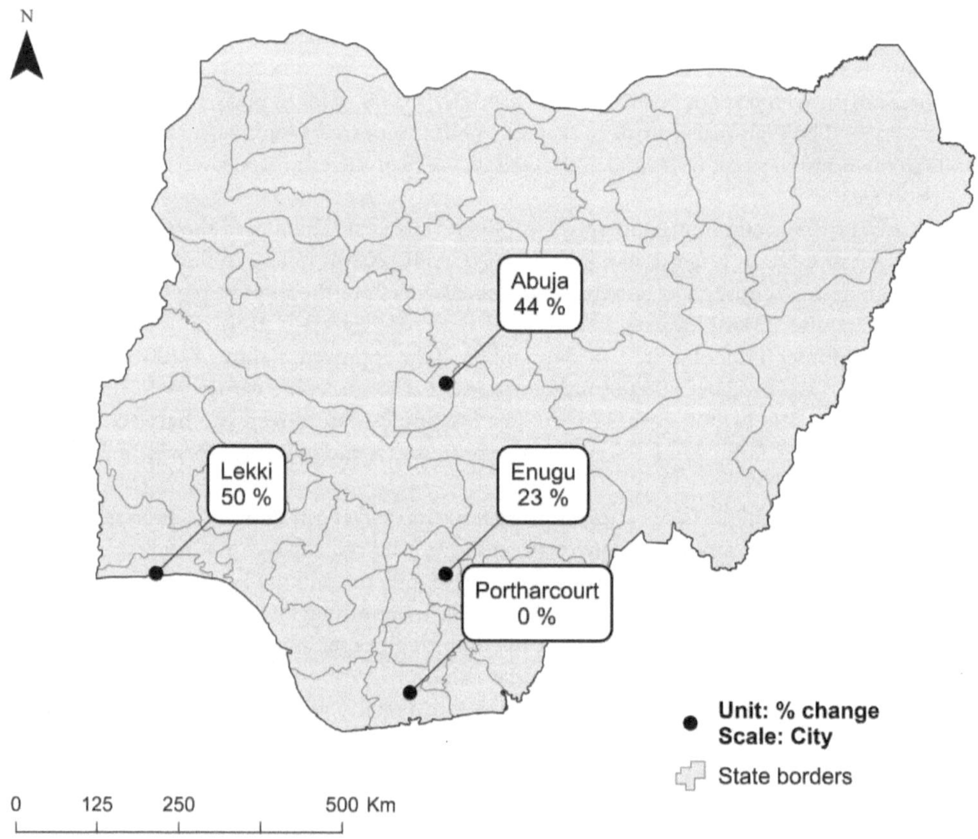

Figure 28.2 Price change of 3-to-4-bedroom – semi-detached (in percent), 2015–2020

Source: Author

from 2009 to 2020. The data cover major cities in Nigeria such as Abuja, Lagos, and Port-Harcourt.

Recently, the rental values in major cities have witnessed a reduction in percentage rentals based on 2020 figures of all housing markets observed including Lagos High Income, Abuja High Income, Abuja Medium Income, Abuja Low Income, and Port-Harcourt. This could be because of the COVID-19 pandemic, which may have led to change in rental values being lower than the change in percentage values of inflation. We observed that average rentals in Nigeria between 2009 and 2020 were volatile. This could also be attributed to the singular source of data.

Due to paucity of data availability, in terms of transacted property prices in Nigeria, we observed 3-to-4-bedroom semi-detached houses from 2015 to 2020[3] as illustrated in Figure 28.2. We employed transaction data from Ubosi Eleh + Co. The results show that there has been a 27.8% increase in the transacted price of 3-to-4-bedroom semi-detached houses in Nigeria between 2015 and 2020. Lagos represented by the Lekki metropolis had the highest increase in transacted prices at 50%,[4] followed by Abuja with an increase of 44.44%, and Enugu with an increase in transacted price of 22.6%. During the period investigated, the transacted price of Port-Harcourt remained unchanged.

2. Nigeria housing regulatory framework

In Nigeria, government, over time, has made the provision of housing an objective of governance and national development. The regulatory framework for housing provision originated from the British rule establishment of the Lagos Executive Development Board (LEDB) in 1928 through the Town Planning Ordinance Cap 95 of 1928. This was a result of the need to have a decent hygienic habitat in response to an outbreak of bubonic plague between 1924 and 1930 in Lagos. The LEDB was charged with responsibilities for housing and infrastructure development. Since then, the Government of Nigeria has been playing a huge role in housing through direct provision, policies and regulation.

In 1972, the LEDB transformed into Lagos State Development and Property Corporation (LSDPC), which is the property development arm of Lagos state with expansion to commercial (Office and Retail) property development in addition to housing provision. The LSDPC was responsible for the outstanding housing development during Governor Lateef Jakande's administration as Governor of Lagos state between 1979 and 1983. In order to make home finance affordable and accessible to civil servants, the Lagos Building Investment Company (LBIC) was established in 1980 to provide mortgages for allottees of the low-income houses developed by the LSDPC.

Until the early 1990s, the Government of Nigeria gave little attention to public housing provision. The focus was on the provision of accommodation for government senior staff through development of staff quarters (Nigeria Railway, Nigeria Airways, National Electric Power Authority [NEPA], Nigerian Television Authority [NTA], Nigerian National Petroleum Corporation [NNPC], Central Bank of Nigeria [CBN], Nigerian Ports Authority [NPA], Nigerian Shippers' Council [NSC], etc.). The early years of the LEDB were devoted to development of living quarters for the officers of the colonial administration as well as senior civil servants. The elites and business partners of the government also benefited from LEDB activities. The federal Government of Nigeria in 1971 established the National Council on Housing under the then National Development Plan. The Employee Housing Scheme Decree 54 of 1979 also focused more on staff housing and housing estates developments. The defunct regional governments also established independent Housing Corporations in 1959 for the purpose of housing provision at the regional levels, which have today become the State Housing Corporation in each successive state in Nigeria.

The Federal Housing Authority (FHA) established in 1973 was charged with making proposals to the federal government on housing and ancillary infrastructural services and implementing those approved by government. The FHA is the main public organ in the provision of housing in Nigeria. Under the national housing programme, festival town (Festac town) and Gowon Estate, Ipaja both in Lagos as well as the first ever low-cost housing estates in eleven state capitals across the country were developed were developed. This marked the first major federal government effort in providing affordable housing to Nigerian citizens on long-term mortgage repayment arrangements.

The regulatory framework from the Nigerian government cuts across every area of importance in ensuring housing for citizens, especially during the era of the various development plans. In 1955, the Nigerian Building Society (NBS) was established to provide finance for homeownership in term of mortgage creation. The Staff Housing Loan Scheme for employees of both public and private sectors was established to provide housing loan for interested employees. The pre-independence NBS later transformed into the Federal Mortgage Bank of Nigeria (FMBN) in 1977. The FMBN serves as the main engine room for public housing delivery with a dual function of both primary and secondary mortgage institution.

The huge deficit in meeting the need for the public housing and the unsustainable cost of improving (management and maintenance of) staff quarters resulted in the federal government resorting to finding ways of meeting the public housing demand. Then in 1990, a new National Housing Policy (NHP) was enacted which resulted in the establishment of National Housing Fund (NHF) in 1992. The housing policy looked into various aspects of housing provision including access to land, finance, building materials, quality, and research. The Mortgage Institutions Act, a precursor to NHF Act, ushered the establishment of Primary Mortgage Institutions (PMI) which now serves as the link/agent between individuals and the FMBN to access the housing fund. The NHF Act provided for mandatory contribution of 2.5% of the salary of every worker earning up to € 6.3 and above per month into the housing fund. The NHP has been reviewed at various times (2006, 2012, 2018) with the latest amendment in 2019 to meet the current realities. The review affected the amount of loan (90% of cost up to € 0.03 million), interest rate (6% fixed), loan term (up to 30 years) and participation of the private sector in housing development and supply. The National Mortgage Refinance Company (NMRC) was established in 2013 to further strengthen the housing finance sector.

Generally, the government housing provision cuts across all income groups but mostly focuses on the low-income group. The NHF is by design targeted at the low-income group with maximum funding amount of € 0.03 million. Most of the private sector developments in the cities are priced beyond € 0.03 million and the FMBN also finances the private sector development under the Estate Development Loan scheme. The housing policy and regulatory framework guarantees the low-cost housing, employees housing, staff housing, and private sector estate development loan. The private sector group of the Real Estate Developers Association of Nigeria (REDAN) also contributes to low-income housing supply.

The provision of housing is two-fold:

1 *Self-development*: Individual construction of buildings and a large percentage is self-constructed by getting materials and artisans to work on-site. Eligible individuals acquire land and then secure the housing loan to construct the building.

2 *Outright purchase*: Both public and private housing developments are available for outright purchase. Individuals can secure mortgage for the purchase of housing units.

3. Regulatory framework

Nigeria operates a National Land Policy anchored on the Land Use Act of 1978 (the Act). The Act vests all land in the territory of each state of the Federation on the governor of the state for purposes of administration and management. The Act empowers the governors to grant title to land and consent to assignments of land from one party to the other. Since the Act has vested ownership and administrative powers over land to the governor of the state, each state and the federal government maintains a separate Land Registry. A basic and uniform requirement for conveyance is that a vesting document for transfer of title must be in writing, signed by the parties and witnessed. It must also contain the sale consideration, full description of the property involved and other basic covenants. The possession of registered title remains the only easily verifiable evidence of title. It must however be stressed that only about 3% of the land area in Nigeria have registered title (Atilola, 2010). Transactions still take place for the huge unregistered portion, but such transactions usually go through a more strenuous ownership verification process.

4. The conveyance process

This process basically involves the agreement by the buyer (assignee) and seller (assignor) to engage in a property transaction and transfer title from one to the other. The buyer and seller can act directly for themselves or through agents. Whilst it is not mandatory that an agent must be involved, especially in private sector transactions, agents have gradually been used. Estate agency practice is not fully regulated, and many quack local bodies exist. The only regulated body of estate agency professionals is the Nigerian Institution of Estate Surveyors and Valuers (NIESV). An agent will be expected to act objectively and with integrity and owes a duty of care to his principal. Agents' fee is usually 5% of the sale price, but this is negotiable. Figure 28.3 shows the conveyancing process in Nigeria.

Marketing phase

A typical conveyance process involving an agent will commence with the agent being instructed by the seller to sell a property or by the buyer to find a suitable property to purchase. There is no mandatory exclusivity, and the agent operates at the discretion of his principal. The agent usually advertises the property or requests amongst his colleagues, in property magazines, newspapers, and online channels. Advertising/marketing/sourcing costs are usually borne by the agent except in rare cases.

The process will usually involve inspections of the property by the purchaser/his agents and most negotiations are done on private treaty basis. The process will also involve several offers and counter-offers until an agreement on price and terms are reached, usually without prejudice. At that point, the buyer's solicitor will conduct a title search on the property. The said solicitor will be paid by the buyer, and the seller may or may not retain a solicitor. Where a mortgage is involved, the bank will usually take over the process once an agreement is reached on price and terms.

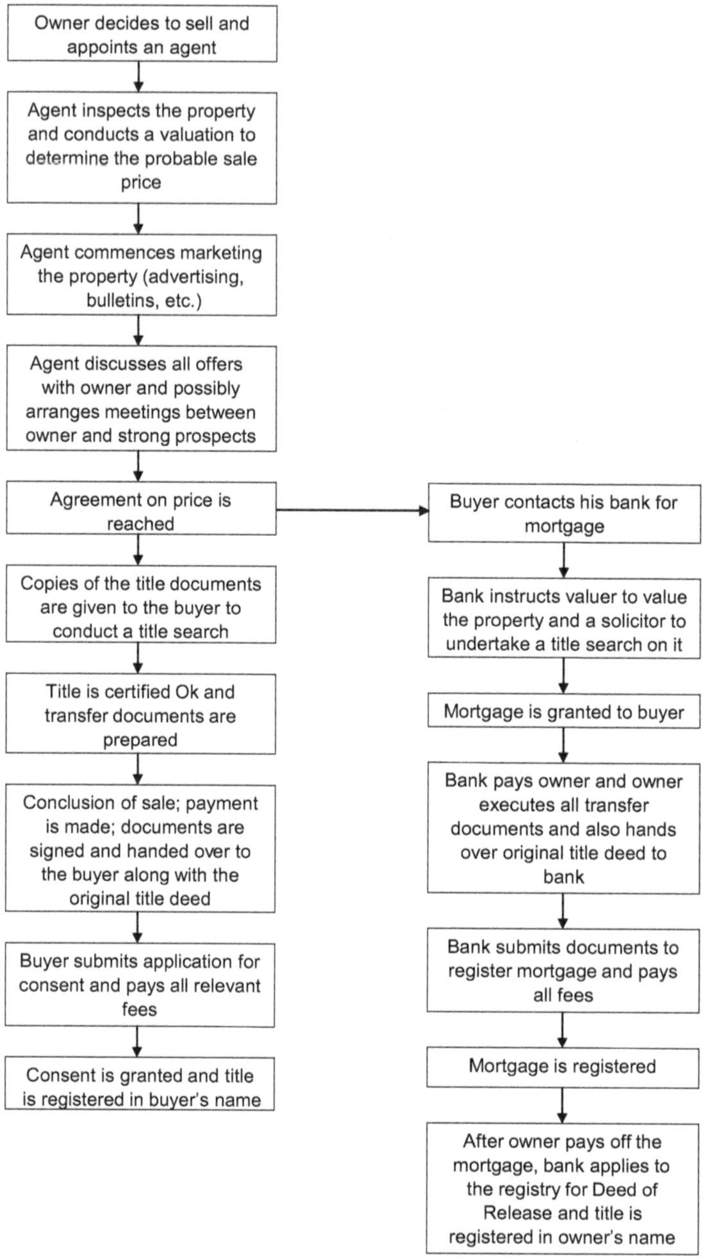

Figure 28.3 The conveyance process in Nigeria

Source: Author

Contract phase

Once the title is verified to be in order, the Deed of Assignment or Power of Attorney, purchase receipt, and consent applications are prepared. At an agreed date, the documents are executed and together with the original title documents handed over to the purchaser upon full payment for the property. Most properties are usually sold on cash basis or agreed instalments. The level of mortgage penetration is quite low and is less than 1% (Siyan, Adegoriola, & Erivwode, 2019). The seller will pay all levies and taxes up to the date of conclusion. Where the property is in a multi-occupied block or estate, the buyer will automatically assume the position of the seller and if the property is occupied by tenants, they are notified of the change of ownership. Upon conclusion of the transaction, the buyer can choose to proceed immediately to seek the governor's consent to the transaction and title is effectively registered in the name of the new owner. The entire process is usually undertaken by solicitors, but there is no law preventing the buyer or his agent from doing so. However, the process of obtaining consent is deemed a separate assignment and remunerated separately. However, there is no time limit for applying for consent.

Registration phase

The land registration system in accordance with the Land Use Act of 1978 takes two similar forms: the acquisition of land through the government (state land) allocation or through private purchase from indigenous landowner (non-state land). Where state land is allocated to a beneficiary, the process will culminate in the grant of statutory right of occupancy and registration of the title in the name of the beneficiary upon fulfilment of all the conditions. Where a non-state land with a registered root of title is acquired, the purchaser will need to apply for the consent of the governor for the title to be registered in his favour. Where non-state land without a registrable root of title is acquired, the purchaser will go through a slightly lengthier process of seeking for a grant of statutory right of occupancy and registration of the title of the new acquisition in his favour. All subsequent transactions in the three instances previously stated will require the application for the grant of governor's consent to the transaction. Across the states in Nigeria, the preceding procedures are similar with slight variations.

The Land Registry Directorate in the Lagos State Lands Bureau has been described as the most developed Land Registry in Nigeria and by extension in West Africa (Lagos State Land Bureau, 2014). Therefore, the registration process in Lagos State is a good representation of the Nigerian market. In 2017, an integrated land administration and automation system called EGIS was adopted and established to reduce the time frame for land rights documentation and registration.

The grant of statutory right of occupancy and registration of the title over state land takes several steps across departments units/section in the State Land Bureau as shown in Figure 28.3, spanning about 90 days but can take much longer.

The process of obtaining the required consent to property assignment follows a similar procedure. The entire land right registration process is quite expensive, time consuming, and tedious and may involve a series of formal and informal payments. The consent process entails the submission of the Deed of Assignment or Power of Attorney or other vesting documents and will culminate in the registration of the title in the name of the new owner or assignee. The process is listed by the government department to last one month but can take much longer. For a typical sales transaction in the state, the fees payable will be a consent fee of 1.5% of the assessed value of the property, registration fee of 0.5%, stamp duty of 0.5%, capital gains tax of 0.5%, and mortgage registration of 0.025% of the mortgage sum (see Table 28.1).

Table 28.1 Fees and taxes during the conveyance process in Nigeria

		Lagos		*FCT-Abuja*		*Port-Harcourt*	
Phase	*Fees and taxes*	*Seller*	*Buyer*	*Seller*	*Buyer*	*Seller*	*Buyer*
Marketing	Real estate agent's fees	5%	5%	-	-	5%	5%
	Marketing fees	-	-	5%	5%	-	-
Contract	Legal fees	5%	5%	5%	5%	5%	5%
Registration	Consent fee	-	1.5%	-	8%	-	10%
	Capital gains tax	0.5%	-	5%	-	10%	-
	Stamp duty	-	0.5%	-	5%	-	0.75%
	Title registration fee	-	0.5%	-		-	3%
	Mortgage registration fee	-	0.025%	-	2%	-	2%
Post-registration	Legal fees	-	5%	-	5%	-	5%
	Sales tax	-	-	-	-	-	-
Total (EUR for the model case price)		**€ 10,000**	**€ 17,525**	**€ 10,000**	**€ 30,000**	**€ 10,000**	**€ 30,750**
Total (% share to the model case price)		**10.0%**	**17.5%**	**10.0%**	**30.0%**	**10.0%**	**30.7%**
Excluded costs from the model case		0.5% CGT	-	5% GCT	-	10% GCT	-

Note. Model case price: € 100,000

Source: Author

5. Summary

Access to land and land documentation is guided by the Land Use Act of 1978 (the Act). The Act vests the right over all land in each federating state in the hands of the governor of the state to hold in trust for all Nigerians. The governor is empowered to allocate land to interested and eligible individuals or entities (S. 5 of the Act) on a leasehold interest of 99 years or fewer evidenced with a statutory right of occupancy certificate called 'Certificate of Occupancy' (C of O), (S. 9 of the Act). The land allocation to individuals for purpose of housing will not extend half a hectare of land. The transaction over a piece of land so granted (assignment, sublease, mortgage) must obtain the consent of the governor to make it valid, legal, and enforceable (S. 22 of the Act). However, the process of documentation or registration (securing the land title by getting C of O, and/or obtaining the governor's consent to subsequent transactions) is bureaucratic and time consuming. The process needs to be revised to ease access to adequate housing and reduce the widening housing deficit in Nigeria.

Notes

1 1 Nigerian Naira = 0.0021 Euro (as of 1 October 2021).
2 Data should be employed with caution, as these data emanate from a single source and may not be a true reflection/aggregate of the rental market in these cities.
3 The observational Data on 3-to 4-bedroom detached houses in Nigeria by Ubosi Eleh + Co after careful consideration is regarded as the most reliable and consistent.
4 We explored further data from Northcourt Nigeria, a real estate consultancy company domiciled in Nigeria. Their data from Northcourt (2021) report shows that between 2015 and 2020, the transacted price increases in Lagos was 8.9% – the report covered Ikoyi, Victoria Island, Lekki Phase 1, Ikeja, 1004, Magodo, Agungi, Sangotendo and Abraham Adesanya. The disparity in results reveals the lack of consistency or the absence of a standardized housing index in Nigeria.

References

Atilola, O. (2010). *Land administration reform in Nigeria, issues and prospects, INV 4, land policies in Africa, FIG congress 2010*. Retrieved from https://www.fig.net/pub/fig2010/papers/inv04/inv04_atilola_4758.pdf

Center for Affordable Housing Finance Africa. (2020). *Housing finance in Africa: A review of Africa's housing finance markets* (pp. 1–300). Retrieved from https://www.fsdafrica.org/wp-content/uploads/2019/08/CAHF_Housing-Finance-in-Africa-Yearbook-2016.09-1.pdf

Ekpo, A. (2019). *Housing deficit in Nigeria: Issues, challenges and prospects. Affordable housing investment summit 2019*. Retrieved from https://www.cbn.gov.ng/out/2020/rsd/efr%20vol%2057%20no%204%20december%202019%20housing%20deficit%20in%20nigeria%20issues,challenges%20and%20prospects.pdf

Lagos State Government Lands Bureau. (2014). *Handbook of activities of the directorate of land services*. Retrieved from https://landsbureau.lagosstate.gov.ng/

Moore, E. A. (2019). *Addressing housing deficit in Nigeria: Issues, challenges and prospects. Affordable housing investment summit 2019*. Retrieved from https://www.cbn.gov.ng/out/2020/rsd/efr%20vol%2057%20no%204%20december%202019%20addressing%20housing%20deficit%20in%20nigeria%20issues,%20challenges%20and%20prospects%20moore.pdf

National Bureau of Statistics. (2015). *Nigerian real estate sector*. Retrieved from https://www.nigerianstat.gov.ng/

Northcourt. (2021). *Nigeria real estate market outlook* (pp. 1–57). Retrieved from https://www.proshareng.com/news/PROPERTY%20&%20REAL%20ESTATE/Northcourt-s-2021-Nigeria-Real-Estate-Market-Outlook/55130

Olisah, C. (2020). CBN approves N200 billion housing loan for 300,000 households at 5% per annum. *Nairametrics*. Retrieved November 20, 2021, from https://nairametrics.com/2020/09/16/cbn-approves-n200-billion-housing-loan-for-300000-households-at-5-per-annum/

Siyan, P., Adegoriola, A. E., & Erivwode, O. M. (2019, March). The development of mortgage finance in Nigeria and its impact on the economy. *International Multilingual journal of Science and Technology*, *4*(3), ISSN.2528-9810.

World Bank. (2020). *The World Bank in Nigeria – overview*. Retrieved November 20, 2021, from www.worldbank.org/en/country/nigeria/overview

29
SOUTH AFRICA

Omokolade Akinsomi, Alison Tshangana and Marimo Taderera

1. The South African housing market

According to the general household survey of 2019, South Africa[1] has approximately 17.2 million households. This number has increased continuously from 11.2 million households as of 2002. Households in South Africa are segmented into formal households which represent 81% of total households, 18.7% of informal households, and 0.2% of traditional households with 0.2% representing other (Stats SA, 2019a). In terms of the ownership structure of households in South Africa, as of 2019, 65% of total households in are owner occupied while 22% are rented and 13% are occupied rent-free (Stats SA, 2019a).

The South African government invests heavily in housing programmes. As of 2019 about 18.7% of residential properties in the total South African households were subsidized, a significant increase from 2002 in which 5.2% of households had received a government subsidy (Stats SA, 2019a). Yearly change in consumer price as well as house prices in South Africa is seen in Figure 29.1.

As of 2019, residential properties registered on South Africa's deed registry was a total of 6.6 million residential properties, which are valued at approximately R5.5 trillion (€ 0.32 trillion) (CAHF, 2020). Figure 29.2 shows the changes in house price index in South Africa from 2001 to 2018.

As of 2019, South Africa residential properties can be separated into five segments by value (CAHF, 2020):

- *The entry market*: residential properties € 17,400 and below (33% of residential households)
- *The affordable market*: residential properties worth € 17,400 to € 34,800 (22% of residential households)
- *The conventional market*: residential properties worth € 34,800 to € 52,200 (15% of residential households)
- *The high-end market*: residential properties worth € 52,200 to € 69,600 (9% of residential households)
- *The luxury market*: residential properties worth more than € 69,600 (20% of residential households)

In terms of new build in South Africa, new residential transactions from the deeds registry office serve as a good proxy for new build; in 2019, there were 56,610 new transactions or new

DOI: 10.1201/9781003130383-33

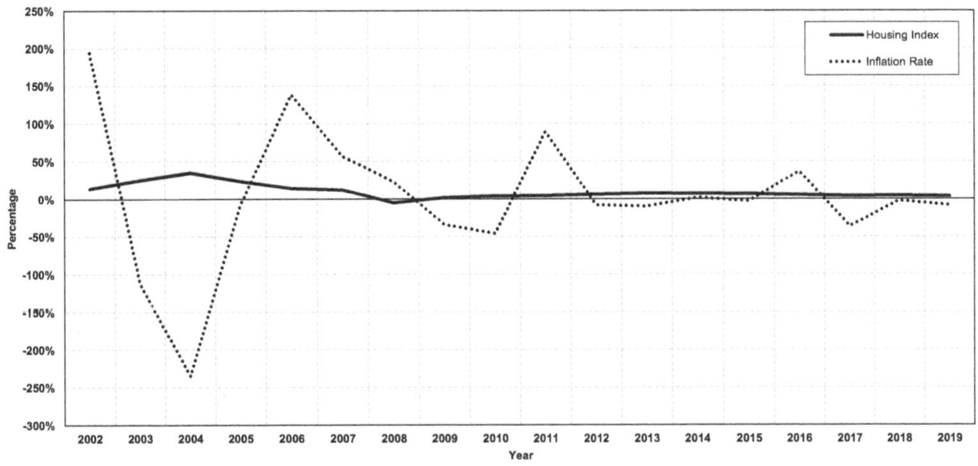

Figure 29.1 Yearly change in consumer price and house prices in South Africa (in percent), 2002–2019

Source: inflation.eu and FNB

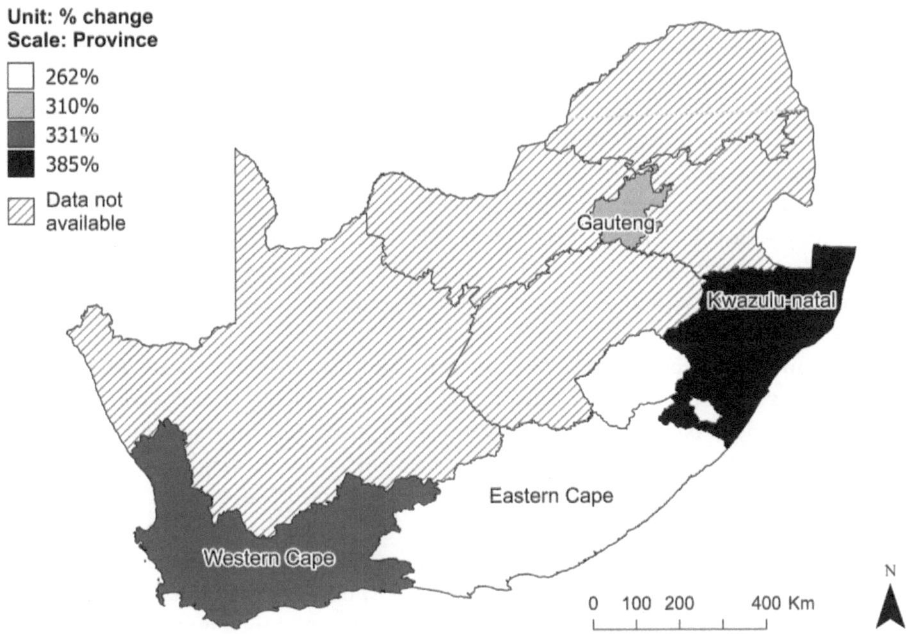

Figure 29.2 House price index change in South Africa (in percent), 2001–2018

Source: Author

builds in South Africa (CAHF, 2020). The availability of housing finance in South Africa has stimulated the housing markets; in 2019, 54% of all residential transactions in South Africa were financed through mortgages (CAHF, 2020).

2. Regulatory framework

Since the advent of democracy in 1994, national government has prioritized its response to the massive housing backlog in South Africa, primarily through the state delivery of subsidized houses for ownership by low-income households, the delivery of subsidized rentals in urban areas, and the upgrading of informal settlements. Since 1994, it has been reported that over 3.2 million houses have been provided through the national government housing programme (City Press, 2019), although annual delivery has decreased from a high of approximately 230,000 in 1999 to fewer than 100,000 houses in recent years (BBC News, 2019).

Despite this massive government investment, the housing backlog persists. UN-Habitat reports that 26% of the urban population of South Africa live in slums. Over half of households own their dwelling and have fully paid it off (52.5%), 14.2% occupy rent-free, and 6.8% own their house but have not yet paid it off. Over a quarter of households (26.5%) rent (StatsSA, 2019b).

Through government's range of housing programmes, households with a monthly income of € 203 or less, who meet other eligibility requirements, may be able to receive a 40 m^2 house. Subsidized rental units are available through the social housing programme, which has delivered approximately 57,000 rental units in urban areas.

Government housing programmes are led by the National Department of Human Settlements, which sets policy and administers the € 1.74 billion conditional grant funding framework which finances the implementation of housing programmes by provinces and municipalities, supported by an array of public entities. The Social Housing Regulatory Authority (SHRA) regulates, invests in, and supports the rental sector, while the Housing Development Agency works with public and private partners to acquire and prepare land for housing and human settlements. Formed by the merger of three public agencies that operated in housing finance, the newly launched Human Settlements Development Bank aims to leverage both public and private sector financing to support large-scale development projects and support housing finance intermediaries (National Treasury, 2019). The Estate Agency Affairs Board is responsible for regulation of estate agents and has focused on transformation of the sector in recent years, while the Community Schemes Ombud Service (CSOS) is responsible for the regulation of sectional title schemes.

The government's direct impact in housing provision has been substantial. The percentage of households that have received some form of government housing subsidy is 13.6% (StatsSA, 2019b). In 2019, 38% of new residential transactions were government-subsidized products compared with homes built and sold by private developers. The primary means of government delivery has been through large-scale projects in urban areas. However, with persistently large housing waiting lists, challenges to access affordable, well-located land, and the reality that the model of delivery is unsustainable, in recent years government policy has been shifting towards rental housing, small-scale landlords in township areas, support to the bonded market, and the provision of serviced sites which then enable households to build incrementally on their own.

Private sector delivery of new homes remains targeted at the high end of the market. Excluding government-subsidized housing, 35% of new transactions in 2019 were in the high-end market over € 69,600, while 20% (only 7,128 new house transactions) were valued under € 34,800.

The bulk of residential properties in South Africa are freehold properties (82%), with 12% sectional title (outside estates) and 6% estate properties. Private estates are common at the upper end of the market, as high-income households opt to reside in estates or complexes with private security. The great majority (70%) of estate properties are valued over € 69,600, while more than 44% of sectional title properties are valued over € 52,200.

The housing delivery scheme benefits from a strong regulatory and institutional framework providing a solid foundation for the business of residential property development and sale. According to the World Bank's *Doing Business 2020* report, South Africa is ranked 84 of 190 countries globally, falling two places from 2018. This ranking is the lowest yet for South Africa, comparing poorly to a ranking of 34 ten years ago. In terms of registering property, South Africa dropped one place in 2020 to a ranking of 108, which is still above-average for the sub-Saharan Africa region. Registering property requires seven procedures and 23 days, at a cost of 8% of the property value. South Africa scores 15.5 (out of 30) on the quality of land administration index, up slightly from 2018 (WB, 2019).

3. The conveyance process

Various legislative acts regulate the process of transferring property from buyer to seller in South Africa. These implications of these acts range from regulating the persons who are allowed to undertake the transfer of property, the taxes/duties applicable, along with conditions that are supposed to be met in order to effect transfer. The most noteworthy acts are discussed next.

The Deeds Registries Act 47 of 1937 consolidates and amends the laws that regulate the transfer of property in South Africa. The act stipulates that the transfer process must be undertaken by a conveyancer and also regulates the processes that the conveyancer must follow in order to effect transfer.

The Transfer Duty Act 40 of 1949 amends and aims to consolidate previous laws in South Africa pertaining to the transfer duty that is payable when transferring property. The Act stipulates that the buyer is required to pay transfer duty to the receiver of revenue, the South African Revenue Service (SARS), within six months of acquisition of the property.

The Housing Act 107 of 1997 stipulates that the recipient of a housing subsidy cannot sell or alienate his or her rights without first offering the property to the relevant provincial housing department at a price not greater than the subsidy which the person received for the property.

The Electronic Deeds Registration Systems Act 19 of 2019 instructs the South African deeds office to develop an electronic deeds registration system that will facilitate the preparation, lodgement, registration, and storage of electronic title deeds. The intention of the Act is to ensure the deeds office leverages technological advancements to create efficiencies.

The Spatial Planning and Land Use Management Act 70 of 2013 forbids the registration of any property resulting from a land development application from occurring unless the local authorities (municipalities) are satisfied and certify that all conditions have been met ensuring that the development is in line with the municipality's spatial planning and land use management system. The flowchart in Figure 29.3 shows the conveyance process in South Africa.

Marketing phase

The first phase of the conveyance process is primarily driven by the estate agent. Once a seller has decided to sell their property, they would appoint a real estate agent. The activities of the real estate agent are regulated by the Estate Agency Affairs Board (EAAB), a regulatory body that has been established to control and monitor the activities of property practitioners, including real estate agents. The EAAB operates in accordance with Act No. 22 of 2019, the Property

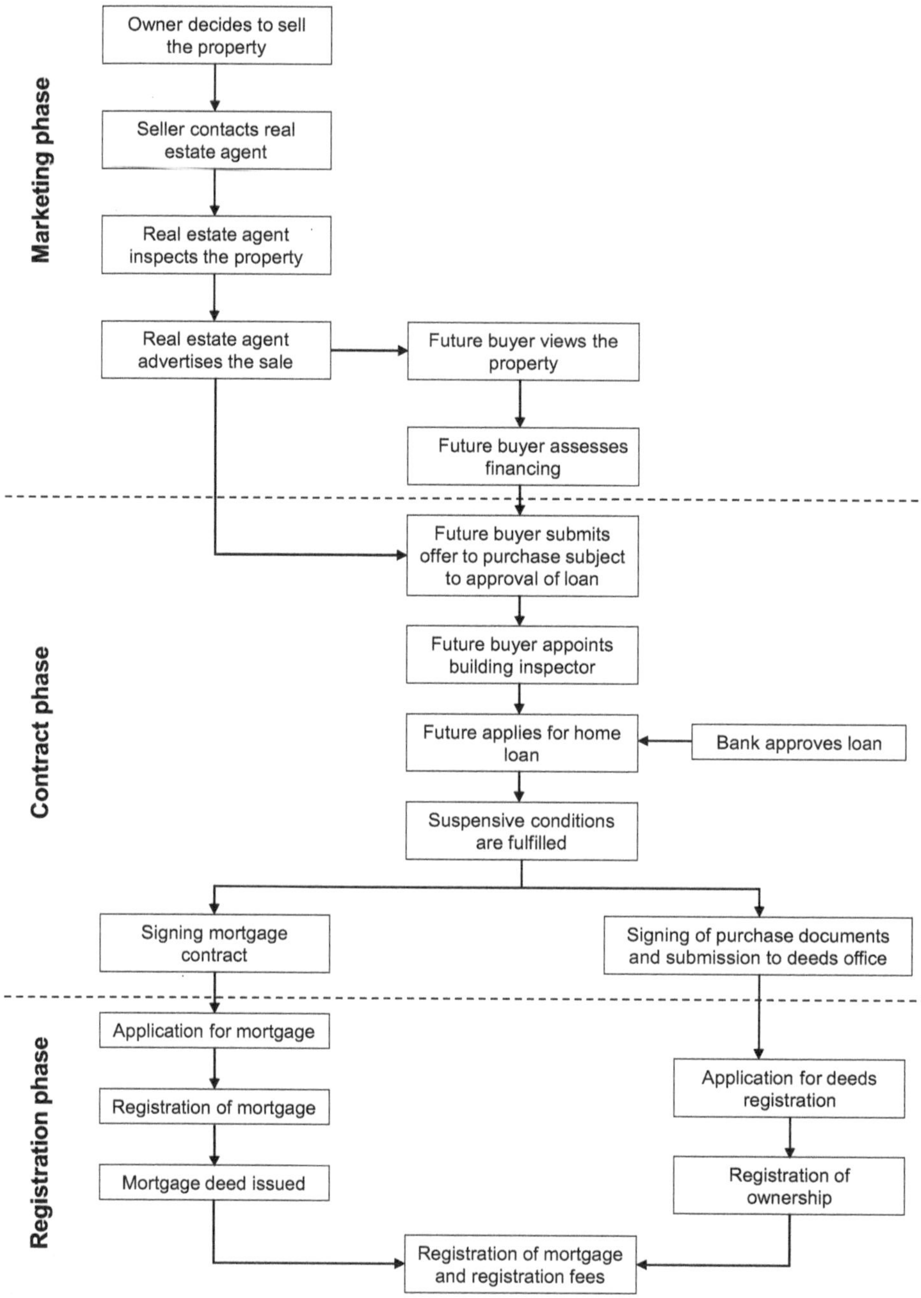

Figure 29.3 The conveyance process in South Africa

Source: Author

Practitioners Act (PPA). All real estate agents must be registered with the EAAB and must have a valid Fidelity Fund Certificate ('FFC') in order to be allowed to practice as a real estate agent and claim commission on any transactions.

The seller of a property can appoint an estate agent by signing an estate agent mandate, a legally binding agreement between seller and agent. There are three main types of mandates that a seller can use to appoint an estate agent:

- *Open mandate*: An open mandate is a mandate that allows multiple estate agents to list the seller's property and works on the basis that the estate agent that manages to introduce the buyer to the seller and serves as the effective cause of the sale of the property gets the commission.
- *Sole mandate*: A sole mandate is a mandate whereby the seller authorizes a single estate agent to market and sell a property. A sole mandate would be issued for a fixed period of time during which the estate agent has the exclusive right to sell the property.
- *Multi-listing mandate*: Some estate agencies have agreements in place to work with each other called a multi-listing mandate. If a sellers opts for a multi-listing mandate, the agencies within the multi-listing group would all market the property and would all share the commission equally between the agent who listed the sale and the agent who referred the buyer to the listing agent.

Once appointed, the estate agent would market the property at their own cost. Marketing activities would typically include directly approaching potential buyers, listing the property on the estate agency's website, listing the property on property listing websites, placing 'for sale' boards outside of the property along with advertising the sale in local newspapers and the agencies' social media platforms. The various methods of advertising attract different costs, and the extent to which an agent would choose to market a particular property would be dependent on the exclusivity of the mandate. The estate agent's marketing efforts would also involve arranging viewings of the property with any interested potential buyers.

The estate agent role would also involve intermediating the negotiating of the sale price between the buyer and seller. This would involve managing the seller's expectations and demonstrating the value to the buyer. The agent's fees are typically paid by the seller and are charged as a percentage of sale price thus incentivizing the agent to seek the best possible price for the seller. The agent's fees are payable only upon transfer of the property.

Contract phase

The second phase of the conveyance process is the contract phase. Once the estate agent has secured a buyer of the property and the buyer and seller have negotiated a price for the property, the buyer and seller would record their agreement in an offer to purchase agreement, a legally binding document that records the buyer's intention to purchase the property and the terms and conditions upon which the purchase is intended to take place. South African law stipulates that a sale agreement for property must be in writing for it to be deemed binding.

The offer to purchase would typically contain some suspensive conditions that would need to be fulfilled in order for the contract to be enforceable. The suspensive conditions would include the successful completion of an inspection of the building by a building inspector to check for any defects in the property and to ensure the property is compliant with statutory regulations which forms part of the buyer's due diligence. The second suspensive condition would be the approval of the buyer's loan application to finance the purchase of the property.

The offer to purchase also stipulates that if all conditions of the offer are met, the buyer and seller would enter into a formal agreement of sale.

Once the offer to purchase is submitted by the buyer and accepted by the seller, the buyer would appoint a building inspector to inspect the building for any defects. The inspector issues a report detailing the findings of their inspection which can be kept by both parties. This would present an opportunity for the buyer to request any repairs to be made. Furthermore, it would serve as a record of the condition in which the property was sold to the buyer. The seller would be liable to remedy any latent ('defects that are not easily identifiable upon inspection') defects for a period up to three years after transfer has taken place.

After the building has been inspected, the buyer would then apply for a home loan from their bank of choice to fund the purchase of the property. If the bank approves the buyer's home loan application and any other suspensive conditions are met, the offer to purchase becomes enforceable.

The seller then appoints a transfer attorney to whom the transfer process is handed over to. The transfer attorney draws up documents that the buyer and seller must sign and submits them to the deeds office along with the required supporting documents.

Transfer phase

The third phase of the conveyance process is the transfer phase. For transfer to take place, the seller is required to cancel any bond they may have on the property that is being sold. The seller would typically be required to give the bank 90 days' notice of their intention to cancel the existing bond. If the bond is cancelled in less than 90s days of notice being given to the bank. The bank charges the seller a bond cancellation fee, which is calculated as a percentage of the outstanding loan amount. The transfer attorney would inform the seller's bank that the property is being sold, and the bank would appoint a cancellation attorney to cancel the bond. The cancellation attorney will then advise the conveyancing attorney on the amount outstanding on the bond. The cancellation attorney is requested to cancel the bond upon receipt of a guarantee of the amount owing.

Once the transfer attorney receives the title deed and cancellation figures from the seller's bank, the buyer and seller will sign transfer documents and the buyer pays all transfer costs to the transfer attorney who then pays the transfer duty and pays the rates and taxes to the relevant local authorities.

The buyer appoints a bond attorney who prepares the necessary bond documents for signature by the buyer. The buyer then pays the bond registration fees so that the bond attorney can prepare documents to be lodged with the deeds office and issues guarantees that are sent to the transfer attorney.

The cancellation attorney will then cancel the bond upon receipt of the guarantees from the transfer attorney. The transfer attorney then lodges the cancellation of the seller's bond and the registration of the buyer's bond with the deeds office which then executes the transfer of the property from the seller to the buyer. Upon transfer, the transfer attorney settles the outstanding bond amount, pays the agent his commission, and releases the balance of the money to the seller's account.

4. Summarizing the conveyance process and its transaction costs

In South Africa, the conveyance process involves three main phases. The first phase is the marketing phase, which is driven by the seller's estate agent. The second phase is the contract phase, which involves the buyer presenting the seller with an offer to purchase the property and the seller accepting the offer. The offer to purchase typically includes several suspensive conditions which allow the buyer to conduct their due diligence and apply for a loan to purchase the property. Once the suspensive conditions are met, the seller appoints a conveyancing attorney who then drives the conveyance process from there on. The buyer appoints a bond registration attorney and the seller's bank appoints a bond cancellation attorney. The final phase of the conveyance process is the transfer process whereby the transfer attorney lodges the transfer of the property with the deeds office while simultaneously coordinating with the cancellation attorney and the bond registration attorney to cancel the seller's bond on the property and register the buyer's bond with the deeds office. Upon transfer of the property at the deeds office, the agent receives their commission and the seller receives the proceeds from sale. The process in South Africa is fairly simple and is in line with global practices.

Table 29.1 summarizes the transaction costs that are applicable to the conveyance process.

Table 29.1 Fees and taxes during the conveyance process in South Africa

Phase	*Fees and taxes*	*Real estate*	
		Seller	*Buyer*
Marketing	Real estate Agent's fees	5–8%	-
	Marketing fees	Included in agent's commission	-
Contract	Bond cancellation fees	1% of the amount owing	-
	Bond registration fees	-	0.5–1.5%
	Bond initiation fee	-	€ 348–406
	Bank administration fee	-	€ 3–6 per month
	Legal fees (conveyancer)	-	€ 302–2,958
Registration	Transfer duty	-	0–13% of value of property
	Deeds office registry fee	-	€ 25–415 (based on value of mortgage)
	Transfer registration fee	-	€ 2–332 (based on value of the property or purchase price, whichever is greater)
Post-registration	Capital gains tax	18% for individuals 22.4% for companies 36% for other trusts (calculated on taxable gain on sale)	-
	VAT	-	15% of sale price
	Sales tax	-	-
Total (EUR for the model case price)		**€ 5,000–8,000**	**€ 15,677–32,110**
Total (% share to the model case price)		**5–8%**	**15.7–32.1%**
Excluded costs from the model case		Bond cancellation fee 1% of the amount owing, 13% CGT for individuals, 20% GCT for organizations, 36% for other trusts	Bond registration fees: 0.5–1.5%, bank administration fee: € 3–6 per month

Note. Model case price: € 100,000

Source: Author

Note

1 1 South African Rand = 0.058 Euro (as of 1 October 2021).

References

Centre for Affordable Housing Finance in Africa. (2020). *South Africa housing market report*. Retrieved November 12, 2021, from https://housingfinanceafrica.org/app/uploads/2021/03/Citymark-National-Property-Market-Report-3-March-2021-FINAL.pdf

National Treasury. (2019). *2019/2020 National treasury annual report Vote* 7. Retrieved February 12, 2022, from http://www.treasury.gov.za/publications/annual%20reports/national%20treasury/nt%20annual%20report%202019-20.pdf

Pretorius, L. (2019). Housing in SA. *City Press*. Retrieved January 26, 2021, from www.news24.com/citypress/news/housing-in-sa-3-faqs-answered-20190410

Reality Check Team. (2019). South Africa elections: Has the ANC built enough homes? *BBC*. Retrieved January 26, 2021, from www.bbc.com/news/world-africa-48093711

Stats SA.(2019a). *General household survey* (pp. 1–192). Retrieved from https://www.statssa.gov.za/publications/P0318/P03182019.pdf

Stats SA. (2019b). *Statistical release P0318: 2018 general household survey* (pp. 25–32). Retrieved December 15, 2020, from www.statssa.gov.za/publications/P0318/P03182018.pdf

World Bank. (2019). *Economy profile South Africa: Doing business 2020* (pp. 23–32). Retrieved December 15, 2020, from www.doingbusiness.org/content/dam/doingBusiness/country/s/south-africa/ZAF.pdf

30
THAILAND

Nattapong Puttanapong and Magnus Andersson

1. The Thai housing market

The housing market in Thailand[1] has attracted attention during several different periods: the collapse of the financial and real estate markets during the Asian financial crisis in 1997, the government's efforts to solve the country's urban low-income shortage, and today's focus on the developing expansion of the mass transit network spanning Bangkok and greater Bangkok, which will make commuting within the city convenient.

This section summarizes the main characteristics of Thailand's real estate market based on several categories. First, Table 30.1 shows the distribution of real estate properties based on registered transaction data. Obtained from the Real Estate Information Center (REIC) at the research unit of the Government Housing Bank (GHB), these data show that the transaction of residential property and that of land are the first- and second-largest types. Reflecting Thailand's economic expansion, commercial and office real estate and industrial real estate are the third- and the fourth-largest transactions, in addition to the concentration of residential construction.

Conventionally, all business transactions and legal documents related to real estate in Thailand use the domestically defined units of land. Therefore, in Thailand, land area is uniquely measured in the domestically defined units of Wah, Ngan, and Rai.[2]

Table 30.1 Number of annual transactions (in millions/US dollars)

In Millions EUR						
Year	*2015*	*2016*	*2017*	*2018*	*2019*	*2020*
Residential	18,835	16,981	17,102	24,935	25,774	25,549
Office & Commercial	675	161	173	402	563	292
Hotel & Resort	50	22	46	135	172	48
Industrial	232	222	214	287	317	201
Land	19,635	14,941	12,250	19,373	18,453	13,772
Total	**39,427**	**32,327**	**29,785**	**45,132**	**45,279**	**39,862**

Source: REIC

DOI: 10.1201/9781003130383-34

Influenced by economic fluctuations, these real estate transactions categories are not stable. As shown in Table 30.1, residential real estate numbers have been increasing, while those of the other categories are declining.

Figure 30.1 and Figure 30.2 give an alternative perspective of the market, showing the proportions of real estate classifications based on the numbers of granted construction permits. Based on this dataset, low-rise residential buildings have dominated roughly half of the total construction permits, whereas high-rise condominiums, commercial buildings, and industrial estates are the second-, third-, and fourth-largest proportions, respectively.

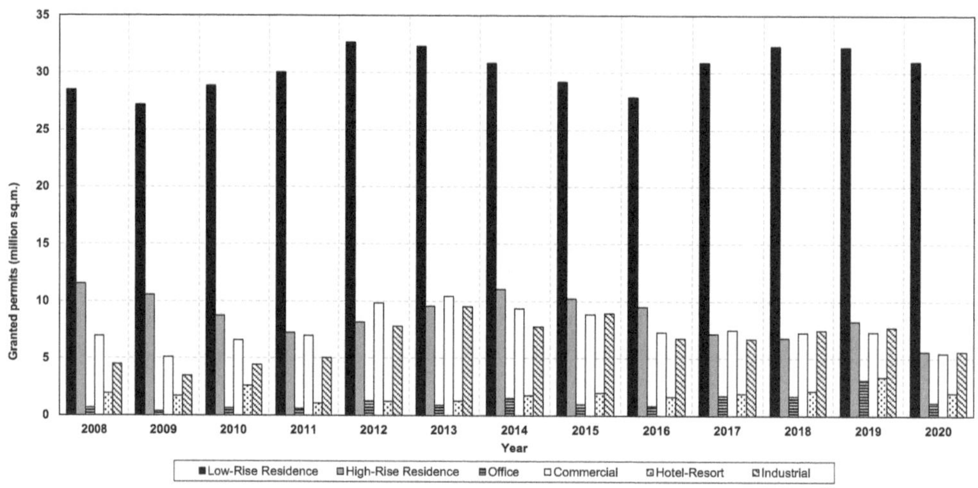

Figure 30.1 Real estate development in Thailand (granted permits – total m^2)

Source: REIC

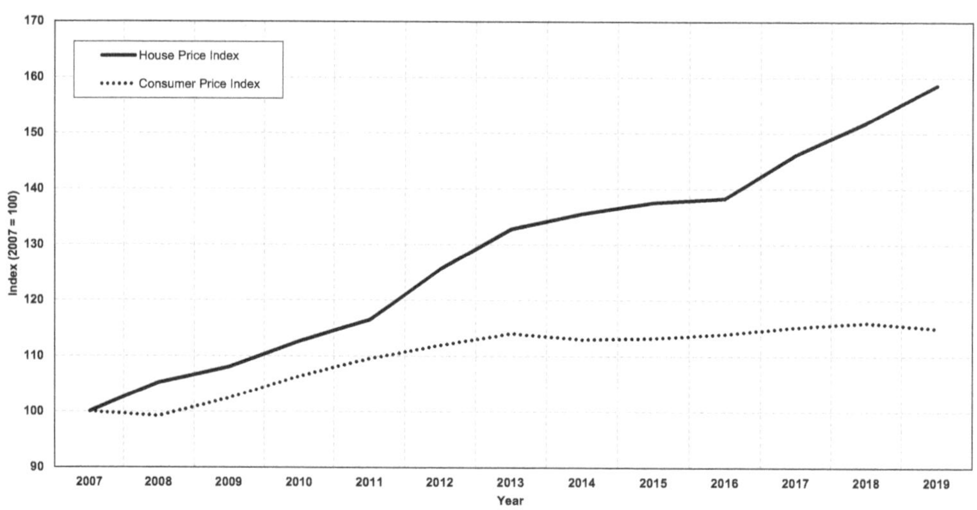

Figure 30.2 House price index and consumer price index

Note. 2007 = 100

Source: Bank of Thailand

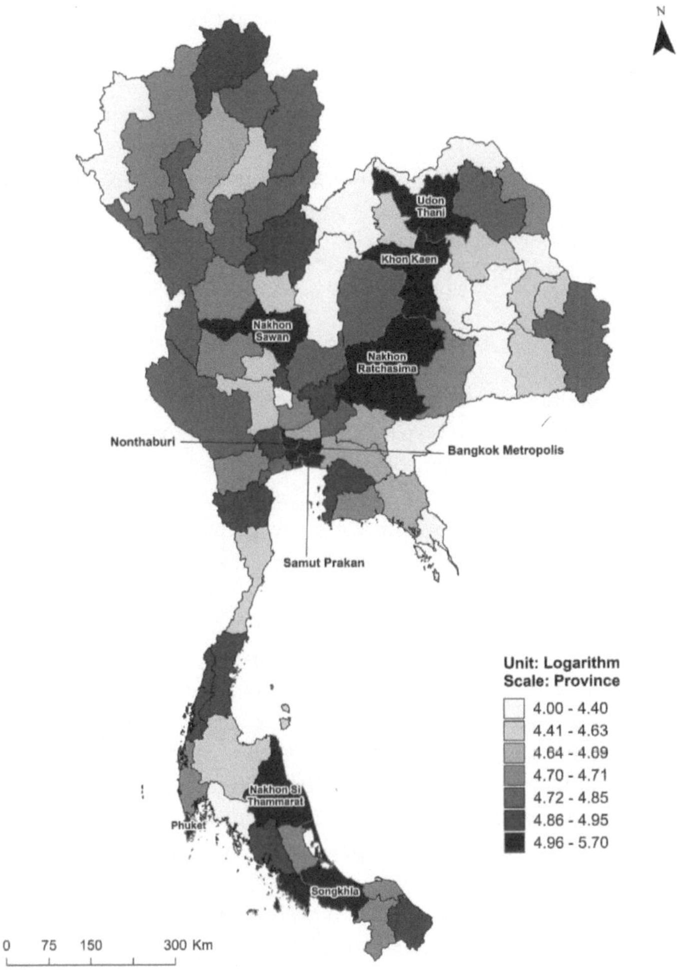

Figure 30.3 Map of sale prices

Source: Real Estate Information Center, Thailand's Government Housing Bank (2019)

These statistics indicate that residential properties are the main real estate construction in Thailand. Likewise, the demand for buildings and facilities serving the manufacturing and service sectors is clear by the large numbers of granted construction permits.

Figure 30.2 depicts the increasing price of houses according to the House Price Index 2007–2019. This shows that the average house price has been domestically outperforming inflation.

The map in Figure 30.3 shows that most real estate activities are spatially clustered around Bangkok and its vicinity. Also, the real estate development is concentrated in the eastern economic corridor (EEC), which are the three coastal provinces (Chachoengsao, Chonburi, and Rayong) of hosting the high density of industrial productions and tourism destinations.

2. Regulatory framework

Thailand's property law (i.e. real estate law) mainly defines the rights of possession, classifications of ownership, and the legal uses of immovable property (land, house, condominium). The main laws and regulations that directly govern real estate properties and related activities are the following:

- Civil Code Book III – 'Specific Contracts' (including laws governing property leases and mortgages)
- Civil Code Book IV – 'Property' (including laws governing ownership, possession, usufruct, superficies)
- The Thailand Land Code Act
- The Thailand Condominium Act

Land titles and deeds

The most important document indicating the ownership of land is the title deed. This official ownership document is issued by the Department of Lands, Ministry of Interior. The title deed includes a map identifying the boundary of the land plot, the survey status, and details about its geographical location. In addition, it also states the obligations and mortgages attached to the property. In addition, this legal document states the obligations and mortgages attached to the property and a list of previous owners and historical transactions related to the property, including all previous disposals and encumbrances.

A title deed in Thailand has various forms and classifications. The conventional one is *Chanott ti din*, which authorizes the incontestable possessions and the full legal rights to the owner. This form of title deed is issued after the land plot has been accurately surveyed. The title deeds of all urbanized and some agricultural lands are issued as *Chanott ti din*.

The second category of title deed is *Nor Sor Sam* or *Nor Sor Sam Kor*, which is the title deed of most rural and agricultural areas. This classification of title deed allows the land to be sold or leased. However, the boundary of the land plot is not as accurately surveyed as with the *Chanott ti din* titles. Therefore, the less clearly defined physical boundaries can lead to legal disputes with neighbouring landowners.

The third category is *Sor Kor Nung*, *Tor Bor Tor Hoc*, and *Tor Bor Tor*. These documents pertain to squatter's rights. The documents are issued based on the registration of rights at the local district administration office. These classifications of land cannot legally be sold.

The last category is *Sor Bor Kor*. This is a type of official title deed which identifies the accurately surveyed boundary of land. Its main propose is to authorize landownership to farmers, and this category of land can be mortgaged. However, it cannot be leased, sold, or transferred.

In addition to the abovementioned title deeds, there is also the condominium title. Governed by the Condominium Act of 1979, it is an official document identifying the ownership of a particular part of a building jointly with multiple owners, and fractional ownership of the land, other common assets, and common parts of the building. The condominium title specifies the floor area, the ground area of the common land, and the percentage of fractional interest in the common property. Also, this percentage is proportionate to the voting interest in the owners' association.

Foreign ownership

Strictly controlled by law, foreign nationals cannot purchase land in Thailand. However, foreign nationals are permitted to own condominium units. Nevertheless, there is a legal restriction that allows foreign nationals to own only 49% of the total units. Recently, Thai law has enabled more options for foreign nationals; for example, non-residents can obtain temporary ownership through the 30-year renewable lease. This option requires both the buyer and seller to register the lease contract at the land office. However, the lessee cannot sell, sublease, or transfer their interest.

Alternatively, foreign nationals can partially own land through the establishment of a private company. However, Thai law stipulates that the maximum share a foreign shareholder can own is 49%. Therefore, full ownership is still not possible.

The last option is to obtain land through permission granted by the Board of Investment (BOI), with the condition of contributing at least THB 40 million (€ 1 million) in the approved investment project. However, this permission allows an ownership of the land up to a total area of 1 Rai (1,600 m^2). In addition, there are rights defined by Thai law that are related to real estate ownership and transactions.

Thailand right of usufruct

Both sections 1417 and 1428 of the Civil and Commercial Code specify the usufruct as the right to use or occupy another person's real property for one's life or up to 30 years. The right of usufruct has to be established and registered at the Land Office.

Thailand right of superficies

Sections 1410 to 1416 of the Civil and Commercial Code legally define the right of superficies, which is a civil law real estate right. In particular, a person can grant to another person the right to construct and own buildings, structures, or plantations even while they remain the owner. Therefore, by registration at the Land Office, the superficies agreement legally separates the ownership of the land and everything on the land.

3. The conveyance process

In this section, we outline the conveyance process for real estate transactions in Thailand, as presented in Figure 30.4.

Marketing phase

This first phase of the conveyance process usually involves an estate agent. In terms of residential property, the buyer should visit and inspect the property. For commercial and industrial properties, due diligence should be undertaken. Also, it is highly recommended that a caveat search and ownership verification be conducted.

When interested in purchasing a property, the buyer submits the offer to the seller. Once the seller agrees with the offer, the purchase agreement can be signed. The details of the real estate property, together with the agreed price and other legal and financial conditions, shall be clearly identified. In addition, the purchase agreement shall identify which party will be responsible for the legal fees, transfer fees, and related taxes.

Contract phase

A deposit is paid to the seller upon signing the purchase agreement. In general, the amount of the deposit is at least 10% of the property's agreed price, and it is non-refundable. Moreover, deposits in escrow are not conventionally conducted. Therefore, the purchase agreement is practically the main document enforcing the legal binding of the subsequent conveyance process.

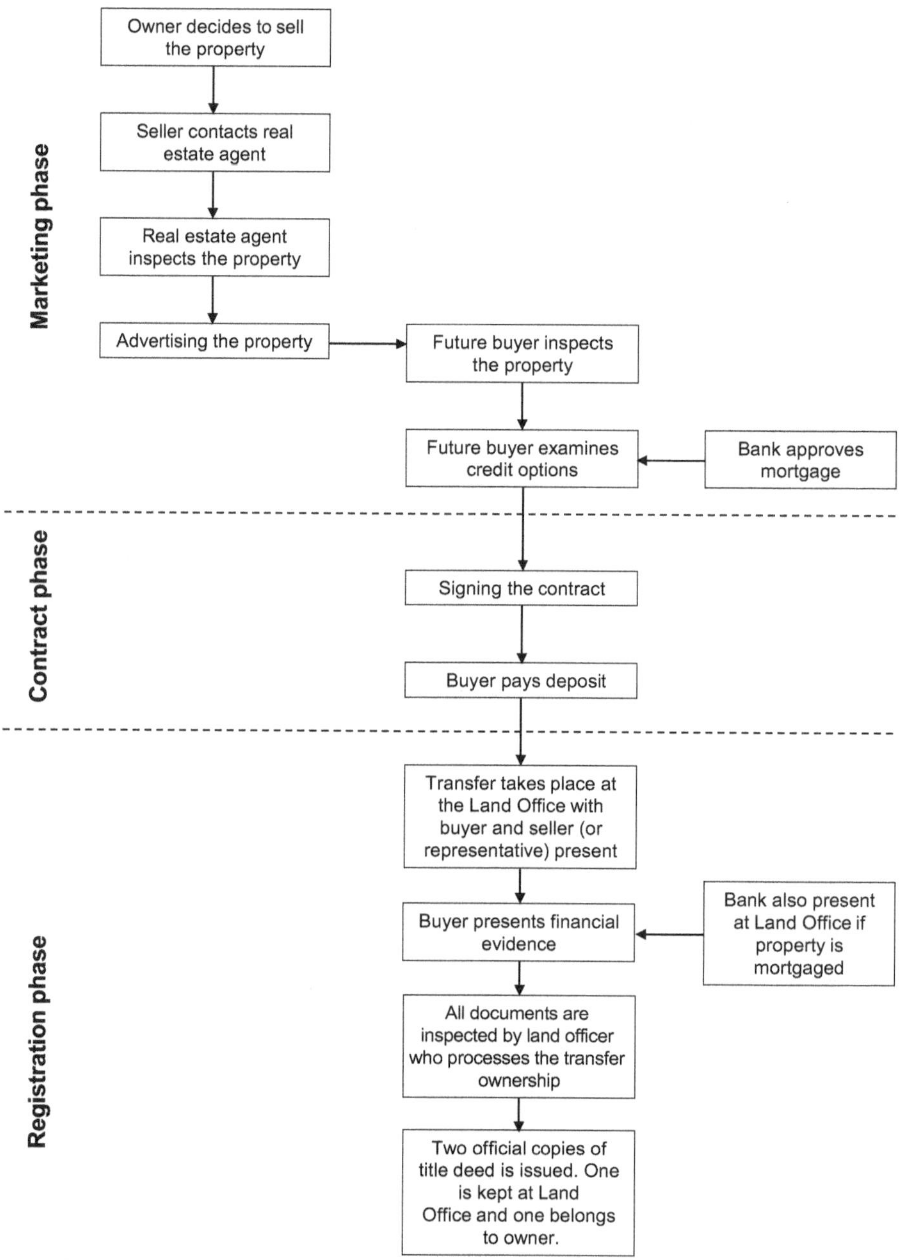

Figure 30.4 The conveyance process for real estate, respectively condominium ownership, in Thailand

Source: Author

Registration phase

The buyer and the seller agree to meet at the Land Office on the transfer date, although each party can legally authorize a representative to conduct the legal process of ownership transfer and registration. Also, if the property is still mortgaged, the bank will authorize a representative to cooperate in this process. As legally defined, if any person is absent, the process cannot be conducted.

Initially, to reassure all parties, the buyer (or their representative) should show the original cashier cheque (or the other financial evidence) to the other parties.

Thereafter, the seller and the buyer (or their representatives) complete and sign the sales contract and submit it with any other required documents to the land officer.

After verifying the authenticity of all the documents, the land officer will certify the transaction. Also, the land officer will register the transfer of ownership. Specifically, there are two official copies of title deed. One is kept in the database of the Land Office and the other copy belongs to the landowner. With the transfer of ownership, a new line will be added to the second page of both copies of the title deed listing the name of the current owner.

The expenditures on taxes and transaction fees to the Land Office are based on the statement identified in the purchase agreement. It is the buyer's responsibility to contact all utility providers (e.g. electricity, running water, fixed-line internet) to transfer the subscriptions of services.

4. Summarizing the conveyance process and its transaction costs

For the newly developed real estate property, as specified by law, the developer can pass on up to 50% of the transfer fee to the buyer. Also, the law specifies that fees and other taxes are the responsibility of the developer. However, for other types of transactions, the responsibility for payment is based on agreement between the buyer and seller.

Table 30.2 shows all transaction costs involved in the purchase and conveyance process.

It should be noted that the transfer fee is conventionally fixed at the rate of 2% of the property's value. However, this rate is subject to government decree, which usually lowers this rate to stimulate the economy during times of economic recession. For example, from November 2019 to December 2020, this fee was reduced to 0.01% of the property's value lower than BHT 3,000,000 (€ 77,630). In addition, in the conveyance process, the property value is estimated by the land officer, incorporating the depreciation of the building. As a result, this depreciation allowance equivalently gives a further reduction of 0.08–0.5% to the original 2% rate.

Table 30.2 Details of transaction costs

Phase	*Fees and taxes*	*Real estate*	
		Seller	*Buyer*
Marketing	Real estate agent's fees	3–5%	-
Contract	Transfer fee	-	0.01–2%
	Stamp tax	0.5%	-
	Land registration fee	2%	-
Registration	Legal fees	-	-
Post-registration	Sales tax	3.3%	-
Total (EUR for the model case price)		**€ 8,800–10,800**	**€ 10–2,000**
Total (% share to the model case price)		**8.8–10.8%**	**0.01–2.0%**
Excluded costs from the model case		-	-

Note. Model case price: € 100,000

Source: The Land Department, Thailand

Notes

1 1 Thai Baht = 0.0259 Euro (as of 1 October 2021).

2 1 Wah = 4 m^2 = 42.7 ft^2.
1 Ngan = 100 Wah = 400 m^2 = 4,277 ft^2
1 Rai = 400 Wah = 1,600 m^2 = 17,109 ft^2
1 Acre = 2.5 Rai = 4,000 m^2 = 42,772 ft^2
1 hectare = 6.25 Rai.

References

Civil Code Book III.
Civil Code Book IV.
Real Estate Information Center, Thailand's Government Housing Bank. (2019). *Price database*. Retrieved from https://www.reic.or.th/?REICPage=1
The Thailand Condominium Act.
The Thailand Land Code Act.

PART III

31

THE CHALLENGES OF COMPARING LEGAL REGIMES AND CONVEYANCE ON THE HOUSING MARKET

Ola Jingryd

1. Introduction

Any researcher who has ever undertaken a comparative study can attest that they entail challenges, pitfalls, and obstacles from conception to completion. Some obstacles are easy to foresee, such as language barriers and the (un)availability of easily accessible sources. Indeed, the feasibility of any intended comparative study hinges on the researcher's ability to access and understand the right sources. However, other, perhaps less obvious obstacles must also be overcome in order to succeed with the undertaking and produce valid findings. Suppose, for instance, that a researcher sets out to compare state subsidies of some kind between countries. In their home country, the subsidy takes the form of a cash payment, as in, a monthly allowance. Suppose further that the search for an equivalent allowance in the country or countries of comparison yields nothing, leading the researcher to draw the conclusion that the country or countries in question do not have any subsidies of the studied kind. However, the conclusion may well be a mistake, as the subsidy could take the form of a tax break instead of an allowance (cf. Bogdan, 2003, p. 46; Dannemann, 2006; Strömholm, 1971). The researcher thus runs the risk of looking in the wrong place and drawing invalid conclusions as a result.

In addition to the aforementioned obstacles and other pitfalls related to the researcher's contextual limitations and their effect on the researcher's understanding of the gathered information (Öcuru & Nelken, 2007; Jingryd, 2012, pp. 38–45), one challenge seems to tower over the rest: the diverging institutional contexts. The word 'institutional' is used here in a fairly broad sense, to denote legal, economic, cultural, and other related factors that give particular meaning to data from a particular country.

For instance, the Swedish Land Code does not prescribe any mandatory intervention by third parties for the validity of a real estate conveyance. The buyer is obliged to seek title registration at the Land Registry, but its function is merely declaratory, not constitutive. This may lead a researcher to conclude that buyers and sellers in Sweden are not afforded legal assistance, as buyers and sellers in civil law countries have a notary intervene to draw up the deeds and advise the parties. Again, this is a mistaken conclusion, as more than 90% of all residential real estate conveyances in Sweden are accomplished through real estate brokers who are legally required, *inter alia*, to draw up the deeds and give advice. By the same token, the lack of

DOI: 10.1201/9781003130383-36

mandatory intervention would seem to invite the conclusion that there are no transaction costs of that kind; however, given that the market voluntarily involves estate agents to such a large extent, the typical Swedish conveyance must be said to entail such costs.

Palm, Jingryd, and Kordic (2020) examined and compared the transaction processes and transaction costs in residential real estate conveyances in Sweden and Croatia. Here, the challenges in making viable comparative analyses are clear. For instance, the Croatian process involves a pre-contract phase, which is ultimately because a binding sale is accomplished only upon notarization. In Sweden, a binding sale is accomplished once both parties have signed the contract of sale, making pre-contracting redundant. The example illustrates how the comparative researcher must take the legal and institutional framework in each studied country into account.

This aim of this chapter is to highlight the inherent challenges facing the comparative researcher and to provide a caveat for the reader when comparing the contributions of the various countries. To that end, we will focus on the theoretical and methodological aspects of comparative law. As the format does not allow for too deep a plunge into the research field, a brief overview of the main schools of thought will have to suffice. Following that, a very brief introduction to comparative institutional analysis is given to broaden the perspective. The chapter then ends with a few concluding remarks.

2. The theory and method of making comparisons

2.1 Comparative law – the art of comparing laws

The inherent difficulty in making viable comparisons between legal orders is by no means unknown to the legal academic community. Indeed, merely establishing the position of the law in another legal order than one's own is a process fraught with obstacles and pitfalls. First, what are the recognized sources of law, and what importance does the legal community in the jurisdiction of interest place in those sources? For instance, are judgements from courts of first instance relevant, and if so, to what degree? Second, the language is a barrier of its own. The researcher may well be proficient enough in the other language to handle day-to-day conversations, but is that enough to make legal interpretations? Legal terms are not always straightforward to translate, as they may carry different meanings in different legal orders.

While the aforementioned examples are largely of a practical nature, they are nonetheless fundamentally important to the researcher's chances of carrying out the intended study and therefore not to be overlooked. Indeed, approaching the subject from the individual researcher's perspective is arguably the most fruitful point of entry, as the purpose of the endeavour is ultimately to understand how the researcher can conduct comparative studies and produce valid and reliable results.

On a more theoretical – and arguably more problematic – level, it is impossible to escape the importance of the clash between the two great schools of thought in comparative law: functionalism and culturalism. A third perspective, which is perhaps more closely related to functionalism, is the idea of legal transplants.

2.2 Functionalism

The basic idea of functionalism is that laws and other public institutions exist to fulfil functions in society (Zweigert & Kötz, 1998, pp. 32–47; Michaels, 2006; Husa, 2009). While laws vary between countries, the needs and functions they are meant to fulfil are assumed to be the same or similar. For instance, the manner in which binding contracts are accomplished varies between legal orders, but the need to provide a regulatory framework to facilitate commerce remains the same. And given that the basic conditions for the transacting parties are the same or similar across nations, the considerations while enacting and applying the law are assumed to be the same or similar; for example, the buyer wants the goods, the seller wants to get paid, and neither party wants to unilaterally perform their part of the bargain only for the other party to default. This gives rise to two central, but contested, tenets of legal functionalism: the *praesumptio similitudinis* and the assumed existence of the functional equivalent. The former is a rule of thumb, amounting to a hypothesis of sorts, which posits that 'as a general rule developed nations answer the needs of legal business in the same or in a very similar way' (Zweigert & Kötz, 1998, p. 40).

The *functional equivalent* is the institution in the studied country or legal order that fulfils the same function as an institution in another country or legal order of interest. For instance, as demonstrated in Chapter 20, Swedish estate agents are legally obliged, *inter alia*, to advise the transacting parties on legal issues pertaining to the sale and to draw up the necessary deeds in a manner that is suitable for the transaction at hand (see further Jingryd, 2021, pp. 255–342). In civil law countries, that function is typically performed by public notaries. In this regard, then, the Swedish estate agent and the civil law notary would seem to be each other's functional equivalent.

2.3 Culturalism

Culturalism, in the context of comparative law, is based on the fundamental understanding that the law is not isolated from the society in which it operates. On the contrary, the law – understood not only as black letter rules in statutes and court cases but also the methods of interpretation that give the legal sources meaning as well as the legal system as a whole – is embedded in the culture (Legrand, 1997; Cotterrell, 2006). For this reason, the culturalist holds that one cannot assume that similarity in appearance bespeaks similarity in substance.

A simple example from Sweden serves to demonstrate that the idea does not lack merit. As stated, legal sources such as statutes and court cases must be interpreted to have meaning. Hence, the method or methods of interpretation applied by lawyers and civil servants come to the fore. An important factor here is what sources are used and deemed to have some bearing on the law. In this, Sweden stands out due to the emphasis on *travaux préparatoires*. While statutes and rulings from the Supreme Court and other high courts are considered to outrank *travaux préparatoires* (although there is no statute that stipulates this), the latter are generally deemed to be so important that it is not uncommon for statutes to be deliberately vague, with the implicit understanding that the more specific interpretation must be sought in the *travaux préparatoires*.

Again, there is no statute that stipulates the relevant sources of law, although the topic is treated in numerous textbooks, theses, and articles which are themselves considered sources of law, albeit mostly of a secondary nature. Rather, the sources that are considered relevant, and their relative weights, is a matter of convention among legal scholars and practitioners. It is an understanding that is passed on to law students at universities and practised by lawyers wherever they operate. It can aptly be called an institution in the legal culture in Sweden.

The idea of law as an inseparable part of society was famously brought forward by Montesquieu in his seminal work *The Spirit of the Laws* (1748). It was later developed by Savigny, whose *Volksgeist* doctrine posited that the law originated in the 'spirit of the people' (de Montmorency, 1910; Kutner, 1972). Savigny stated, among other things, that 'law is first developed by custom and popular faith, next by judicial decisions – everywhere, therefore, by internal silently operating powers, not by the arbitrary will of a law-giver'.

Today, Legrand is one of the most prominent culturalist legal comparatists – one who has most vehemently opposed functionalism, which he brands as 'positivist'. Legrand's view rests on the assertion that the law cannot be understood as rules that can be readily interpreted by any and all but rather as something embedded in the culture of the country where it is conceived and applied, as well as the assumption of the singularity and 'radical contingence' of each culture. This singularity means that the culture – and hence the law embedded therein – cannot be fully known by those who do not belong to the same culture (Legrand, 1996, 1997, 2003). Legrand relies heavily on the critical theory of Derrida and the relation between the 'self' and the 'other' (Derrida, 1967a, 1967b, 1972). It remains unclear, however, to what extent Legrand's positions are compatible with those of Derrida (Miller, 2017).

Curran (1998) has suggested that 'cultural immersion' is indispensable in comparative studies. Despite declaring comparison central in legal analysis, Curran emphasizes 'the impossibility of perfect comparison, the inevitability of some loss in the comparative act, as well as the irreducible nature of some cultural differences', suggesting somewhat counterintuitively both that 'comparison in an ideal form is unattainable' and that 'comparison is indispensable to understanding'. To reconcile these seemingly opposing views, Curran asserts that differences among individuals and groups, even within societies and legal cultures, render all communication and understanding imperfect – but not impossible. Words do not carry the same meaning to all individuals, and by the same token, 'perfect communication by means of human language would require that all interlocutors have identical cultural backgrounds and physical makeups'. Curran thus holds that 'a valid examination of another legal culture requires immersion into the political, historical, economic and linguistic contexts that molded *[sic]* the legal system, and in which the legal system operates'.

Curran's view is visibly more optimistic than Legrand's as to the possibility of comparison. By immersing themselves in the legal culture they wish to study, the comparatist can achieve knowledge. The greater the comparatist's fluency in the target culture's language and the greater their cultural proximity, the greater the degree of cultural immersion. Curran would also seem to accept that it is possible to draw correct conclusions about a foreign legal culture in spite of one's imperfect immersion in it, stating that 'one needs to distinguish between verification and validity, and . . . recognize that unverifiability does not imply invalidity, but only uncertainty about validity'. Nevertheless, her stance remains that legal cultures are ultimately, at least partially, incommensurable. Curran's views are similar to those expressed by van Hoecke (1998).

2.4 Transplants

Watson (1974) has contributed with an alternative approach to comparative law; one that attributes the development of laws to transplants. Using an illustrative collection of samples from antiquity to modern times, Watson traces the origins of the laws of nations throughout the world to transplants from neighbouring and/or influential nations. One intriguing example compares the provisions concerning the liability for injuries caused by aggressive oxen in three historical codes: the Laws of Eshnunna (eighteenth century BC), the Babylonian Code of Hammurabi (seventeenth century BC), and the Book of Exodus (600–400 BC). All three examples address the situation where an ox gores another ox or a human. In the Laws of Eshnunna and the Code of Hammurabi, the owner was liable for damages in various degrees if he had been notified by the authorities that the ox was aggressive but failed to take measures to prevent it from injuring others. In the Book of Exodus, the ox was to be stoned to death, and if it had been testified to the owner that the ox was a threat, and it killed a man or a woman, both ox and owner were to be put to death.

Watson, however, does not claim that all laws are the result of transplants or that transplants are always complete and without alterations. As to the former, Watson gives the example of Roman law, which is not to be confused with Greek law, despite that the Roman Republic had its most dynamic period of development during its last two centuries (approx. 200–27 BC) when there was a particularly strong Greek influence on Roman culture. As for the latter, he names the adoption in Western Europe of the Roman *Lex Aquilia* (287 BC), which concerned the liability for damages to property, as an example of incomplete transplants, and in some cases, transplants in name only (ibid. pp. 26–30, 79–81).

Yet Watson makes the claim that the transplanting of laws – whether it be individual rules or large parts of a legal system – is common and 'the most fertile source of development', as most changes in most systems are the result of borrowing. He further claims, *inter alia*, that the transplanting of legal rules is 'socially easy', in the sense that the adopted rules are accepted without much difficulty (ibid. pp. 95–96).

Watson's writings have drawn criticism from Pierre Legrand (1997). Legrand – who, as mentioned, ranks among the top culturalists in the field – dismisses the idea of transplants as an explanation to legal change, on the grounds that the law cannot be understood as segregated from the society in which it operates and the culture that permeates that society. Therefore, Legrand argues, even if a law is borrowed, it will never have the same content. Watson has answered the criticism (Watson, 2000). It would seem Watson has the right to, given that he stated from the beginning that the transplanting of rules would seldom be without adaptations. Indeed, the aforementioned aggressive ox example would appear to prove his point.

It seems fair to posit that transplants have indeed played a large part in legal development all over the world. To name but one example, the numerous countries that have a civil code modelled after the French *Code civil* clearly attest to this. Again, this is not to say that it is the only contributing factor or that transplants have been accomplished without alterations or adaptions. Even where a law is 'imported' without the express intention of changing it, the way it is practised and interpreted will to varying degrees be affected by the culture in the host jurisdiction. However, just as ideas and innovation spread unrestricted by political boundaries, so too can and do legal rules and institutions proliferate. Understood in this fashion, as an instance of the proliferation of ideas, transplants ought not be too unpalatable even for staunch supporters of culturalism.

The existence of legal transplants would seem to lend credence to the functionalist idea that different societies often have the same, or similar, problems to be solved and functions to be fulfilled. After all, if it were true that each society is singular and insular, there would be no need to borrow legal rules or institutions. Indeed, the very idea would be moot. Yet borrowing has been going on since far antiquity. At the very least, it refutes the extreme culturalist idea of singularity (cf. Grazidei, 2006).

2.5 Institutional analysis

What Legrand and other culturalists define as culture bears kinship with comparative institutional analysis. While there is no room in this chapter to delve deeply into that field, the reader is urged to bear in mind that it can be a fruitful perspective in comparative studies.

Douglass C. North (1991) has defined *institutions* as

> the humanly devised constraints that structure political, economic and social interaction. They consist of both informal constraints (sanctions, taboos, customs, traditions, and codes of conduct), and formal rules (constitutions, laws, property rights).

This rather broad definition of institutions can be categorized in innumerable ways. For the purpose of comparative studies in real estate, the following three categories would seem particularly relevant.

- *Legal institutions* are all the rules, processes, and actors that have a bearing on the law. Rules comprise both legislation and other types of rules, such as case law, legal principles, or extra-legal norms that are recognized by the legal community, usually known as 'soft law'. Processes are also aptly defined in a broad sense, for example, by encompassing the manner prescribed by law in which legally binding contracts are entered into or how title registration is obtained. By the same token, all actors on the legal arena are of interest, such as courts, authorities, dispute resolution bodies, and similar.
- *Economic institutions* involve all the factors of an economic nature that affect human behaviour through incentives of different kinds. Economic incentives play an important part in shaping people's choices and the way institutions function.
- *Cultural institutions* are not as straightforward to define as the former two categories. This is in large part due to the elusive nature of the concept of 'culture'. A pragmatic approach would see cultural institutions defined as any norms or patterns of behaviour that tend to come into play among people in a given geographic area.

As is readily seen and appreciated, it is difficult to draw clear lines between these three categories, as they are intertwined – sometimes intimately so. For instance, a tax, prescribed by law, is at once a legal and an economic institution. Similarly, transaction processes have legal, economic, and cultural aspects and origins, of which the pre-contract prevalent in many countries serves as a fine example.

Moreover, the nature of the relationship between institutions is oftentimes complex. This is particularly the case in the matter of causality, at times giving rise to chicken-and-egg questions with no clear answers, as one institution may shape another, which in turn my affect and shape the former, *ad infinitum*. For instance, the substantive contents of a particular piece of legislation may be a codification of norms in the marketplace which are in turn rooted in the culture as well as in economic institutions. At the same time, once a law is enacted, it shapes people's patterns of behaviour – and thus both economic and cultural institutions – with its normative reach.

In real estate conveyances, the civil law notary and the Swedish estate agent – and the respective regimes regulating them – stand out as central institutions that shape the way in which transactions are conducted.

3. Concluding remarks

The three perspectives on comparative legal studies that have been presented all seem to have some degree of merit. While it is clear that not all laws in all legal orders have inspired or resulted from legal transplants, it is equally evident that many such transplants have taken place throughout history. In this regard, it can be noted that a large part – often the lion's share – of new legislation in EU member states are in fact those member states' implementation of EU directives. And while EU legislation is not mandated by a particular member state, the legal regimes and solutions prescribed in EU directives nonetheless often bear the hallmark of the legal order of one or a few member states. In such instances, legal regimes are indeed transplanted from one member state to another via EU legislation.

Functionalism, for its part, appears sound insofar as laws and institutions are by definition meant to serve some kind of purpose; hence, fulfilling a function is their very *raison d'être*. It is also evident that these functions are usually relevant in more than one country or legal order. In the case of real estate conveyances, land is and has always been an important asset. While the theoretical foundations of the proprietor's claim to the land may vary between legal orders, the need for clarity and certainty with respect to the chain of title remains much the same. Hence, regulating the way in which valid and binding property conveyances are accomplished is a function that is common to at least most legal orders. By the same token, there is a need to regulate it in such a manner that ensures clear chains of title; hence, formality requirements of some kind are to be expected in most legal orders. The specific traits of those requirements may vary, giving rise to different legal regimes, but the function they fulfil remains the same. In that, functionalism is a sound and useful perspective in comparative studies, and the search for functional equivalents a worthwhile endeavour.

It is important, however, not to mistake similarity in function with similarity in execution. In this book, for the purpose of comparison, the conveyance process has been given a framework of different transaction phases: marketing, pre-contract, contract, registration, and post-transaction. However, not all regimes are easily accommodated in such a model. For instance, the existence of a pre-contract phase is evident in legal orders where buyer and seller can enter into a mutually binding agreement as between themselves, whereas the agreement becomes binding only towards third parties upon notarization. The same applies where the validity of the conveyance itself is contingent on notarization; under such a legal regime, it appears natural that buyers and sellers will seek to bind each other to the conveyance, at least given the threat of liability. However, where the sale itself becomes binding both as between the contracting parties and towards third parties once the sales contract is signed by the buyer and the seller – which is the case in Sweden – there is no need for a pre-contract. Hence, there is typically no such phase in the Swedish transaction process. Brand (2007) has addressed this type of problem and has suggested conceptual comparisons as a useful approach.

As is often the case, the three perspectives in part pose and answer different questions, with none being universal. The relative merits of the perspectives to the researcher would therefore appear to vary depending on the nature of the task at hand. If one seeks to explain how laws evolve over time, then transplants is not a bad place to start. Just as ideas in general proliferate, so do ideas of legal solutions proliferate, albeit perhaps often at a slower rate. At the same time, culturalism and institutions can help explain why a particular legal regime did *not*, or will *not*, take root.

The functionalist approach, meanwhile, seems excellently suited to help the researcher to learn about different legal orders. The assumption of functional similarity serves as a working hypothesis, which may be either confirmed or falsified. It helps the researcher to pose the right questions: How are binding contracts accomplished? Who draws up the necessary documents? Do the parties receive legal advice, and if so, from whom? and so forth. The answers may be found in different places in different legal orders; as stated, in Sweden, several tasks performed by civil law notaries are entrusted to estate agents. In that instance, the assumption of similarity is at least in part confirmed. The researcher must never forget, however, that sometimes one has to cast aside one's assumptions and embrace the evidence. And that is more difficult than it sounds.

References

Bogdan, M. (2003). *Komparativ rättskunskap*. Stockholm: Norstedts Juridik.

Brand, O. (2007). Conceptual comparisons: Towards a coherent methodology of comparative legal studies. *Brooklyn Journal of International Law, 32*(2), 405–466.

Cotterrell, R. (2006). Comparative law and legal culture. In M. Reinmann & R. Zimmermann (Eds.), *The Oxford handbook of comparative law* (pp. 709–737). Oxford: Oxford University Press.

Curran, V. (1998). Cultural immersion, difference and categories in U.S. comparative law. *The American Journal of Comparative Law*, *46*(1), 43–92.
Dannemann, G. (2006). Similarities or differences. In M. Reinmann & R. Zimmermann (Eds.), *The Oxford handbook of comparative law* (pp. 383–419). Oxford: Oxford University Press.
de Montmorency, J. E. G. (1910). The great jurists of the world: Friedrich Carl von Savigny. *Journal of the Society of Comparative Legislation*, *11*(1), 32–54.
Derrida, J. (1967a). *L'écriture et la différence*. Paris: Éditions du Seuil.
Derrida, J. (1967b). *De la grammatologie*. Paris: L'Éditions de Minuit.
Derrida, J. (1972). *La dissemination*. Paris: Éditions du Seuil.
de Secondat, C. (1748). *The spirit of the laws*. Baron of Montesquieu.
Grazidei, M. (2006). Transplants and receptions. In M. Reimann & R. Zimmerman (Eds.), *The Oxford handbook of comparative law* (pp. 467–470). Oxford: Oxford University Press.
Husa, J. (2009). Farewell to functionalism or methodological tolerance? *Rabels Zeitschrift für ausländisches und internationales Privatrecht*, *67*(3), 419–447.
Jingryd, O. (2012). *Impartial legal counsel in real estate conveyances: The Swedish broker and the Latin notary* (Doctoral thesis). Royal Institute of Technology, Stockholm.
Jingryd, O. (2021). *Fastighetsförmedling* (6th ed.). Lund: Studentlitteratur.
Kutner, L. (1972). Legal philosophers: Savigny: German lawgiver. *Marquette Law Review*, *55*(2), 280–295.
Legrand, P. (1996). European legal cultures are not converging. *The International and Comparative Law Quarterly*, *45*(1), 52–81.
Legrand, P. (1997). The impossibility of legal transplants. *Maastricht Journal of European and Comparative Law*, 111–124.
Legrand, P. (2003). The same and the different. In P. Legrand & R. Munday (Eds.), *Comparative legal studies: Traditions and transitions* (pp. 240–311). Cambridge: Cambridge University Press.
Michaels, R. (2006). The functional method of comparative law. In M. Reimann & R. Zimmermann (Eds.), *The Oxford handbook of comparative law*. Oxford: Oxford University Press.
Miller, R. A. (2017, July). On hostility and hospitality: Othering Pierre Legrand. *The American Journal of Comparative Law*, *65*(suppl_1), 191–206.
North, D. C. (1991). Institutions. *Journal of Economic Perspectives*, *5*(1), 97–112.
Öcuru, E., & Nelken, D. (Eds.). (2007). *Comparative law: A handbook*. Oxford: Hart Publishing.
Palm, P., Jingryd, O., & Kordic, L. (2020). Housing and transaction costs. In P. Håkansson & H. Bohman (Eds.), *Investigating spatial inequalities: Mobility, housing and employment in Scandinavia and South-East Europe (SEE)*. Bingley: Emerald.
Strömholm, S. (1971) Integritetsskyddet. *Svensk Juristtidning*, *56*, 695–736.
van Hoecke, M. (1998). Legal paradigms and legal doctrine: Towards a new model for comparative law. *The International and Comparative Law Quarterly*, *47*(3), 495–536.
Watson, A. (1974). *Legal transplants: An approach to comparative law*. Edinburgh; Scottish Academic Press (2nd ed., University of Georgia Press in 1993).
Watson, A. (2000). Legal transplants and European private law. *Electronic Journal of Comparative Law*, *4*(4).
Zweigert, K., & Kötz, H. (1998). *Introduction to comparative law* (Vol. 3). Oxford: Clarendon Press.

32

REFLECTIONS AND CONCLUSIONS

Magnus Andersson and Peter Palm

From Chapter 2, we learned the importance of institutions and how New Institutional Economics emphasizes the role of institutions for a market economy to function. Here, property rights are considered particularly pivotal, not only in terms of property law but also within a property system. A property system consists of both property law, rules in conjunction with properties (as, for example, real estate agent regulations), systems (as, for example, land registration), and organizations involved in the real estate market. All institutions and actors need to work jointly to enable a well-functioning real estate market.

From the country chapters, we learn that traditions, particularly legal traditions, have a great impact on the conveyance process. The different legal families, as outlined in Chapter 3, more or less set not only the legal framework of the conveyance process but also to a large extent the process in itself. This is an example of how these norms spread to the culture, in terms of the transfer of norms, values, and ideas.

More or less all countries that have participated in the book use the services of real estate agents. From a theoretical perspective, and to a large extent also a practical perspective, the arguments are to ease the conveyance process bridging the information asymmetry. By using a real estate agent, the seller gains full access to information about how to sell the property and the formal paperwork following a transaction. However, it is not mandatory to use a real estate agent, but we learn from the country chapters that their services are used to a very high degree. This is interpreted as how culture and norms shape how the actors in practice navigate the institutional setting to minimize the transaction costs. And although we can see that real estate agents charge between 1.5% and 10%, the market seems to nevertheless value their involvement, as it lowers the total transaction cost of the conveyance process, which would be higher than the agent's commission.

This common practice of using real estate agents has also made its mark in the legislation. Most countries, particularly with common law, have clear regulations regarding real estate agents, which can be interpreted as a recognition of the high importance of their involvement in the conveyance process. With the legislation in Sweden as an example, it takes a clear consumer perspective, aiming to protect both the seller and buyer from misleading counselling. From this perspective, the institutional framework aims to ensure correct information and good counselling in the conveyance process without the involvement of lawyers or notaries, as is commonly the case in many other countries. This is a clear example of how institutions evolve to ensure,

DOI: 10.1201/9781003130383-37

in this case, a secure conveyance process to enable a strong property system where not only property law but also the whole system of laws, rules, and norms interplay.

1. Summing up the transaction costs

This section summarizes the transaction costs from the country chapters by comparing our three legal families: civil law, Nordic law, and common law, as shown in Figure 32.1.

The first category, civil law, comprises Napoleonic and Germanic law systems, in which both include the notary as an important actor. The countries with Napoleonic law exhibit rather high transaction costs, particularly for the buyer. Spain, Brazil, and Dubai (UAE) report transactions in the higher range of the sample or the buyer of the property. An interesting finding within this category is that the buyer of property in Cambodia pays much fewer transaction costs than the seller. Also, there are quite large differences between the states in the United Arab Emirates, with Abu Dhabi and Northern Emirates exhibiting low levels of transaction costs.

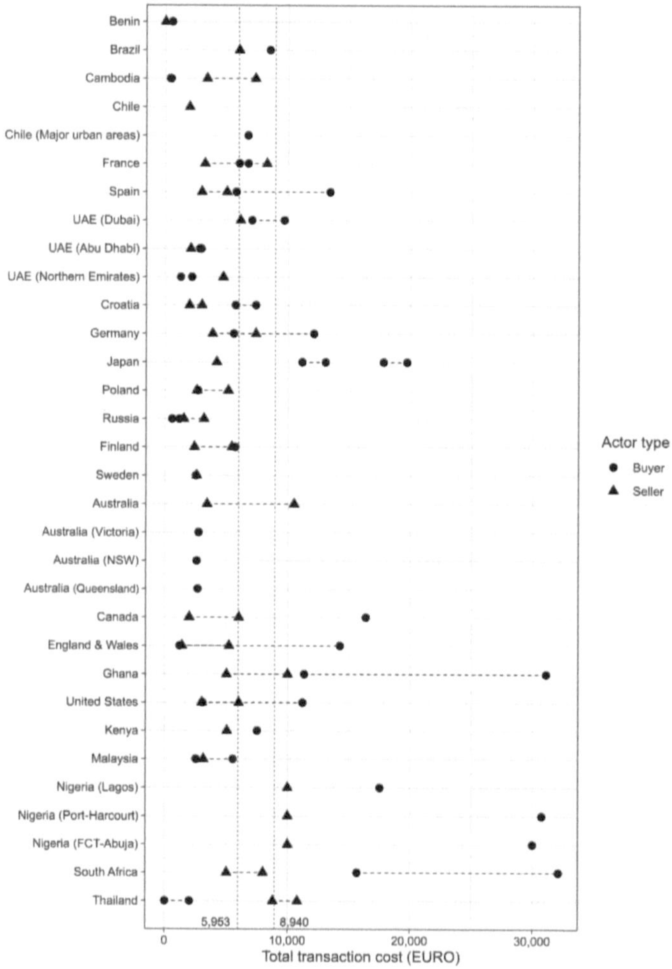

Figure 32.1 Transaction costs

Source: Author

In the Germanic law system, the transactions costs are in the lower range, with Croatia, Germany, and Japan in this group. These countries have similar levels for both sellers and buyers. Poland and Russia have law systems with a mix of Germanic law, and there, we can observe very low transaction costs.

The second category, Nordic law, represented here by Finland and Sweden, presents very low transaction costs for both the buyer and seller.

In the third category, with common law systems and mixed systems using common law, we observe the largest differences within the group of countries. For example, South Africa, Nigeria, and Ghana exhibit high transaction costs, particularly for the buyer. The costs are three times higher than, for example, in Kenya, Malaysia, Australia, the United States, and Thailand. It is in this third category that the highest transaction costs within all the countries in the book can be observed, with Nigeria, South Africa, and Ghana particularly standing out.

2. What we cannot see in the numbers

From Figure 32.1, we can see the differing amounts of fees and taxes that together encompass the transaction cost for the conveyance process of private homes in the specific country. However, this is not the full truth when it comes to transaction costs. The chart displays only fees and taxes, and transaction costs involving aspects such as time, information, complexity, and insecurity are not considered. These components of transaction costs are harder to compile, as they are often of a more intangible nature.

Time comes into play in the conveyance process in more than one dimension and in different ways for the seller and buyer. In the marketing phase, the future buyer can be affected depending on how the search is conducted and the information about homes for sale are displayed. In the Swedish example, 'all' homes for sale are marketed on the same website. From the buyer perspective, this is time-efficient, as you do not have to conduct more than one search to reach the whole supply of the market. The same is accomplished by the seller, who will be able to reach the whole market by advertising their home on one platform. In other countries, such as England, we can see that there is no independent e-marketplace where all, or at least the majority, of the homes for sale are marketed regardless of real estate agent firm. Instead, there are several platforms advertising homes for sale, but this makes the search take longer and the potential buyer risks missing out when a home comes up for sale. The same goes for the seller, who, first of all, must consider which platform to advertise on or to advertise on more than one. Second, given that the whole market is not covered by one platform, the seller also risks missing out on a potential buyer. This makes the matching process more time-consuming for both the seller and future buyer, and thereby, also adds to the transaction costs for both of them.

During the pre-contract and contract phase, both the buyer and seller will have to invest time, in terms of not only meeting for the signing of the contracts but also the amount of work time involved, in terms of the finding right officials to process the documents. When comparing the countries with the notary system with those without, we expect that the notary system would be more time-consuming, and therefore, also add on to the transaction costs, time-wise.

The registration phase is to a large degree dependent on the effectiveness of the government officials. From the World Bank's Worldwide Governance Indicators (WGI), we can see a grading, among other information, of how efficient a country's public administration is.

Countries like, for example, Finland, have a shorter process in general, with a government administration characterized by independence and impartiality speaking for an efficient process. Other countries with a less transparent property system can be less effective in the conveyance process as a result. Countries with a less efficient government administration cause inefficiency and a longer conveyance process, which adds on to transaction costs.

Information, both in terms of availability and quality, also influences the transaction costs. As mentioned, the search time will be affected during the marketing phase depending on how the information is available, but the availability of information also influences the transaction cost during the conveyance process if certain information is not publicly available or the quality of the information is poor. For example, in Sweden, most information regarding landownership is public and available for everyone, sometimes for a small fee. However, in Germany, this information is not public and available for everyone, which leads to a more time-consuming process that also could add to insecurity.

The pre-contract and contract phase is also affected by the availability of information and the quality of the information. For example, the lack of availability of information about ownership will add to transaction costs, both in terms of a more time-consuming process and a higher degree of insecurity for the buyer. In some countries, such as Egypt or Russia, the public records are not complete or accessible, which can lead to a great deal of insecurity, and the conveyance process will be affected to a large degree by the quality or accessibility of public records. In other countries, disputes over land are not uncommon, and property registration remains difficult. Although the state recognizes property rights, the difficulties remain, especially on the local level outside major cities. However, steps are being taken and new legislation for new construction has been enforced, making the process more expensive but at the same time more standardized and foreseeable, reducing the transaction cost of insecurity.

Poor public records will add to insecurity in different ways. For the buyer, records are important for knowing that the seller truly is the rightful owner and that the property bought will remain private property, with no one else having better claim to it. However, for the seller, records can be problematic, as they will need to prove their ownership, which can be a timely process.

The complexity of the conveyance process is also a source of transaction costs. How many persons/institutions that are involved will have consequences for the transaction costs of the conveyance process. The number of institutions involved in the conveyance process differs between the countries participating in the book. Thailand, Benin, Chile, Sweden, and Brazil are examples of countries with few institutions involved in the conveyance process. In contrast, Japan, Germany, South Africa, Poland, Kenya, Nigeria, and Spain involve quite a few different institutions when selling and buying a property. When combining the rate of the real estate agent's commission with the number of institutions involved, one can see how the overall costs increase with the number of actors.

Taking the Croatian process as an example, it is more complex than it is in Sweden. This is to some extent explained through the different legal traditions that the countries have inherited. But it also comes from how the Croatian transaction process features at least one additional party to be remunerated. The Swedish regulatory regime, where the agent has a mandatory university education comprising courses in relevant fields of law, handles the legal aspects of the transaction, and this lowers the transaction costs, as no legal counselling (or notary) are involved in the process. However, this system (in Sweden) can also be criticized because the agent is selected by the seller and paid by the seller but nevertheless is expected to act in an impartial way. To add to this, agents are employed by a private company and thus not officials, although they do act under official regulations. This might render a sense of insecurity by the buyer to not have the same kind of support or control as the seller, who will naturally have a closer relationship with the agent. In this case, the reduced fees together with less complexity in the process may result in insecurity for the buyer.

The number of persons involved are not only due to the legal regime but also to tradition. For example, both France and Spain come from the civil law family with a Napoleonic tradition that uses a notary system for the conveyance process. In the French tradition, the notary draws up all the documents and explains to the parties their meaning and importance before notarizing the transaction. A meeting with the notary can take up to several hours and can be rather costly. In the Spanish tradition, the notary expects all documents to be in order before the meeting and only notarizes the transaction. The meeting will last only a couple of minutes. However, in this case, the parties are required to have had the documents drawn up by a lawyer or similar before the meeting with the notary. This add to the complexity, as at least one more additional party has to be involved. This also adds to a kind of insecurity, as the notary will notarize the transaction only if all the documents are in place. Otherwise, the parties will have to make amendments and reconvene at a later time.

Insecurity is not only a question of good official records and a process that is foreseeable without risk of corruption and so on, but also a consideration of to whom the fee is paid. If the fee is to a private actor, such as a bank, instead of the state, we can expect less transparency and the risk of changing rates to the fee. The Croatian example involves several fees to the bank, whereas the French simply pay fees to officials. This comes with a risk, both in terms of less foreseeability regarding the rates and a risk in terms of how willing the parties are to include the bank in the transaction. Less trust in the banks and the financial system could restrain the market and make the process less efficient. If we take Russia as an example, where paying cash seems to be the norm, this could hinder the mobility of the inhabitants, as one may not willing to use the banks to take out a mortgage. This may also mean that people are less inclined to buy a new home and thus hindered to move. At the same time, an overconfidence in the financial system might lead to an overheated market with rapidly increasing prices. However, trust in the financial system is not the same as a willingness to take new mortgages. Taking Germany as an example, the loan-to-value ratio is low, although people in Germany tend to have a high degree of trust in banks and the financial system. Nevertheless, by tradition, taking out a mortgage, even for homes, is not that common. In contrast, in the case of Australia, there is a possibility to have very high loan-to-value ratio, leading to high risk, but these loans also come with a guarantor for the mortgage's payments.

Another source of insecurity and information quality is when taxes and fees are differentiated in regard to the price of the property. This gives the seller and buyer an incentive not to disclose the true sales price. Instead, we could expect that one part of the price is paid on the side so that the price will end up in a lower category of the fee. Several countries have this system with differential rates. Poland is one example, but they have taken measures towards payments on the side. However, as with other regulations where both sides can gain by not disclosing the true price (i.e. the seller does not have to pay as much in capital gains tax and the buyer pays a lower notary fee), creativity will be high. Given that trust in the financial system and bank mortgages is less common, as with Poland in the case of apartments, we end up with less control mechanisms and the risk for misleading information on the market as well as less transparency.

3. Summing up

The complexity of a given conveyance process determines the level of people's involvement and also the transaction costs. If it is complex from the start, this is a sign that the process will be time-consuming, costly, and hard to understand for the common buyer and/or seller.

However, as stated in Chapter 2 and Chapter 3, transaction costs as such are not always something negative. In some circumstances, they counteract to other, perhaps more harmful, transaction costs. One example is the tradition of notaries drawing up the contracts. It is a time-consuming process, but nevertheless a process that adds legitimacy and a kind of security, as all contracts will have the same standard. These clear rules of the game – that, on one hand, are time-consuming, but on the other hand, reduce uncertainty and insecurity – must instead be considered as involving transaction costs that on the whole reduce the amount of overall transaction costs in the conveyance process.

4. Looking ahead

Time, and its relation to transaction cost, has only been touched upon in this book. In the future, it would be interesting to also include the time dimension in this kind of overview, as time plays a significant role in the conveyance process as well as being a vital component of the transaction costs.

In addition, the perspective of how the fees and taxes are measured and collected would be interesting to investigate further. Is it, for example, more beneficial to have a fixed amount or a percentage rate, and should the rates be differentiated depending on the price/value of the real estate that is sold? How these different ways of determining the fees and taxes related to different questions of justice and fairness is also a perspective that would be important to look further into in the future.

Last, the question of trust in relation to the conveyance process and its transaction costs would be highly interesting to investigate further, given that trust comes into play in several dimensions within the conveyance process. For example, how does it relate to the complexity of the contracts? Or will a higher degree of trust for institutions such as banks imply a real estate market with higher debts but more secure transactions, in terms of less physical money-handling?

INDEX

Note: Page numbers in *italics* indicate a figure and page numbers in **bold** indicate a table on the corresponding page.

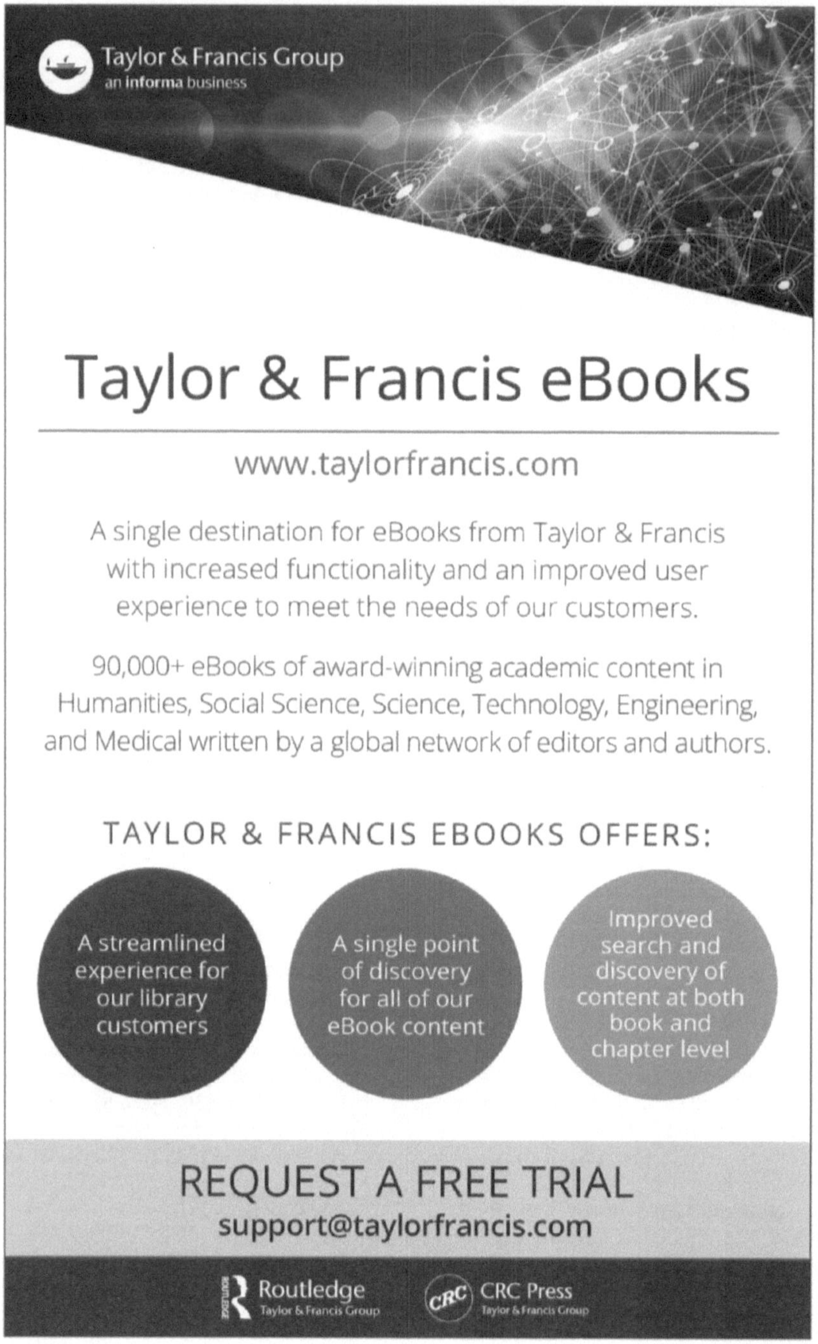
Taylor & Francis Group
an informa business
Taylor & Francis eBooks
www.taylorfrancis.com
A single destination for eBooks from Taylor & Francis with increased functionality and an improved user experience to meet the needs of our customers.
90,000+ eBooks of award-winning academic content in Humanities, Social Science, Science, Technology, Engineering, and Medical written by a global network of editors and authors.
TAYLOR & FRANCIS EBOOKS OFFERS:
A streamlined experience for our library customers
A single point of discovery for all of our eBook content
Improved search and discovery of content at both book and chapter level
REQUEST A FREE TRIAL
support@taylorfrancis.com
Routledge
Taylor & Francis Group
CRC Press
Taylor & Francis Group